Innovative Perspectives on Interactive Communication Systems and Technologies

Muhammad Sarfraz
Kuwait University, Kuwait

IGI Global
PUBLISHER of TIMELY KNOWLEDGE

A volume in the Advances in Wireless Technologies and Telecommunication (AWTT) Book Series

Published in the United States of America by
IGI Global
Information Science Reference (an imprint of IGI Global)
701 E. Chocolate Avenue
Hershey PA, USA 17033
Tel: 717-533-8845
Fax: 717-533-8661
E-mail: cust@igi-global.com
Web site: http://www.igi-global.com

Library of Congress Cataloging-in-Publication Data

Library of Congress Cataloging-in-Publication Data

Names: Sarfraz, Muhammad, editor.
Title: Innovative perspectives on interactive communication systems and technologies / Muhammad Sarfraz, editor.
Description: Hershey, PA : Information Science Reference, 2020. | Includes bibliographical references and index. | Summary: "This book explores theories, approaches, methods, and tools on the use of interactive communication systems and technologies in interdisciplinary and multidisciplinary areas"-- Provided by publisher.
Identifiers: LCCN 2019051979 (print) | LCCN 2019051980 (ebook) | ISBN 9781799833550 (h/c) | ISBN 9781799833567 (s/c) | ISBN 9781799833574 (eISBN)
Subjects: LCSH: Interactive computer systems. | Multimedia communications. | Computer networks.
Classification: LCC QA76.9.I58 I559 2020 (print) | LCC QA76.9.I58 (ebook) | DDC 302.23/1--dc23
LC record available at https://lccn.loc.gov/2019051979
LC ebook record available at https://lccn.loc.gov/2019051980

This book is published in the IGI Global book series Advances in Wireless Technologies and Telecommunication (AWTT) (ISSN: 2327-3305; eISSN: 2327-3313)

British Cataloguing in Publication Data
A Cataloguing in Publication record for this book is available from the British Library.

All work contributed to this book is new, previously-unpublished material. The views expressed in this book are those of the authors, but not necessarily of the publisher.

For electronic access to this publication, please contact: eresources@igi-global.com.

Advances in Wireless Technologies and Telecommunication (AWTT) Book Series

Xiaoge Xu
University of Nottingham Ningbo China, China

ISSN:2327-3305
EISSN:2327-3313

Mission

The wireless computing industry is constantly evolving, redesigning the ways in which individuals share information. Wireless technology and telecommunication remain one of the most important technologies in business organizations. The utilization of these technologies has enhanced business efficiency by enabling dynamic resources in all aspects of society.

The **Advances in Wireless Technologies and Telecommunication Book Series** aims to provide researchers and academic communities with quality research on the concepts and developments in the wireless technology fields. Developers, engineers, students, research strategists, and IT managers will find this series useful to gain insight into next generation wireless technologies and telecommunication.

Coverage

- Cellular Networks
- Wireless Sensor Networks
- Global Telecommunications
- Telecommunications
- Mobile Web Services
- Radio Communication
- Broadcasting
- Grid Communications
- Virtual Network Operations
- Network Management

IGI Global is currently accepting manuscripts for publication within this series. To submit a proposal for a volume in this series, please contact our Acquisition Editors at Acquisitions@igi-global.com or visit: http://www.igi-global.com/publish/.

The Advances in Wireless Technologies and Telecommunication (AWTT) Book Series (ISSN 2327-3305) is published by IGI Global, 701 E. Chocolate Avenue, Hershey, PA 17033-1240, USA, www.igi-global.com. This series is composed of titles available for purchase individually; each title is edited to be contextually exclusive from any other title within the series. For pricing and ordering information please visit http://www.igi-global.com/book-series/advances-wireless-technologies-telecommunication/73684. Postmaster: Send all address changes to above address.

Titles in this Series

For a list of additional titles in this series, please visit:
https://www.igi-global.com/book-series/advances-wireless-technologies-telecommunication/73684

Handbook of Research on the Political Economy of Communications and Media
Serpil Karlidag (Baskent University, Turkey) and Selda Bulut (Ankara Haci Bayram Veli University, Turkey)
Information Science Reference • © 2020 • 404pp • H/C (ISBN: 9781799832706) • US $195.00

Forensic Investigations and Risk Management in Mobile and Wireless Communications
Kavita Sharma (National Institute of Technology Kurukshetra, India) Mitsunori Makino (Chuo University, Japan) Gulshan Shrivastava (National Institute of Technology Patna, India) and Basant Agarwal (Indian Institute of Information Technology Kota, India)
Information Science Reference • © 2020 • 314pp • H/C (ISBN: 9781522595540) • US $195.00

IoT and WSN Applications for Modern Agricultural Advancements Emerging Research and Opportunities
Proshikshya Mukherjee (KIIT University (Deemed), India) Prasant Kumar Pattnaik (KIIT University (Deemed), India) and Surya Narayan Panda (Chitkara University, India)
Engineering Science Reference • © 2020 • 145pp • H/C (ISBN: 9781522590040) • US $175.00

Design, Implementation, and Analysis of Next Generation Optical Networks Emerging Research and Opportunities
Waqas Ahmed Imtiaz (University of Engineering and Technology Peshawar – Jalozai, Pakistan) and Rastislav Róka (Slovak University of Technology in Bratislava, Slovakia)
Information Science Reference • © 2020 • 221pp • H/C (ISBN: 9781522597674) • US $195.00

Ambient Intelligence Services in IoT Environments Emerging Research and Opportunities
Dmitry Korzun (Petrozavodsk State University (PetrSU), Russia) Ekaterina Balandina (Tampere University of Technology (TUT), Finland) Alexey Kashevnik (St. Petersburg Institute for Informatics and Automation of the Russian Academy of Sciences (SPIIRAS), Russia) Sergey Balandin (FRUCT Oy, Finland) and Fabio Viola (University of Bologna, Italy)
Engineering Science Reference • © 2019 • 199pp • H/C (ISBN: 9781522589730) • US $195.00

Handbook of Research on the IoT, Cloud Computing, and Wireless Network Optimization
Surjit Singh (National Institute of Technology Kurukshetra, India) and Rajeev Mohan Sharma (National Institute of Technology Kurukshetra, India)
Engineering Science Reference • © 2019 • 563pp • H/C (ISBN: 9781522573357) • US $425.00

IGI Global
PUBLISHER of TIMELY KNOWLEDGE

701 East Chocolate Avenue, Hershey, PA 17033, USA
Tel: 717-533-8845 x100 • Fax: 717-533-8661
E-Mail: cust@igi-global.com • www.igi-global.com

Table of Contents

Detailed Table of Contents

The chapter analyzes three types of computer systems from the hardware and software point of view, slot machines, respectively. The owners of these devices, by different methods of illegal manipulation, can have total control of the chances of winning and losing, and on the other hand, they can determine at will the amounts to be recorded in accounting. The investigation techniques used in the analysis of electronic gaming machines with winnings show that they present particular problems of cyber security. As a result, solutions have been proposed to solve these problems, such as those concerning the verifications that the legal metrology offices must carry out.

Pedagogic conversational agents are interactive systems that allow students to dialogue with them about a certain domain to learn. PCAs have been used in multiple domains from pre-primary education to university, in roles such as teacher, student, or companion. In this chapter, Alcody, a PCA to teach programming to children, is enhanced with a new proposal to manage emotions in the dialogue with students. The goal is that when children are learning to program, Alcody can help them with the emotions associated to the learning. Six emotions have been integrated into Alcody: happiness, anger, sadness, fear, surprise, and disgust. A description of how a PCA to teach programming can modify its face and verbal expressions according to the emotion detected in the student. This is given for any other researcher that would like to incorporate emotions in dialogues between PCAs and students.

OAuth is an open security standard that enables users to provide specific and time-bound rights to an application to access protected user resources. It stored on some external resource servers without needing them to share their credentials with the application. Unlike websites, for locally installed packaged web applications, the main security challenge is to handle the redirect response. The OAuth flow initiated from packaged web apps is similar to the OAuth flows explained in the current literature. However, for packaged web apps, it is difficult to define an HTTP endpoint as redirection endpoint since these apps are locally installed. The authors have proposed a novel method to execute OAuth flow from such

The simultaneous multiple data transmission can improve the use of the network. Unlike existing solutions in the literature, in this chapter, the authors propose a solution to the network resource allocation problem under the selfish behavior of mobile device with multiple connections to several available network interfaces simultaneously, to resolve the conflict of interest in network. They analyze the impact of interactions between users based on two conflicting factors (i.e., throughput and monetary cost). Also, a diverse set of user service types is taken into consideration, which makes the proposed approach suitable for an integrated service network. Analytical and numerical results demonstrate the validity of the proposed approach, which show that the non-cooperative game has an equilibrium point that depends on all parameters of the system, and they show that this situation between mobile devices is much more beneficial in terms of the performance of mobiles, cost, and the data transfer rate.

Optimization is an essential tool in the field of decision support. In this chapter, the authors study an inverse problem applied in the telecommunication networks. Indeed, in the telecommunication networks, service providers have subscription offers to customers. Since competition is strong in this sector, most of these advertising offerings, totally or partially ambiguous, are prepared to attract the attention of consumers. For this reason, customers face problems in making decisions about the choice of the operators that gives them a better report price/QoS. Mathematical modeling of this decision support problem led to the resolution of an inverse problem. More precisely, the inverse problem is to find the function of the QoS real knowing the QoS theoretical or advertising. This model will help customers who seek to know the degree of sincerity of their operators, and it is an opportunity for operators who want to maintain their resources so that they gain the trust of customers.

A typical activity of some university studies is to make visits to places outside the university where students must observe certain elements and take notes of what has been observed. Normally these visits are carried out the instructions of a guide that has been made by the teacher where it is indicated in detail how to carry out the visit: what should be seen and observed, what type of information should be retrieved, and the type of report that should be done of the visit. Likewise, these activities have some evaluation mechanism associated. The creation of these activities consumes a lot of time for the teacher. This chapter describes a proposal to automate the creation of such activities using an application that would act as an added value service that would process the information available in open data repositories and linked data in order to offer an editor/publisher of activities of visits.

Interactive communication systems and technologies are big motivation for e-learning nowadays. This chapter is dedicated for studying the effect of e-learning worldwide. The state of Kuwait is considered as a special case for this study. Two research methods have been used in this study. One method uses a questionnaire for an input of 104 students to study their level of acceptance and their attitudes towards e-learning. Other method has been set to collect interviews with 21 instructors who are involved in e-learning at undergraduate and graduate levels. The participants among students and instructors have been randomly chosen from different Kuwait institutions. A detailed analysis of the input has triggered the best practices to change the people's behavior towards e-learning and find a possible solution to bridge the gap between system makers and the users.

Preface

Interactive Communication Systems and Technologies (ICST) is an active and important area of study, research and practice today. It is an interdisciplinary area which covers a broad spectrum of existing and emerging topics and trends. Specifically, it covers Internet-based social interaction technologies which may include their applications, functions, and impact on people's digital communication. Due to a high need of the time, it is desired to explore innovations and methods in ICST. A compilation of latest advances in ICST research may be a source revolutionizing for facilitating and enhancing the exchange of information among researchers involved in both the theoretical and practical aspects.

This book consists of original and innovative findings in the field of ICST. It is intended to focus on theories and methods from various dimensions. It would look at trends and needs in the modern perspective of present and future. It is also intended to address a variety of disciplines including communication science, psychology, human-computer interaction, information systems, computer science, knowledge management, business, economics, media studies, public relations, advertising, marketing, education, law, anthropology, and social work. The book would integrate interdisciplinary and multidisciplinary needs to the significant issues of ICST in this age of technology and communication. It would focus on theories, approaches, methods, tools, experiences and state-of-the-art technologies for solving real life issues and problems.

The book is targeted to internet-based community, communication experts, system engineers, scientists, researchers, practitioners, academicians, and related professionals. It will aim at the latest advances for an in-depth discussion on ICST and its applications, functions, and conceptualizations. The chapters in this comprehensive reference explore the latest developments, methods, approaches and applications of ICST in a wide variety of fields and endeavors. This book is compiled with a view to provide readers of backgrounds and methods with an in-depth discussion of the latest advances. It consists of fourteen chapters from different disciplines of life.

Ye-Sho Chen begins the book with a discussion upon Innovative Perspectives on an Interactive Platform for Sustainable Supply Chains. He describes that United Nations setup seventeen sustainable development goals in 2015 to transform our world. Sustainable supply chains play a key role in achieving the global goals. In this chapter, he proposes an innovative platform with interactive communication systems and technologies to empower sustainable development participants actively engaging in managing sustainable supply chains and developing global entrepreneurship through international soft landings. Specifically, he illustrates how to develop an interactive platform of "Flying High, Landing Soft" for wetland innovation and entrepreneurship to address the growing global problem of wetland losses. The interactive platform, grounded in the theories of strategic entrepreneurship, distributed cognition, digital nervous systems and equipped with social interaction technologies, allows student entrepreneurs to participate in exploring and developing innovative solutions and business ventures to help build and maintain healthy wetlands to help achieve sustainable development goals.

Delello and Mokhtari, in Chapter 2 of the book, follow with a discussion of "Student and Faculty Perceptions of Social Media Use and Relationships Inside and Outside the Higher Education Classroom." Their work aims at examining student and faculty perceptions of social media use inside and outside the classroom. Three hundred and ninety-six students and fifty faculty members, at a regional university campus in the south central United States, voluntarily completed an online survey soliciting quantitative and qualitative data about their perceptions of social media use. Results revealed important findings highlighting similarities, differences, and insights among student and faculty perceptions of social media use in the classroom. It presents their views about whether social media use constitutes a distraction and how each group views social media relationships in and out of the classroom. These findings are quite consistent with prior and emerging research about social media use and have implications for how institutions of higher education can explore meaningful ways of incorporating social media in the classroom with the goal of strengthening teaching and learning.

This follows the chapter on "Security Issues Detected in the Computer Slot Machine Systems" introduced by Popescu and Naaji. It analyzes three types of computer systems, from the hardware and software point of view, for the slot machines. This is done from nowadays gambling market as a result of the detection of some criminal activities by some malicious persons. They can intervene with the help of some special keys on the electromechanical meters from their construction/equipment. Thus, the owners of these devices, by different methods of illegal manipulation which is contrary to the legal provisions, can have the total control of the chances of winning and losing. On the other hand, they can determine at will the amounts to be recorded in accounting as well as those they may obtain. The investigation techniques used in the analysis of electronic gaming machines with winnings show that they present particular problems of cyber security. As a result, in this chapter, future solutions have been proposed to solve the problems such as those concerning the verifications that the legal metrology offices must carry out.

Pedagogic Conversational Agents are interactive systems that allow students to dialogue with them about a certain domain to learn. PCAs have been used in multiple domains from Pre-Primary Education to University, in roles such as teacher, student or companion. Morales-Urrutia et al., in Chapter 4 of the book, follow with a discussion of "How to Integrate Emotions in dialogues with Pedagogic Conversational Agents to teach programming to children." In this chapter, Alcody, a PCA to teach programming to children, is enhanced with a new proposal to manage emotions in the dialogue with students. The goal is that when children are learning to program, Alcody can help them with the emotions associated to the learning. Six emotions have been integrated into Alcody which include happiness, anger, sadness, fear, surprise, and disgust. It provides a description of how a PCA, to teach programming, can modify its face and verbal expressions according to the emotion detected in the student. This will help the community that would like to incorporate emotions in dialogues between PCAs and students.

OAuth is an open security standard that enables users to provide specific and time-bound rights to an application to access protected user resources. It stored on some external resource servers without needing them to share their credentials with the application. Unlike websites, for locally installed packaged web applications the main security challenge is to handle the redirect response. The OAuth flow initiated from packaged web apps is similar to the OAuth flows explained in the current literature. However, for packaged web apps it is difficult to define an HTTP endpoint as redirection endpoint since these apps are locally installed. Shawon Rahman et al., in Chapter 5 of the book, follow with a discussion of "OAuth 2.0

A Framework to Secure the OAuth-based Service for Packaged Web Application." They have proposed a novel method to execute OAuth flow from such applications with the help of a web runtime framework that manages the life cycle of these applications. They have compared their proposed approach with another two existing approaches. After conducting experiments, they have found the proposed approach blocking all illegal OAuth flow executions. It also delivers better OAuth response handling time and power consumption performance.

Chapter 6 is presented by Alkhaldi and Muhammad Sarfraz, it is about "A Technological Solution for Communication with Deaf Students in the Institutions of Kuwait." The Deaf category represents a significant class at the global level, and it is therefore necessary to pay attention to the provision of prominent educational programs and services to help them achieve self-realization and full integration into society. The study, in this chapter, proposes a new translation system for deaf students in faculties, universities and higher institutes in Kuwait. It can help deaf students to obtain better educational services and improve their communications with their hearing students and teachers in the State of Kuwait. The authors have named such a system as Video Relay Service (VRS), it can also be adopted in institutions in other parts of the worldwide. The overall goal of this research is to propose a new translation system that can help students obtain better educational services and improve their communication with their hearing peers and teachers in the State of Kuwait. The proposed translation system, for deaf students, is a detailed study and methodology of the targeted research and has been explained with all its essential elements.

The data sponsored scheme allows the content provider to cover parts of the network subscription fees for end-users. As sponsored data gains traction in the industry, it is important to understand its implications. Garmani et al., in Chapter 7, have contributed for "Strategic Interaction Among Content Providers in Sponsored Content Market." Their contribution considers content providers (CP) choice of how much content to sponsor and the implications for end-users and CPs. In particular, they have modelled the interactions of CPs as a non-cooperative game in terms of pricing, the credibility of content and the number of sponsored content. They have proved it through a detailed analysis of the existence and uniqueness of the Nash equilibrium. Based on the game properties, they implemented a learning scheme using best-response dynamics that allows CPs to learn their strategies in a fast, accurate yet completely distributed fashion. Extensive simulations have also been provided for attractive insights on how the sponsoring content may increase the CPs payoff.

Amrani et al. in "A Pricing Model to Optimize Transmission Strategies for Mobile Devices" describe that the simultaneous multiple data transmission can improve the use of the network. Unlike existing solutions in the literature, in Chapter 8, they propose a solution to the network resource allocation problem. This is done under the selfish behavior of mobile device with multiple connections to a several available network interfaces simultaneously. This helps to resolve the conflict of interest in network. They have analyzed the impact of interactions between users based on two conflicting factors (i.e., throughput and monetary cost). They have also taken a diverse set of user's service types into consideration. This makes the proposed approach suitable for an integrated service network. Analytical and Numerical results, in this chapter, demonstrate the validity of the proposed approach. It shows that the non-cooperative game has an equilibrium point which depends on all parameters of the system. This situation between mobile devices is much more beneficial in terms of the performance of mobiles, cost and the data transfer rate.

The following chapter, "Decision Choice Optimization With Genetic Algorithm in Communication Networks," by Omar et al., shows that optimization is an essential tool in the field of decision support. In this chapter, authors have studied an inverse problem applied in the telecommunication networks. Indeed, in the telecommunication networks, service providers have subscription offers to customers.

Since competition is strong in this sector, most of these advertising offerings, totally or partially ambiguous, are prepared to attract the attention of the consumers. For this reason, customers face problems in making decisions about the choice of the operators that gives them a better report price and/or Quality of Service (QoS). Mathematical modeling of this decision support problem led to the resolution of an inverse problem. More precisely, the inverse problem is to find the function of the QoS real knowing the QoS theoretical or advertising. This model will help customers who seek to know the degree of sincerity of their operators. This will also provide an opportunity for operators who want to maintain their resources so that they gain the trust of customers.

Ahmad has contributed Chapter 10 for "Online Self-Presentation Strategies Among Social Networking Sites' Users in Emirates." This study explored the strategies of self-presentation (ingratiation, supplication, and enhancement) among United Arab Emirates users (n=230) of popular social networking sites (SNS). The size of social networks, degree of network connectivity, and perceptions of self-presentation success have been examined here. The results of the study indicate a significant positive correlation between the frequency of SNS use and ingratiation and enhancement strategies. Greater diversity of online friends among the respondents is positively associated with the perception of online self-presentation success. Males and females differed in the size of the online social network they interacted with, diversity of online friends, and preferred self-presentation strategies. However, no significant gender differences have been found in the levels of network connectivity and perceptions of self-presentation success.

Next chapter is searching for answers to Hybrid Approaches in Communication and Learning Environments. In this chapter, Brown Sr. examines the continued growth of computer-mediated communication as a starting point in working toward a hybrid approach to combining face-to-face and online activities. This examination focuses on hybrid approaches, and highlights two in particular: blended learning and the hybrid shift. Blended learning can support online and face-to-face students equally by providing similar learning experiences. Incorporating synchronicity is recommended as a way to address the limitations of asynchronous learning resident in the proposed hybrid approach in this chapter. Synchronicity is connecting two or more people in related educational events at the same time. Another approach, the "hybrid shift," features faculty design content and structure with a focus on being thoughtful, engaging, and intentional in promoting learning, effective collaboration, critical thinking, and communication skills. This examination looks closely at hybrid approaches and offers some answers for educational implications to overcome communication limitations of geographical distance and time.

Academic interest in parasocial relationships has increased dramatically in the past decade. A subset of this broader topic concerns the relationships between sports fans and a host of athletes, coaches, and teams. An overwhelming majority of those studies examine fan reactions following a singular event rather than over an extended period of time. Rodriguez is motivated in this direction and proposes an article " 'It's a Lot Like Ending Up on the Losing Side of a Relationship': Sports Fans, Grief, and Parasocial Divorce," in Chapter 12. This article unpacks the ways in which college basketball fans reacted to the departure of a beloved coach over a period of nine years. Relevant reader comments were analyzed thematically and coded by an expressed grief state using the Kubler-Ross model. The term "parasocial divorce" is introduced to describe the depth of parasocial relationships for highly motivated fans. The findings reveal how fans can come to loathe a particular persona, and the conditions under which that parasocial relationship may be rehabilitated.

A typical activity of some university studies is to make visits to places outside the University where students must observe certain elements and take notes of what has been observed. Normally, these visits are carried out by the instructions of a guide that has been made by the teacher where it is indicated in detail how to carry out the visit, what should be seen and observed, what type of information should be retrieved, and the type of report that should be done on the visit. Likewise, these activities have some evaluation mechanism associated. The creation of these activities consumes a lot of time for the teacher. This has motivated to the Chapter 13, "A Model for the Creation of Academic Activities Based on Visits", by Sarasa-Cabezuelo and Fernández-Vindel. It describes a proposal to automate the creation of such activities using an application that would act as an added value service that would process the information available in open data repositories and linked data in order to offer an editor/publisher of activities of visits.

Final chapter of the book is on "Interactive Communication Systems and Technologies for Effective E-Learning," which is contributed by Al-Ajeel et al. Interactive Communication Systems and Technologies are big motivation for e-learning nowadays. This chapter is dedicated for studying the effect of e-learning worldwide. The state of Kuwait is considered as a special case for this study. Two research methods have been used in this study. One method uses a questionnaire for an input of 104 students to study their level of acceptance and their attitudes towards e-learning. Other method has been set to collect interviews with 21 instructors who are involved in e-Learning at undergraduate and graduate levels. The participants among students and instructors have been randomly chosen from different Kuwait institutions. A detailed analysis of the input has triggered the best practices to change the people's behavior towards e-Learning and finds a possible solution to bridge the gap between system makers and the users.

Muhammad Sarfraz
Kuwait University, Kuwait

Chapter 1
Innovative Perspectives on an Interactive Platform for Sustainable Supply Chains

Ye-Sho Chen
https://orcid.org/0000-0002-7459-9891
Louisiana State University, USA

ABSTRACT

The United Nations set up 17 Sustainable Development Goals in 2015 to transform our world. Sustainable supply chains play a key role in achieving the global goals. In this chapter, the authors propose an innovative platform with interactive communication systems and technologies to empower sustainable development participants actively engaging in managing sustainable supply chains and developing global entrepreneurship through international soft landings. Specifically, they illustrate how to develop an interactive platform of "Flying High, Landing Soft" for wetland innovation and entrepreneurship to address the growing global problem of wetland losses. The interactive platform, grounded in the theories of strategic entrepreneurship, distributed cognition, digital nervous systems, and equipped with social interaction technologies, allows student entrepreneurs to participate in exploring and developing innovative solutions and business ventures to help build and maintain healthy wetlands to help achieve the Sustainable Development Goals.

INTRODUCTION

In the era of global hypercompetition, dealing with the "hidden dynamics" of diplomacy, espionage, and geopolitics has been a major challenge in international business. The rise of climate change and global terrorism, however, has brought nations together to seek for collaborative solutions. For example, in the November 2015 Paris Climate Change Conference, more than 150 Head of State and Government had the political will to come to a new, legal framework of climate change agreement (UNFCCC, 2015). Another example is the unanimous call by the 15-member body of United Nations Security Council for eradicating ISIL safe havens in Syria and Iraq (UNSC, 2015). Furthermore, on September 25th, 2015,

DOI: 10.4018/978-1-7998-3355-0.ch001

United Nations adopted a set of seventeen global sustainability goals to transform our world (UNSDG, 2015). Sustainable supply chains play a key role in achieving the goals (Busse, Schleper, Niu, & Wagner, 2016). For example, consider Goal 9: Build resilient infrastructure, promote sustainable industrialization and foster innovation. Making cities resilient is a campaign for Goal 9, started in 2010 by the United Nations International Strategy for Disaster Reduction (UNISDR, 2018), encouraging cities to exchange their experiences with each other to help enhance their resiliency.

One example is the New Orleans / Gothenburg Exchange in 2015 (NOGE, 2015) during the ten-year anniversary of the devastating Hurricane Katrina in Louisiana and five years after the disastrous oil spill in 2010. According to Louisiana Economic Development (LED, 2018), Louisiana has attracted, since 2008, more than $60 billion in Foreign Direct Investment that enables the state to be number one in the U.S. for per capita FDI attraction. The recent New York Times three-part special report (Baurick, 2018), entitled "The Drowning Coast," indicates a long way to go to make Louisiana coastal cities resilient. The question for multinationals then is: As a part of corporate social responsibility, how can companies participate in making cities resilient while investing in Louisiana? The Louisiana experience indicates that it takes three waves for multinationals to actively engage in making cities resilient. First, individual entrepreneurs with proven business solutions shall be encouraged with their stories well told to inspire students to develop innovative business solutions. Second, resources in the startup ecosystem for making coastal cities resilient shall be effectively connected to cultivate new generations of student entrepreneurs developing business solutions grounded in sound science and engineering foundations. Third, multinationals play a key role in scaling up the startups nationally and globally if the solutions developed by the student entrepreneurs have significant value propositions in advancing global sustainable developments of multinationals.

Coopetition (Dagnino and Padula, 2009), combining competition and cooperation, is thus a strategy needed for business executives and managers engaging in strategic planning and operations of their international businesses in the 21st century (Castaldo and Dagnino, 2010). In this paper, we propose a two-stage process of coopetition. The first stage is to develop a strategy of cooperation in sustainable supply chains, which delivers value by using ethical, economic, social, and environmental levers throughout the supply chains (PwC, 2014). The second stage is to expand the strategy globally through an interactive platform of international soft landings (Chen, et. al, 2013; Chen, et.al, 2016) to mitigate the risks associated with the "hidden dynamics" of global hypercompetition and the extreme politically uncertain environment (Akash, 2015). We illustrate how the two-stage process is used to develop an interactive platform with communication systems and technologies to empower sustainability development participants actively engaging in managing sustainable supply chains and international soft landings. Specifically, we illustrate how to develop an interactive "Flying High, Landing Soft" platform of wetland entrepreneurship to address the growing global problem of wetland losses. The interactive platform, grounded in the theory of digital nervous systems and equipped with social interaction technologies, allows student entrepreneurs to participate in exploring and developing business solutions to maintain a healthy wetland and have positive impact on global sustainability development.

SUSTAINABLE SUPPLY CHAINS

Sustainable supply chain is an emerging and important research topic in the current context of climate change. Consider global food supply chain management as an example (Paloviita and Järvelä, 2015). Early literature of food supply chain focused on "farm to fork" (Bourlakis and Weightman, 2004; Eastham, et al, 2007), including consumer, procurement, livestock systems & crop production, food manufacturers, organic foods, retailing & supermarket distribution networks, wholesaling, catering. Corporate social responsibility (CSR) emerged later as a research focus in the global food supply chain (Piacentini, et al, 2000), including motivations, CSR drivers, rural economy development, reputation for quality and reliability, CSR within food stores, impact on consumer trust, CSR in emerging markets, cross-cultural comparison, and nutrition information disclosure. More recent research focus of food supply chain management consists of addressing growing sustainability issues such as food waste and sustainability (Leal Filho and Kovaleva, 2015), design for sustainability through social practice approaches (Niimi, et al, 2014), sustainable intensification strategies for food security (Godfray and Garnett, 2014), climate change impacts on food availability (Shackleton, 2014) and threat to future global food security (Tai, et al, 2014).

In responding to the growing demand for sustainable supply chain research and practices, PwC and APICS Foundation (PwC, 2014) recently surveyed 500 supply chain executives and developed a collaborative framework of sustainable supply chains to help companies prioritize their sustainability strategies and unlock value in the end-to-end supply chain. The framework consists of an end-to-end plan identifying environmental, social, economic, and ethical priorities for each of the following five processes: design and plan, source, make, deliver, and use and return. Such a holistic framework, when appropriately implemented, can help reduce the risks in the "hidden dynamics" of local politics.

Global entrepreneurship in sustainable supply chains is believed to be an effective solution to address the growing global issues in climate change (Jaggernath, 2015). Consider sustainable supply chains in global food management as an example (Tripathi and Agarwal, 2014; Kline, et al, 2014). To cultivate food entrepreneurs to address the growing global demand and issues in food sustainability, there is an emerging trend in integrating vital players in food supply chain to form food clusters (Lee and Wall, 2012; Woods, 2014; Cooperhouse and Surgi, 2014), including government agencies, existing food companies, food service companies, food logistic companies, value-added technology companies, agricultural associations & cooperatives, agribusinesses people, faculty and students in regional universities, and start-up food entrepreneurs. Based on the food clusters and their existing networked resources, various food incubators (Edward and Policy, 2012; Khanduja, 2013; Salinger, 2013) are created to cultivate new food entrepreneurs (Tarr, 2011; McFadden and Marshall, 2014).

Sustainable supply chain from a global entrepreneur's perspective has many needs today (Hall and Stelvia, 2010; Peters, et al., 2011). Consider sustainable supply chain in global food management as an example again. The needs (Cooperhouse, 2014; Dakup, et al., 2014; Okello and Were, 2014) include information on product markets; expertise to run the food business; the expertise to develop competitive food products; certified food production facility; networked resources with value added to the food supply chain; cost competitiveness; access to capital; scale up to grow; and finally the ability to go global.

INNOVATIVE PERSPECTIVES ON AN INTERACTIVE PLATFORM FOR SUSTAINABLE SUPPLY CHAINS

In this section, we propose a "Flying High, Landing Soft" platform of global entrepreneurship in sustainable supply chains. The platform helps develop a strategy of cooperation in sustainable supply chains, which delivers value by using ethical, economic, social, and environmental levers throughout the supply chains (PwC, 2014). The platform also expands the strategy globally through international soft landings to mitigate the risks associated with the "hidden dynamics" of global hypercompetition. International soft landings (Chen, et. al., 2011), originally developed by the International Business Innovation Association (InBIA, 2015), is a process to help a company from one country land softly – without crashing – into the market of another country through a designated incubator. The purpose is to help the international soft landings companies identify business opportunities and succeed in the new foreign market with least risks and costs (Chen, et al., 2010). There are five steps in international soft landings (Chen, et. al, 2013; Chen, et.al, 2016): (1) attending a going global workshop; (2) attending an international trade trip; (3) developing a business plan for going abroad; (4) exporting the product or service through networks of international soft landings incubators; and (5) surviving and growing in the international business.

We illustrate how such the platform can be developed using the example of wetland entrepreneurship to address the growing global problem of wetland losses. With rising seas as a global phenomenon (Pilkey and Young, 2011), developing sustainable coastal cities with interdisciplinary cooperation of science, engineering, architecture, technology, socio-economics, etc. is becoming a research agenda of high priority (Rodriguez and Brebbia, 2015). Consider the state of Louisiana as an example. Before Hurricane Katrina hit Louisiana gulf coast in 2005, Louisiana already had its coastal wetlands restoration plan (LCWRP, 1993). Ten years after the devastating flood (Rivlin, 2015), several significant projects for protecting the Louisiana coastal cities were implemented which includes learning from Netherlands to design barriers to protect New Orleans from high storm surge (Folger, 2013) and the establishment of the Water Institute of the Gulf to develop innovative science and engineering in the realms of coasts & deltas, communities, and water resources (WIotG, 2015). The Gulf of Mexico Research Initiative (GoMRI, 2015), established after the disastrous oil spill in 2010 (Freudenburg and Gramling, 2012), adds another stream of research efforts to restore and improve the long-term environmental health of the coastal cities in the Gulf of Mexico.

One of the effective sustainable developments of coastal cities is creating and restoring wetlands to protect land losses (Craft, 2015). In responding to the two complex crises of Hurricane Katrina and Oil Spill (Miller, et. al., 2014), Louisiana people have risen up to develop wetland entrepreneurship businesses to provide solutions to help address the issues of wetland losses and maintain healthy wetland ecosystems. For example, Louisiana coastal wetlands have been significantly damaged by nutria (LDWF, 2007). Marsh Dog (MD, 2015) turns nutria meat into dog treats and Righteous Fur (RF, 2015) designs clothing for the contemporary fashion market using nutria fur. Both businesses help save the wetlands. Martin Ecosystems (ME, 2015) is another Louisiana startup that developed floating islands to clean polluted lakes, support a wildlife habitat, protect wetland erosion, and facilitate wetland restoration. More examples of Louisiana entrepreneurs for wetland recovery and restoration are shown in Table 1.

LSU has a rich history of entrepreneurial outreach activities. Consider Louisiana Business & Technology Center (LBTC, 2015) as an example. Established in 1988 to enhance economic development in Louisiana, LBTC has received numerous awards including the 2005 Randall M. Whaley Incubator of the Year by the National Business Incubation Association and the 2009 Economic Development Administration Excellence in Economic Development Award for Rural Economic Development, recognizing the tremendous service the Mobile Classroom (MC, 2015) provides. LSU also gained a national reputation for its rapid and effective response to Hurricane Katrina and emerged as a leader in a time of extreme crisis through the establishment of Stephenson Disaster Management Institute (SDMI, 2015; Mitchell, 2015). Both LBTC and SDMI also played significant roles in providing effective solutions to address the environmental disaster due to oil spill in 2010. As a strategy of capacity building for extension and

Table 1. Exemplar Louisiana entrepreneurs for wetland recovery and restoration

Solutions	Examples
Turning invasive species into business solutions	· Nutria (DWF, 2007) is an invasive species that significantly damages Louisiana wetlands. Louisiana entrepreneurs turn this invasive species into value-added products such as Marsh Dog (MD, 2015) dog treats made of nutria meat and Righteous Fur (RF, 2015) for the contemporary fashion market made of nutria fur and teeth. The New York Times report "The Drowning Coast" (Baurick, 2018) also pointed out new invasive species, such as small sap-sucking insects and feral hogs, are on the rise and altogether they are destroying the vegetation essential for restoring wetlands. More entrepreneurial solutions are very much needed. · Asian carp is another invasive species damaging Louisiana wetlands. Can't Beat 'EM, Eat 'EM (CBEEE, 2015) turns Asian carp into value-added natural protein food products for human consumption.
Constructing wetland ecosystems using oysters	· Seventy percent of the oysters caught in the USA are from the Coast of Gulf of Mexico and Louisiana's oyster industry contributes US$317 million annually to the state economy. In addition, oyster reefs along with other bioengineered reefs are being built by entrepreneurial businesses such as ORA Estuaries (ORAE, 2015) and Oyster Bed (OB, 2015) to help protect Louisiana's shoreline and coastal wetlands.
Helping save Louisiana's coastal wetlands through alligators	· Alligators are common in wetlands and Louisiana accounts for 80 percent of American alligators' production (Millar, 2012) with a total annual economic impact of $104 million (LS, 2015). To make the alligator economy environmentally sustainable, Louisiana State developed rules and regulations in the alligator industry (AP, 2015). This balanced approach empowers the landowners to maintain their healthy wetlands and helps save Louisiana's coastal wetlands **(Lewis and Lallo, 2013).**
Investing in wetland forests for carbon financing	· Tierra Resources (TR, 2015) focuses on innovative carbon financing that allows companies to invest in wetland restoration projects in Louisiana to offset their greenhouse gas emissions.
Cleaning up oil spills with floating islands	· Martin Ecosystems (ME, 2015) is another Louisiana startup that developed floating islands to clean polluted lakes, recycle water bottles (PN, 2016), support a wildlife habitat, protect wetland erosion, and facilitate wetland restoration. · The SHORE|LINKS (SL, 2015) system provides a unique, cost- effective technology that prevents and repairs erosion damage through the creation of living shorelines.
Scaling up the proven solutions through the investments of multinationals	· Multinationals are also investing in making Louisiana cities more resilient. For example, energy companies such as Conoco Phillips and Entergy (Abrams, 2015) supported Tierra Resources to plant mangroves using crop duster planes to spread the seed through the air. · Similarly, through the support of Shell and Entergy (RT, 2015), Martin Ecosystems work with Coastal Conservation Association's Building Conservation Trust and 100 Lafourche Parish school students to help build the Floating Islands Restoration Project in Grand Isle, a frontier city in Louisiana fighting the eroding wetlands. · On the other hand, Dow (2015) helped students along the shores of Louisiana's Lake Pontchartrain explore wetland conservation and encouraged students in the Netherlands to build their own wetland chain reactions.

community outreach, LSU Student Incubator (SI, 2015) enables entrepreneurial students to leverage the networked resources of LBTC and SDMI to create effective entrepreneurial solutions to address the growing problems in the coastal regions.

Working together with those units related to entrepreneurship at LSU and in Louisiana, we have developed a "Flying High, Landing Soft" platform to bring our local entrepreneurial solutions to go global with impact. The platform is grounded in the theories of strategic entrepreneurship and docility-based distributed cognition. Strategic entrepreneurship (Hitt et al. 2011), rooted in the resource-based view of the firm (Barney 1991), creates value for shareholders in three dimensions: resource inputs, consisting of individual resources, organizational resources, and environmental factors; resource orchestration processes (Sirmon et al. 2011), consisting of structuring the firm's resource inputs into portfolios, bundling resource portfolios into capabilities, and leveraging the capabilities to create value for shareholders; and outputs, consisting of creating individual benefits, organizational benefits, and societal benefits. Docility is "*the tendency to depend on suggestions, recommendations, persuasion, and information obtained through social channels as a major basis of choice*" (Simon 1993, p.156). Furthermore, "because of docility, social evolution often induces altruistic behavior in individuals that has net advantage for average fitness in the society. Altruism includes influencing others to behave altruistically" (Simon 1993, p.157). The docility-based distributed cognition (Secchi, 2010) has three dimensions: developing learning communities with communication standards and docile people who are willing to share resources and networks to help others in advancing the common interests of the communities; developing coaching procedures for advice giving and taking through accessible social channels; and telling success stories with bandwagon effects as the feedbacks to the communities.

Specifically, the platform consists of three major components (Parker, Van Alstyne and Choudary, 2016) as is shown on the left in Figure 1. First, the participants as the resource inputs, docility-based learning communities with communication standards enabled by communication systems & technologies and docile people who are willing to share networked resources to help wetland entrepreneurs survive and grow the global businesses. Docility-based learning communities with shareable networked resources include Louisiana Business and Technology Center, Mobile Classroom, Student Incubator, and Stephenson Disaster Management Institute. Second, the value unit with processes of resource orchestration, with accessible social channels and effective coaching, including structuring the networked resources into portfolios, bundling resource portfolios into capabilities of developing robust business plans, and leveraging the capabilities to effectively execute the business plans. Third, the filter of cultivating storytellers with effective storytelling of how they developed their businesses successfully and fostering a virtuous cycle of continuous improvement of the platform by feeding back to the communities and enriching the networked resources.

RESOURCE INPUTS: THE PARTICIPANTS

The resources include (1) individual resources: docile people, typically successful alumni and their human capital, social capital, and financial capital such as wetland owners, wetland ecologists, wetland engineers, wetland landscape architects, and wetland entrepreneurs; (2) environmental factors: learning and adapting to changes in context specifics, dynamism, and uncertainty with leading experts such as global trade services, e.g., ports for logistics and transportations, chambers of commerce, U.S. Commercial Services, and World Trade Centers; and (3) organizational resources: wetland-related organizations with tools and resources and the culture of cultivating docile and entrepreneurial leadership such as wetland research institutes with coastal and environmental scientists, architects, and engineers; wetland-related businesses; entrepreneurship education organizations like Stephenson Entrepreneurship Institute at LSU, Louisiana Business and Technology Center, Mobile Classroom, Student Incubator, LSU AgCenter Food Incubator, Specialty Incubators, LSU Innovation Park; and International Programs for global outreach and engagement.

LSU has developed resource inputs with an action-oriented program cultivating entrepreneurs and helping companies to grow through the collaboration of the following entities. The program also cooperates with other Louisiana entities (e.g., Baton Rouge Area Chamber, Louisiana Economic Development, U.S. Commercial Service in New Orleans, and World Trade Center in New Orleans) to help Louisiana companies go abroad.

Figure 1. An interactive platform, empowered by the Digital Nervous System (DNS), for global sustainability development in wetland entrepreneurship

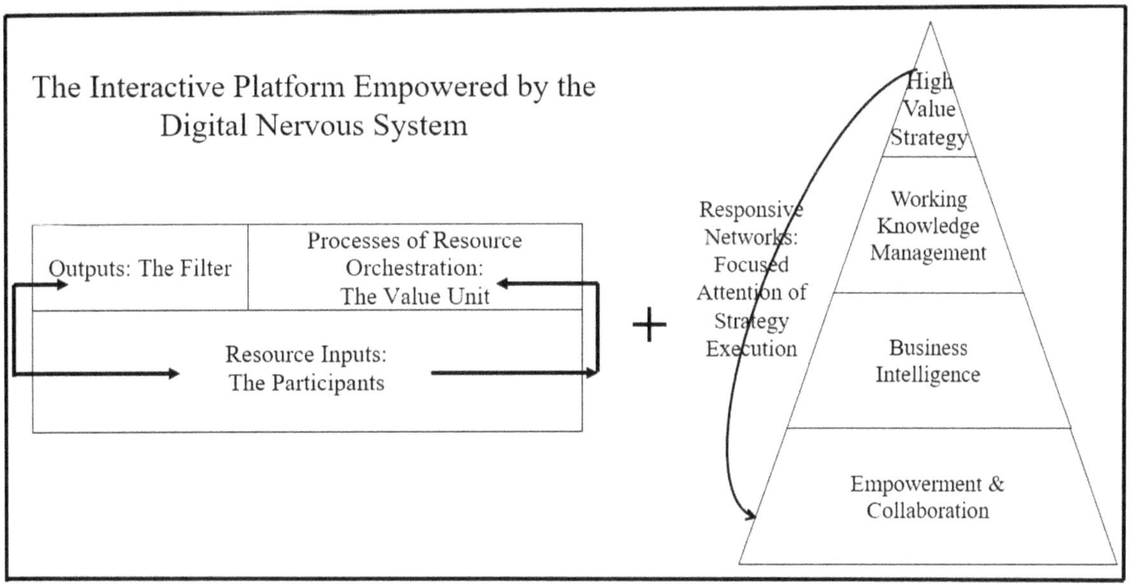

1. Stephenson Entrepreneurship Institute (SEI). SEI (2015) offers entrepreneurship education, organizes seminars, and provides business plan development advice.
2. Louisiana Business and Technology Center (LBTC). LBTC (2015), designated the 2005 National Business Incubator of the Year and the Soft Landings International Incubator in 2011 by International Business Innovation Association (InBIA, 2015), provides 3-D prototyping services, technology transfer, and incubating facilities that assure the successful implementation of business plans developed by entrepreneurs. LBTC also provides space and services to companies from international markets or joint ventures with U.S. firms to give them a starting place in which to develop their businesses in U.S. For example, LBTC was recently awarded to host a two-week agribusiness session to help entrepreneurs in Latin America access international markets (USDS, 2014).
3. Mobile Classroom (MC). MC (2015), dubbed the "Driving Louisiana's Economy" initiative, is a community outreach vehicle of LBTC used to house seminars and workshops for small business owners. LBTC received the 2009 Economic Development Administration Excellence in Economic Development Award for Rural Economic Development, recognizing the tremendous service the Mobile Classroom provides.
4. Student Incubator (SI). SI (2015) allows student entrepreneurs to develop their businesses by leveraging the resources and networks of LBTC and LSU.
5. LSU AgCenter Food Incubator (LAFI). LAFI (2015) helps food entrepreneurs survive and grow their start-up businesses by providing resources, business support, and access to scientists and services within the Food Science and Nutrition Department.
6. Specialty Incubators. In addition to LAFI, there are other specialty incubators such as Louisiana Emerging Technology Center (LETC, 2015), focusing on the biotechnology, life sciences, agricultural and environmental industries; and Pennington Biotech Initiative (PBTI, 2015), focusing on impacting patient health and care through technological advances.

7. LSU Innovation Park (LIP). LIP (2015) is a collaborative community offering businesses and entrepreneurs to develop, foster, and enhance innovation utilizing vast networked resources of LSU.
8. International Programs at LSU (IP). IP (2015) leverages the networked resources of LSU international students from international markets. IP also provides the linkages to the international business programs in the world.

Louisiana also has developed some useful entrepreneurial and sustainable models in saving Louisiana's coastal wetlands. For example, alligators are common in wetland and Louisiana accounts for 80 percent of American alligators' production (Millar, 2012). Entrepreneurial opportunities related to Louisiana alligators include egg hunting, incubation, farming and its supply chain, skin processing, high-end merchandise production (e.g., purses, belts, watchstraps, and shoes), social media marketing for meat consumption, and recycling. To make the alligator economy environmentally sustainable, Louisiana State developed rules and regulations in the alligator industry. This balanced approach helps save Louisiana's coastal wetlands (Lewis and Lallo, 2013). In addition, documenting and filming the role each species plays in developing healthy wetland ecology is itself a fertile ground for creative digital media business cultivation. This is to take advantage of Louisiana's growing film industry to benefit K-16 students in sustainability education including enhancement of environmental science education for all local schools; eco-tourism and economic development; enhancing options for Louisiana tourism and providing opportunities for small business development in conjunction with LBTC and SDMI.

These resource inputs, along with wetland entrepreneurs and networked resources of research organizations, form docility-based learning communities with communication standards and procedures for information sharing. The communities have docile individuals who are willing to share their networked resources to help wetland entrepreneurs survive and grow the businesses. For example, LBTC has a strong advisory board which provides advices and resources to empower entrepreneurs to succeed. The docile individuals and their behaviors cultivate a culture of docility and entrepreneurial leaders. For example, SEI's Entrepreneurship Fellows Program helps produce future leaders of entrepreneurship. Both docile culture and entrepreneurial leadership are vital elements of our organizational resources. The individual resources and organizational resources allow us to tap into the established wetland networks and adapt to changing environmental factors. For examples, the International Programs at LSU help us tap into the diverse groups of international students and visitors when the local communities have the context specific needs of foreign countries; SEI's LSU 100 – Fastest Growing Tiger Businesses identify rising new business ventures responding to the market dynamics; and SEI's Entrepreneurship Bootcamp for Veterans help veterans address the uncertainty of life and find business opportunities after they retire from the services.

PROCESSES OF RESOURCE ORCHESTRATION: THE VALUE UNIT

The processes of resource orchestration in Figure 1, drawing upon the recent work of Sirmon et al. (2011), consists of structuring the resource inputs into portfolios, bundling resource portfolios into capabilities, and leveraging the capabilities to create value for shareholders. There are five steps for "Flying High" to develop and grow the wetland related business. The first step is to attend the useful informational and networking presentations. For example, interested wetland entrepreneurs with food products in mind, e.g., Marsh Dog (MD, 2015) and alligator meat entrepreneurs, may inquire the Food Incubator online and schedule to attend the quarterly informational presentation. The second step is to develop a short wetland entrepreneurship business plan. For example, a food-related wetland entrepreneur may submit a short business plan form and a cash flow analysis form. The Food Incubator, partnering with Louisiana Business and Technology Center, can help the entrepreneur develop a professional business plan. The third step is to start the business within a related incubator. For example, a food-related wetland startup at the Food Incubator is guided through each step of food production, including permitting, licensing, safety, product liability insurance, accessing a commercial kitchen with dry, cold and pallet storage, specialized processing equipment, analytical and testing services by the food scientist on staff. The fourth step is to bring products to local markets. For example, the Food Incubator helps food-related wetland entrepreneurs bring their culinary products to store shelves in the food markets and network with other tenants to grow the business.

Developing interactive, value-added coaching/mentoring procedures for advice giving and taking through accessible social channels with five steps of "Flying High" and five steps of "Landing Soft":

- "Flying High" Steps
    - Step 1: Attend useful informational and networking presentations and events
    - Step 2: Develop a short wetland entrepreneurship business plan
    - Step 3: Start the business within a related incubator
    - Step 4: Bring products to local markets
    - Step 5: Grow the local and domestic business, e.g. though franchising
- "Landing Soft" Steps
    - Step 1: Attend going global workshops and networking
    - Step 2: Attend international trade and business-matching trips for market access
    - Step 3: Develop a business plan with yearly strategies for going abroad
    - Step 4: Expand or export the products or services through soft landings networks
    - Step 5: Survive and thrive in the global market, e.g. through international franchising

The most important step of "Flying High", Step 5, is to grow the business in the local and domestic market. For example, if a food-related wetland entrepreneur succeeds in the local market, the LBTC can help develop a strategic plan to grow in the regional and national markets through established food networks such as Whole Food (NLAFI, 2014). Franchising is another effective approach to growth (Chen, 2018). Franchising is "a business opportunity by which the owner (producer or distributor) of a service or a trademarked product grants exclusive rights to an individual for the local distribution and/or sale of the service or product, and in return receives a payment or royalty and conformance to quality standards. The individual or business granting the business rights is called the *franchisor*, and the individual or business granted the right to operate in accordance with the chosen method to produce or sell the product or service is called the *franchisee*." (Justis and Judd, 2002, pp. 1-3) Developing a good relationship between the franchisor and the franchisee is the key for a successful franchise (Justis and Judd, 2002). The step-by-step process of "Flying High" helps the entrepreneur develop relationship-building skills, which are vital for growing the business through franchising.

There are also five steps for "Landing Soft" in Figure 1. The purpose is to help the soft landings companies identify business opportunities and succeed in accessing the new market with least risks and costs (Chen, et al., 2010). The first step is to attend going global workshops. There are several organizations specializing in helping Louisiana companies go abroad, e.g., U.S. Commercial Service in

New Orleans, World Trade Center in New Orleans, and Port of New Orleans. For food-related wetland entrepreneurs, attending food export education workshops will help them understand how to bring the businesses abroad. The second step is to attend international trade and market access trips. U.S. Commercial Service in New Orleans constantly takes SMEs in Louisiana to visit foreign countries with market access opportunities. The LBTC assists interested Louisiana SMEs in the process. The third step is to develop a business plan for going abroad. For example, the U.S. Commercial Service has a one-on-one gold key matching service for SMEs interested in going abroad (USCS, 2009). The fourth step is to expand or export the products/services through soft landings networks. The LBTC is one of thirty-one international soft landings incubators designated by International Business Innovation Association (InBIA, 2015). For food-related wetland entrepreneurs, Rutgers Food Innovation Center, one of the designated international soft ladings incubator, has the outreach capacity to reach food businesses throughout the world (RFIC, 2014).

The most important step of "Landing Soft", Step 5, is to survive and grow in the global market. An objective of soft landings is to help the soft landings company secure the market access in the new country, a key success factor of soft landings (Mencin & Erikson 2009). Another objective is to provide revenue generation services for the soft landings company, another key success factor of soft landings (Mencin & Erikson 2009). It takes typically two or three years for the SMEs to survive and grow in the new foreign market through well-planed international soft landings. International franchising is an effective approach to global growth (Chen, 2016). Global markets, such as BRICS (Brazil, Russia, India, China, and South Africa) commonly known as the emerging markets, have growing opportunities for international franchising. Consider the following statistics estimated by the U.S. Commercial Service: Brazil has 1,643 franchising chains and 79,988 franchising units, ranking the Brazilian franchising market as the 6th largest in the world (in number of units) and the 4th largest (in number of franchise chains) (U.S. Commercial Service, 2011). Russia formally adopted franchise legislation in 1994 and the number of franchise operations has grown steadily ever since (U.S. Commercial Service, 2012). India has over 70 international franchise operations successfully operating with an investment of $1.1 billion and sales turnover of $2.7 billion. Franchising is considered to be the second fastest-growing industry in India with an impressive growth rate of 30% (U.S. Commercial Service, 2009). China has over 2,600 brands with 200,000 franchised retail stores in over 80 sectors and is now the largest franchise market in the world (U.S. Commercial Service, 2008). South Africa has 531 franchise systems with 28,620 units contributing 12.5% to the country's GDP in 2008 (U.S. Commercial Service, 2010).

OUTPUTS: THE FILTER

The outputs in Figure 1 focuses on cultivating storytellers and empowering them to tell their stories in each of the five steps of "Flying High" and "Landing Soft" through networking, team building, and coaching. We tell the stories and feature the storytellers via website news, monthly newsletters, social media platforms such as Facebook and Twitter, and annual top-performer awards to gain bandwagon effects.

Those storytellers and the networked resources behind the success stories become the new resources to the resource inputs. In his study of distributed cognition, Secchi (2010) reported that pro-social behaviors, such as philanthropy and social responsibility, are the byproducts of docility; and the more people are docile, the stronger the intensity of the byproducts. Our experiences confirm these. The success stories and storytellers, derived from the resource inputs of docility-based communities, are added as feedbacks to the communities so that the resource inputs are enlarged and expanded. The arrows in Table 1 show this feedback loop in accordance with continuous learning and improving performance principles. For example, food-related wetland startups cultivated from the Food Incubator generate individual benefits, organizational benefits, and societal benefits as the outputs. Individual benefits include the startup companies and the employees they hire. Organizational benefits include creating partnership opportunities for clients at the incubators to produce new innovation and new technology in food innovation and production. Societal benefits include creating wealth and increasing economic growth for the local communities.

INTERACTIVE COMMUNICATION SYSTEMS AND TECHNOLOGIES

Shown on the right of Figure 1 are the communication systems and technologies which empower the interactive platform. They consist of five layers: empowerment and collaboration, business intelligence, working knowledge management, high business-value venturing strategy, and strategy execution. They are adapted from Gates' Digital Nervous Systems (1999), communication systems and technologies which can detect the changes in the environment and empower the decision makers to response timely to the changes.

The Empowerment and Collaboration layer is used to support the participants in sharing their networked resources to help wetland entrepreneurs to identify business opportunities in addressing the needs in wetland recovery and restoration. For example, there are effective solutions grounded with science-based, engineering-design, and architect-sound principles to mitigate wetland losses. Exemplar solutions are marsh creations, sediment diversion, surge barriers, and shoreline restorations. Technology tools are needed to develop the solutions such as The Nature Conservancy Climate Wizard; Climate Change Vulnerability Assessment Tool for Coastal Habitats; NOAA Sea Level Rise Viewer; Social Vulnerability Index Mapping Dashboard; Climate Change Adaptation Toolkit: Wetland Management, Creation, and Restoration; EPA's Climate Resilience Evaluation and Awareness Tool; Coastal Resilience Decision-Support Tool; Coastal Emergency Risks Assessment; International River Interface Cooperative Software; and Microsoft AI for Earth. Detailed description of each technology tool and its web address are shown in Table 2.

The Business Intelligence layer consists of streaming of data from various sources and formats (such as numeric, text, image, voice, and video), data warehousing of data streams, data marts of the streamed data, data mining techniques used to convert volumes of data into reports and benchmarks for the participants to glean business intelligence. Both Empowerment and Collaboration and Business Intelligence layers are enabled by communication systems and technologies with the objective of identifying the good, the bad, and the ugly in wetland management to gain the attention of the participants for opportunity identifications.

The Working Knowledge Management layer disseminates best practices in sustainability and wetland entrepreneurship. Examples in the Louisiana context include turning invasive species into business solutions, building wetland ecosystems using oysters, how alligators are helping save Louisiana's coastal wetlands, investing in wetland forests for carbon financing, cleaning up oil spills with floating islands, and scaling up the proven solutions through the investments of multinationals. The Working Knowledge Management layer also has Intranet-based systems consisting of the skills of coaching/influencing others and working knowledge profiles in wetland entrepreneurship are implemented for knowledge sharing

Table 2. Exemplar technology tools for wetland recovery and restoration

Tools	Descriptions	Web Sites
The Nature Conservancy Climate Wizard	A tool enabling people to access climate change information in the history and visualize the impacts anywhere on the earth	http://www.climatewizard.org/index.html
Climate Change Vulnerability Assessment Tool for Coastal Habitats	A decision support tool for integrating local data and regional climate change predictions to provide an assessment of potential habitat vulnerabilities	https://toolkit.climate.gov/tool/climate-change-vulnerability-assessment-tool-coastal-habitats-ccvatch
NOAA Sea Level Rise Viewer	A tool allowing people to view sea level rise and potential coastal flooding impact areas and relative depth	https://coast.noaa.gov/slr
Social Vulnerability Index Mapping Dashboard	A tool providing specific information to help public health officials and local planners better prepare communities to respond to emergency events	https://svi.cdc.gov/map.aspx?txtzipcode=70806&btnzipcode=Submit
Climate Change Adaptation Toolkit: Wetland Management, Creation, and Restoration	A tool for identifying and controlling invasive species in wetlands, maintaining and investing in stormwater wetlands, maintaining and investing in wetland restoration, and landscape level wetland assessments	https://www.watershedcouncil.org/wetland-management-creation-and-restoration-tools.html
EPA's Climate Resilience Evaluation and Awareness	A risk assessment tool helping utilities to adapt to extreme weather events by discovering, assessing, and sharing current and long-term weather conditions	https://www.epa.gov/crwu/creat-risk-assessment-application-water-utilities
Coastal Resilience Decision-Support	A science-based computer tool enabling the five U.S. Gulf states to visualize the coastal hazards risks and examine ways to increase coastal resilience through conservation and restoration activities	http://maps.coastalresilience.org/gulfmex/#
Coastal Emergency Risks Assessment	A useful storm surge guidance with interactive maps, enabled by a science-based computer model, for first responders, emergency managers and decision makers	https://cera.coastalrisk.live
International River Interface Cooperative Software	A tool for river flow, morphodynamics, and flood analysis calculations	https://i-ric.org/en/
Microsoft AI for Earth	Cloud and AI tools educating and training people to develop innovative, scalable solutions to address environmental challenges across the globe	https://www.microsoft.com/en-us/aiforearth

and learning for the five steps of "Flying High" and five steps of "Landing Soft". An online learning curriculum of working knowledge modules can also be deployed. The working knowledge management is empowered by social networks with the objective of perfecting the student entrepreneurs.

The High Value Strategy layer is developed with the goal of leveraging working knowledge profiles and applying effectual reasoning principles to create high value strategy for business solutions and its resilience. Examples of the High Value Strategy in the Louisiana context include wetland scientists use predictive modeling tools to develop new management strategies that have the potential to hinder the growth of invasive species, wetland engineers mitigate wetland losses while wetland conservation efforts, wetland architects use big data and analytics to research, evaluate, and design oyster reefs in the Gulf of Mexico off the coast of Louisiana, wetland ecologists analyze diverse data in an effort to study and protect the ecosystem of Louisiana alligators, wetland conservationists use analytical tools to census endangered wetland species by tracking footprints, and wetland administrators use visualization tools to find and address wetland treats.

Finally, the Strategy Execution layer makes sure that the new strategy is executed by empowering decision makers in their value-added collaborations via the responsive networks in wetland management and entrepreneurship. For example, nutria has significantly damaged Louisiana coastal wetlands, Marsh Dog turns nutria meat into dog treats and Righteous Fur designs clothing for the contemporary fashion market using nutria fur. Both businesses along with other entrepreneurial solutions help save the wetlands, however they are managed under Nutria.com managed by the Louisiana Department of Wildlife and Fisheries as the department is responsible for managing and protecting Louisiana's natural resources.

INNOVATIVE PERSPECTIVE

The platform allows students to participate in exploring and developing sustainable supply chains and international soft landings to maintain a healthy wetland to protect local costal land losses. Another objective of the platform is to cultivate wetland entrepreneurs to bring local solutions abroad for global impact. A useful model for accomplishing this objective may be described as an airplane with the "Flying High, Landing Soft" platform depicted in Figure 1 as the main body of the plane and "other incubators" and "new partners" serving as its two wings. The airplane, powered by the collaborative entities, has the focus of environmental scanning and sensing. The plane's elevation, distance, and direction are controlled by the following six strategies adapted from the process-driven business strategies of Hammer (1996). Figure 2 summarizes the intrinsic essence of the six strategies as it is applied to the "Flying High, Landing Soft" platform. The strategies allow the platform to excel in the current wetland entrepreneurship *vertically* by providing an excellent service to current shareholders and expand the services *horizontally* to new partners, other incubators, and new products/services through process diversification.

Intensification refers to empowering the platform with emerging technologies, such as mobile cloud learning practices, to help cultivate global wetland entrepreneurs more efficiently and effectively. The intensified platform is entitled Platform 1.0. Mobile cloud learning, a combination of both mobile learning and cloud computing, has four types of learner-centered scenarios (Park, 2011): high transactional distance socialized mobile-learning, high transactional distance individualized mobile-learning, low transactional distance socialized mobile-learning, and low transactional distance individualized mobile-learning. Based on our experiences at LSU incubators, we found that the fourth type of mobile learning, low transactional distance individualized mobile-learning, is the most useful for the entrepreneurship platform discussed above. Specifically, such a mobile cloud learning platform will help us understand the learner's attributes; help the learner identify his/her passionate food niche; help the learner manage his/her time efficiently and effectively through the platform; interact and collaborate with online teachers to pursue pedagogical inquiry and creativity for the learner; and cultivate new storytellers and empower them to tell their success stories to inspire new online learners. The empowered platform can have more digitalized features in order to maximize interactivity and time-space independency between teachers and learners through the mobile Internet infrastructure and mobile devices such as smart phones, tablets.

Augmentation refers to expanding the platform of wetland entrepreneurship by including any business ventures which can add value to the platform. The augmented platform is entitled Platform 1.1. For example, alligator processors in Louisiana and neighboring southern states annually produce roughly 2 million pounds of waste, including alligator bones, connective tissues, and associated materials (LAM, 2007). Traditionally, the waste was discarded. Instead, a researcher in the in the LSU School of Nutrition and Food Sciences, where LSU AgCenter Food Incubator is located, has developed a method of extract-

Figure 2. Innovative perspectives on advancing the interactive platform for sustainable supply chains toward sustainable development goals

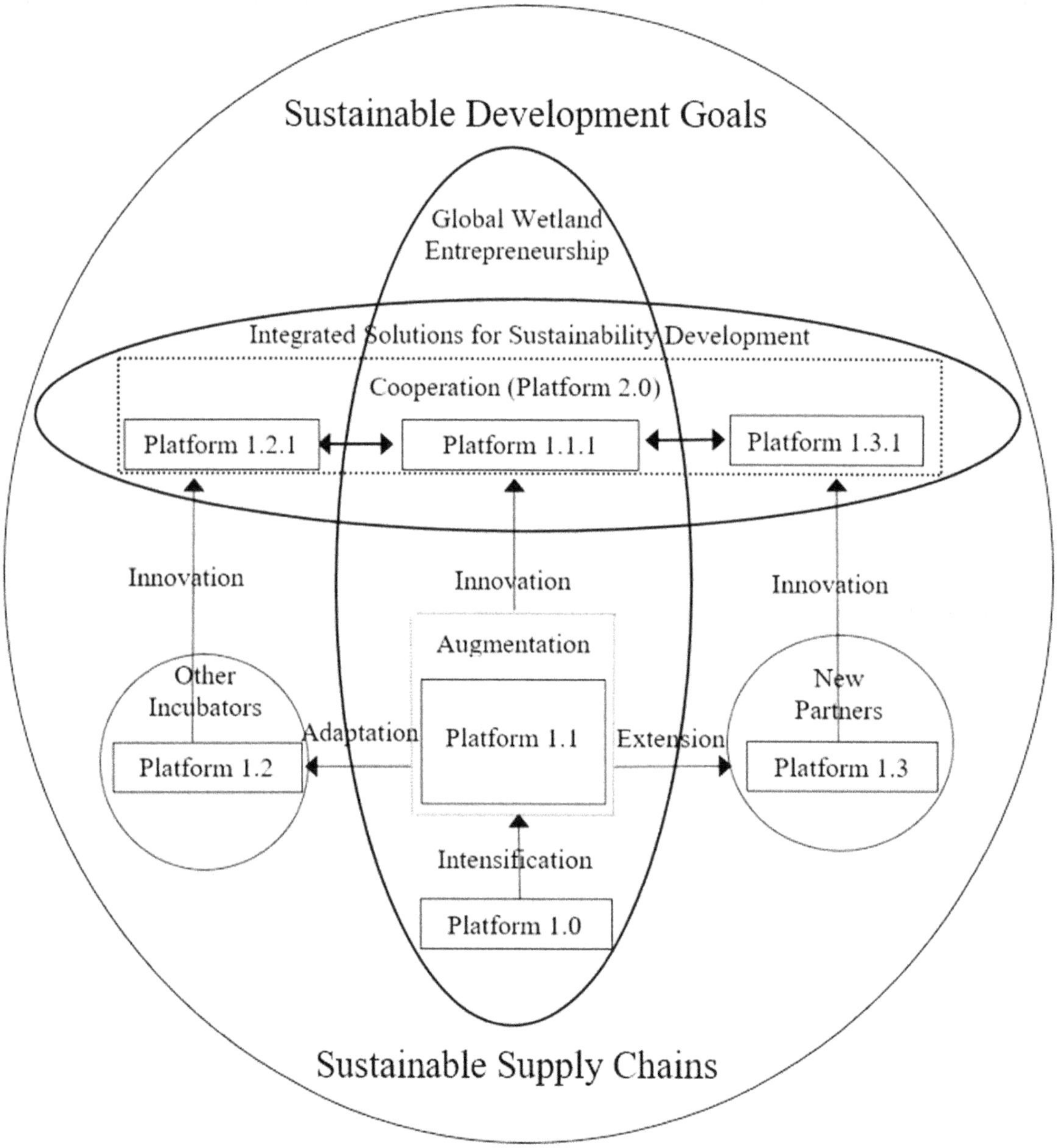

ing collagen from alligator bones which has many potential commercial uses in medicine, cosmetics and foods (LAM, 2007). Another Louisiana wetland species crawfish, 110 million pounds harvested each year and having an annual economic impact of $120 million (LS, 2015), also generates fairly large amount of wastes annually. Value-added products produced from the wastes of crawfish, along with catfish, are being created (LAM, 2012). Furthermore, a project turning invasive species in coastal wetlands into value-added natural protein food products for human consumption (SG, 2015). In turn, they will help save Louisiana's coastal wetlands (Lewis and Lallo, 2013).

Adaptation refers to taking the proven wetland entrepreneurship and incubation as a model for other academic disciplines to develop their entrepreneurship programs. The adapted platform is entitled Platform 1.2. Consider textile entrepreneurship using imperfect alligator skins as an example. With a grant from the Louisiana Alligator Advisory Council, LSU senior design students used imperfect alligator skins to create new designs for the Marsh on the Catwalk competition (Blanchard, 2014). With an abundance of these skins, rejected by large fashion houses, and skills they learn, students can create their fashion businesses and promote the Louisiana alligator products (Blanchard, 2014).

Extension refers to use the proven wetland entrepreneurship and incubation programs to work with new partners to help participative entrepreneurs to gain new market opportunities. The extended platform is entitled Platform 1.3. The new market partners could be identified through the other thirty international soft landings incubators designated by International Business Innovation Association (InBIA, 2015). For example, for those food-related wetland entrepreneurs, Rutgers Food Innovation Center (RFIC, 2014), one of the designated international soft ladings incubator, could be a good partner for international markets.

Innovation refers to using the platform and its networked resources to engage in community outreach activities and create economic value. The participated platforms are entitled Platform 1.1.1, Platform 1.2.1, and Platform 1.3.1. For example, working with local communities, we may hold workshops on Wetland Ecology and Wetland Entrepreneurship allowing wetland ecologists, wetland entrepreneurs, K-16 educators and students to network and develop new venture opportunities. In the long-term, we may develop eco-tourism, living classroom of wetland ecology and living laboratory of wetland entrepreneurship.

Cooperation refers to enriching resource inputs in Figure 1 with sharable resources of other incubators and new partners to advance the platform of cultivating global wetland entrepreneurs. The cooperated platform is entitled Platform 2.0. For example, Office of Research & Economic Development at LSU has seven focal areas as its strategic plan, including costal sustainability & environment and individual behavior & community context (LSUOREC, 2012). A strategic goal is to advance and support the economic and work force development in Louisiana (LSU, 2014). There are cooperative opportunities for the incubators in the Resource Inputs, such as LBTC and LSU AgCenter Food Incubator, to work with the focal areas to enrich and advance the platform.

CONCLUSION

Although dealing with the "hidden dynamics" of diplomacy, espionage, and geopolitics has been a major challenge in international business, the rise of climate change and global terrorism has brought nations together to seek for collaborative solutions for global sustainability. Coopetition is becoming a necessary strategy to survive and thrive in the 21st century. In this paper, we proposed a two-stage process of coopetition sustainable supply chains and international soft landings to mitigate the risks associated with the "hidden dynamics" of global hypercompetition and the extreme politically uncertain environment. Using the two-stage process, we developed an "Flying High, Landing Soft" platform of wetland entrepreneurship to address the growing global problem of wetland losses. This is an interactive platform empowered by communication systems and technologies grounded in Gates' Digital Nervous Systems.

The platform allows students to participate in exploring and developing businesses to help maintain a healthy wetland and have positive impact on public health. For example, alligators are common in wetland and Louisiana accounts for 80 percent of American alligators' production. When looking at the traveling of alligators in the global economy, there are various business opportunities for entrepreneurial students, including egg hunting, incubation, farming and its supply chain, skin processing and high-end merchandise production, social media marketing for meat consumption, and recycling. In addition, documenting and filming the role each species plays in developing healthy wetland ecology is itself a fertile ground for creative digital media business cultivation. This is to take advantage of Louisiana's growing film industry to benefit K-16 students in sustainability education including enhancement of environmental science education for all local schools; eco-tourism and economic development; enhancing options for Louisiana tourism and providing opportunities for small business development.

Finally, the "Flying High, Landing Soft" platform can be used to advance the wetland entrepreneurship globally. Soft landings is a process to help a company from one country land softly – without crashing – into the market of another country through a designated incubator. The purpose is to help the soft landings companies reduce risks/costs and find business opportunities to succeed in the new market. The platform enables the soft landings companies to reduce risks and costs, find business opportunities, and identify talented people for its new business venture. Moreover, the participating students are able to network with key decision makers at both the soft landings companies and the supply chain companies; engage in developing revenue generation models for the soft landings companies; earn income or internships; become the employees or partners of the soft landings companies; and create their businesses by leveraging the networked resources developed in the course of participation. This innovative entrepreneurial platform, useful to both industry and education fields, achieves a win-win situation for everyone involved.

REFERENCES

Abrams, R. (2015). *USA Today Recognizes Tierra Resources in Leading the Charge to Prevent the Next Katrina*. Retrieved September 29, 2019, from https://tierraresourcesllc.com/usa-today-recognizes-tierra-resources-leading-charge-prevent-next-katrina

Akash, D. (2015). *Impact of international terrorism on commodity markets*. LAP Lambert Academic Publishing.

AP. (2015). *Alligator program*. Retrieved September 29, 2019, from https://www.wlf.louisiana.gov/wildlife/alligator-program

Barney, J. B. (1991). Firm resources and sustained competitive advantage. *Journal of Management, 17*(1), 99–120. doi:10.1177/014920639101700108

Baurick, T. (2018). The Drowing coast. *New York Times*. Retrieved September 25, 2019, from https://www.nytimes.com/interactive/2018/02/24/us/louisiana-wetlands-insects.html

Blanchard, T. (2014). Runway show features student designers. *LSU Ag Center News*. Retrieved May 21, 2018, from https://www.lsuagcenter.com/news_archive/2014/may/headline_news/Runway-show-features-student-designers.htm

Bourlakis, M. A., & Weightman, P. W. (Eds.). (2004). *Food supply chain management*. Blackwell Publication.

Busse, C., Schleper, M. C., Niu, M., & Wagner, S. M. (2016). Supplier development for sustainability: Contextual barriers in global supply chains. *International Journal of Physical Distribution & Logistics Management, 46*(5), 442–468. doi:10.1108/IJPDLM-12-2015-0300

Castaldo, S., & Dagnino, B. (Eds.). (2010). *Coopetition: winning strategies for the 21st century*. Edward Elgar Publishing.

CBEEE. (2015). *Can't Beat 'EM, Eat 'EM*. Retrieved September 25, 2019, from http://www.cantbeatemeatem.us

Chen, Y., Watson, E., & Azevedo, R. F. L. (2011). Soft landings curriculum of US-China entrepreneurship. *China Currents, 10*(2).

Chen, Y., Watson, E., Liu, C., Cornachione, E., & Wu, S. (2010). *Soft landing curriculum of entrepreneurship in emerging markets*. In Academy of International Business 2010 Annual Meeting, Rio de Janeiro, Brazil.

Chen, Y. S. (2016). E-Commerce in International Franchising. In Encyclopedia of E-Commerce Development, Implementation, and Management (pp. 578-590). IGI Global. doi:10.4018/978-1-4666-9787-4.ch042

Chen, Y. S. (2018). Utilizing Information Science and Technology in Franchise Organizations. In *Encyclopedia of Information Science and Technology* (4th ed., pp. 4822–4835). IGI Global. doi:10.4018/978-1-5225-2255-3.ch418

Chen, Y. S., Davutyan, N., & Ersoy, İ. (2016). A "Flying High, Landing Soft" platform for supplier diversity. In *Handbook of Research on Global Supply Chain Management* (pp. 72–84). IGI Global. doi:10.4018/978-1-4666-9639-6.ch004

Chen, Y. S., Watson, E., & Azevedo, R. F. (2014). E-strategy and soft landings for franchising in emerging markets. In *Trends in E-Business, E-Services, and E-Commerce: Impact of Technology on Goods, Services, and Business Transactions* (pp. 148–159). IGI Global. doi:10.4018/978-1-4666-4510-3.ch008

Cooperhouse, L., & Surgi, M. L. (2014). *Maximizing your food incubation program*. In National Business Incubation Association 28th Conference, New Orleans, LA.

Craft, C. (2015). *Creating and restoring wetlands: From theory to practice*. Elsevier.

Dagnino, G. B., & Padula, G. (2009). Coopetition strategy. *Coopetition strategy: Theory, experiments and cases*, 25-43.

Dow. (2015). *Going Back to School: Ambassadors for Science*. Retrieved September 29, 2019, from https://corporate.dow.com/en-us/science-and-sustainability/2025-goals/earth-day/stem-outreach

Eastham, J., Sharples, L., & Ball, S. (Eds.). (2007). *Food supply chain management*. Taylor & Francis. doi:10.4324/9780080506678

Edward, J., & Policy, P. (2012). Community food hubs. *The Ralph W. Voorhees Center for Civic Engagement at Rutgers*. Retrieved May 21, 2018, from https://rwv.rutgers.edu/wp-content/uploads/2013/07/FoodHubFinalReport.pdf

Folger, T. (2013). Rising seas. *National Geographic, 224*(3), 30–59.

Freudenburg, W. R., & Gramling, R. (2012). *Blowout in the Gulf: The BP oil spill disaster and the future of energy in America*. MIT.

Gates, W. (1999). *Business @ the speed of thought*. Warner Books. doi:10.1111/1467-8616.00097

Godfray, H. C. J., & Garnett, T. (2014). Food security and sustainable intensification. *Philosophical Transactions of the Royal Society of London. Series B, Biological Sciences*, *369*(1639), 20120273. doi:10.1098/rstb.2012.0273 PMID:24535385

GoMRI. (2015). *Gulf of Mexico research initiative*. Retrieved May 21, 2018, from http://gulfresearchinitiative.org

Hall, J., & Matos, S. (2010). Incorporating impoverished communities in sustainable supply chains. *International Journal of Physical Distribution & Logistics Management*, *40*(1/2), 124–147. doi:10.1108/09600031011020368

Hammer, M. (1996). *Beyond Reengineer*. Harper Business.

Hitt, M. A., Ireland, R. D., Sirmon, D. G., & Trahms, C. A. (2011). Strategic Entrepreneurship: Creating Value for Individuals, Organizations, and Society. *The Academy of Management Perspectives*, *25*(2), 57–75.

InBIA. (2015). *International business innovation association*. Retrieved May 21, 2018, from https://www.inbia.org/resources/directories/soft-landings-members

IP. (2015). *International programs at LSU*. Retrieved May 21, 2018, from http://international.lsu.edu

Jaggernath, R., & Khan, Z. (2015). Green supply chain management. *World Journal of Entrepreneurship, Management and Sustainable Development*, *11*(1), 37–47. doi:10.1108/WJEMSD-06-2014-0018

Justis, R. T., & Judd, R. J. (2002). *Franchising*. DAME Publishing.

Khanduja, J. (2013). *Shared tastes* (Doctoral dissertation). Columbia University, New York, NY.

Kline, C., Shah, N., & Rubright, H. (2014). Applying the positive theory of social entrepreneurship to understand food entrepreneurs and their operations. *Tourism Planning & Development*, *11*(3), 1–13. doi:10.1080/21568316.2014.890126

LAFI. (2015). *LSU AgCenter food incubator*. Retrieved May 21, 2018, from https://www.lsuagcenter.com/en/our_offices/departments/food_science/extension_outreach/incubator

LAM. (2007). Alligator collagen: New source for medical, cosmetic uses. *Louisiana Agriculture Magazine*. Retrieved May 21, 2018, from https://www.lsuagcenter.com/en/communications/publications/agmag/archive/2007/winter/alligator+collagen+new+source+for+medical+cosmetic+uses.htm

LAM. (2012). Value-added from crawfish and catfish. *Louisiana Agriculture Magazine*. Retrieved May 21, 2018, from https://www.lsuagcenter.com/en/communications/publications/agmag/Archive/2002/Fall/Valueadded+from+Crawfish+and+Catfish.htm

LBTC. (2015). *Louisiana business and technology center*. Retrieved May 21, 2018, from http://www.lbtc.lsu.edu

LCWRP. (1993). *Louisiana coastal wetlands restoration plan*. Retrieved May 21, 2018, from https://lacoast.gov/reports/cwcrp/1993/1993lcwrp-1executive_summary.pdf

LDWF. (2007). Nutria control program. *The Louisiana Department of Wildlife and Fisheries*. Retrieved May 21, 2018, from https://nutria.com/site9.php

Leal Filho, W., & Kovaleva, M. (2015). *Food waste and sustainable food waste management in the Baltic Sea Region*. Springer. doi:10.1007/978-3-319-10906-0

LED. (2018). International commerce. *Louisiana Economic Development*. Retrieved May 24, 2018, from https://www.opportunitylouisiana.com/international-commerce

Lee, A., & Wall, G. (2012). *Food clusters: Towards a creative rural economy*. Martin Prosperity Institute, Rotman School of Management, University of Toronto.

LETC. (2015). *Louisiana emerging technology center*. Retrieved May 21, 2018, from https://laetc.com/the-la-emerging-tech-center

Lewis, S., & Lallo, E. (2013). *Alligators are helping save Louisiana's coastal wetlands*. Retrieved May 21, 2018, from https://gulfseafoodnews.com/2013/09/18/the-business-of-alligator-part-four-the-thick-skinned-environmentalists

LIP. (2015). *LSU innovation park*. Retrieved May 21, 2018, from https://sites01.lsu.edu/wp/innovationpark

LS. (2015). *The economy of seafood in Louisiana*. Retrieved May 21, 2018, from https://www.louisianaseafood.com/industry

LSU. (2014). *Gov. Jindal announces funding hike for higher education & new workforce incentive fund*. Retrieved May 21, 2018, from https://sites01.lsu.edu/wp/lovepurple/2014/01/21/gov-jindal-announces-funding-hike-for-higher-education-new-workforce-incentive-fund

LSUOREC. (2012). The constant pursuit of discovery. *LSU office of research & economic development*. Retrieved May 21, 2018, from https://sites01.lsu.edu/wp/lsu2015/files/2013/04/20130409ResearhMtgOREDStrategicPlan.pdf

MC. (2015). *Mobile classroom*. Retrieved May 21, 2018, from https://sites01.lsu.edu/wp/innovationpark/about-lbtc/mobile-classroom

McFadden, D. T., & Marshall, M. I. (2014). Local food systems and interactions with entrepreneurship. *Journal of Food Distribution Research*, *45*(3).

MD. (2015). *Marsh dog*. Retrieved May 21, 2018, from www.marshdog.com

ME. (2015). *Martin ecosystems*. Retrieved May 21, 2018, from www.martinecosystems.com

Mencin, O., & Erikson, C. (2009). Silicon valley's US market access center: The incubator as a soft landing zone. *International Journal of Entrepreneurship and Innovation*, *10*(3), 233–241. doi:10.5367/000000009789067842

Millar, D. (2012). Louisiana alligator industry stages a comeback, thanks to growing demand for skins. *The Times-Picayune*. Retrieved May 21, 2018, from https://www.nola.com/business/index.ssf/2012/06/louisianas_alligator_industry.html

Miller, A., Roberts, S., & LaPoe, V. (2014). *Oil and water: Media lessons from hurricane Katrina and the deepwater hrizon disaster*. Univ. Press of Mississippi.

Mitchell, B. (2015). Making cities resilient: New Orleans/Gothenburg city-to-city exchange. *Stephenson Disaster Management Institute.* Retrieved May 21, 2018, from http://www.sdmi-resilient-cities.com

Niimi, M., Wakes, S. J., & McGuire, M. H. (2014). Design for sustainability: Addressing food waste behaviour through social practice approaches. *Food Design on the Edge*, 29.

NLAFI. (2014). To market, to market…. *News - LSU AgCenter Food Incubator*. Retrieved May 21, 2018, from https://www.lsu.edu/departments/nfs/outreach/incubator.htm

NOGE. (2015). *Making cities resilient: New Orleans / Gothenburg Exchange*. Retrieved May 21, 2018, from http://sdmi-resilient-cities.com

OB. (2015). *Oyster Bed*. Retrieved September 29, 2019, from https://www.theoysterbed.com

OREA. (2015). *ORA Estuaries*. Retrieved September 29, 2019, from https://www.oraestuaries.com

Paloviita, A., & Järvelä, M. (2015). *Climate change adaptation and food supply chain management*. Routledge Advances in Climate Change Research. doi:10.4324/9781315757728

Park, Y. (2011). A pedagogical framework for mobile learning: Categorizing educational applications of mobile technologies into four types. *International Review of Research in Open and Distance Learning*, *12*(2), 78–102. doi:10.19173/irrodl.v12i2.791

Parker, G. G., Van Alstyne, M. W., & Choudary, S. P. (2016). *Platform Revolution: How Networked Markets Are Transforming the Economy and How to Make Them Work for You*. WW Norton & Company.

PBTI. (2013). *Pennington biotech initiative*. Retrieved May 21, 2018, from https://www.businessreport.com/business/budding-ecosystem

Peters, N. J., Hofstetter, J. S., & Hoffmann, V. H. (2011). Institutional entrepreneurship capabilities for interorganizational sustainable supply chain strategies. *International Journal of Logistics Management*, *22*(1), 52–86. doi:10.1108/09574091111127552

Pilkey, O. H., & Young, R. (2011). *The rising seas*. Island Press.

Pivato, S., Misani, N., & Tencati, A. (2008). The impact of corporate social responsibility on consumer trust: The case of organic food. *Business Ethics (Oxford, England)*, *17*(1), 3–12. doi:10.1111/j.1467-8608.2008.00515.x

PN. (2016). *Water Challenge Winner Recycles Water Bottles to Protect Coasts & Reduce Pollutants: Martin Ecosystems wins $15,000 at New Orleans Entrepreneur Week*. Retrieved September 29, 2019, from http://gopropeller.org/news/1583

PwC. (2014). Sustainable supply chains: Making value the priority. *PwC*. Retrieved May 21, 2018, from https://www.pwc.com/us/en/operations-management/publications/sustainable-supply-chain.jhtml

RF. (2015). *Righteous fur*. Retrieved May 21, 2018, from http://www.righteousfur.com

RFIC. (2015). *Rutgers food innovation center*. Retrieved May 21, 2018, from http://foodinnovation.rutgers.edu

Rivlin, G. (2015). *Katrina: After the flood*. Simon & Schuster.

Rodriguez, G. R., & Brebbia, C. A. (2015). *Coastal cities and their sustainable future*. Wit Pr/Computational Mechanics.

RT. (2015). *Rising Tide: Louisiana kids help build Floating Islands Restoration Project*. Retrieved September 29, 2019, from https://risingtide.joincca.org/news3

Salinger, J. H. (2013). *Economic development policies through business incubation and co-working: A study of San Francisco and New York City*. Academic Press.

SDMI. (2015). *Stephenson disaster management institute*. Retrieved May 21, 2018, from http://sdmi.lsu.edu

Secchi, D. (2010). *Extendable rationality: Understanding decision making in oganizations*. Springer.

SEI. (2015). *Stephenson entrepreneurship institute*. Retrieved May 21, 2018, from https://www.business.lsu.edu/Stephenson-Entrepreneurship-Institute/Pages/SEI.aspx

SG. (2015). *The Silverfin™ Group - Our Alternative Solution to the Nation's Asian Carp Crisis*. Retrieved May 21, 2018, from http://cantbeatemeatem.com

Shackleton, S. (2014). Impacts of climate change on food availability: Non-Timber forest products. In *Global Environmental Change* (pp. 695–700). Springer Netherlands. doi:10.1007/978-94-007-5784-4_117

SI. (2015). *LSU student incubator*. Retrieved May 21, 2018, from https://sites01.lsu.edu/wp/innovation-park/about-lbtc/lsu-student-incubator

Simon, H. A. (1993). Altruism and Economics. *The American Economic Review*, *83*(2), 156–161.

Sirmon, D. G., Hitt, M. A., Ireland, R. D., & Gilbert, B. A. (2011). Resource orchestration to create competitive advantage breadth, depth, and life cycle effects. *Journal of Management*, *37*(5), 1390–1412. doi:10.1177/0149206310385695

SL. (2015). *SHORE|LINKS*. Retrieved May 21, 2018, from http://www.shore-links.com

Tai, A. P., Martin, M. V., & Heald, C. L. (2014). Threat to future global food security from climate change and ozone air pollution. *Nature Climate Change*, *4*(9), 817–821. doi:10.1038/nclimate2317

Tarr, N. W. (2011). Food entrepreneurs and food safety regulation. *J. Food L. & Pol'y*, *7*, 35.

TR. (2015). *Tierra resources*. Retrieved September 29, 2019, from https://tierraresourcesllc.com

Tripathi, R., & Agarwal, S. (2014). An empirical study of marketing for guava and its sub-products by farmers in Allahabad: An approach towards agripreneurship through food processing units. *International Journal of Business and Globalisation*, *13*(1), 69–75. doi:10.1504/IJBG.2014.063395

UNFCCC. (2015). Closing address by Ms. Christiana Figueres, *United Nations Framework Convention on Climate Change Executive Secretary*. Retrieved May 21, 2018, from https://unfccc.int/files/press/statements/application/pdf/20151212_closingplenary.pdf

UNISDR. (2018). *Making cities resilient: My city is getting ready*. United Nations International Strategy for Disaster Reduction. Retrieved May 24, 2018, from https://www.unisdr.org/campaign/resilientcities

UNSC. (2015). Security Council calls for eradicating ISIL safe havens in Syria and Iraq. *UN News Centre*. Retrieved May 21, 2018, from http://www.un.org/apps/news/story.asp?NewsID=52623#.VnYWI1Xnbm4

UNSDG. (2015). *Sustainable development goals: 17 goals to transform our world*. Retrieved May 21, 2018 from https://www.un.org/sustainabledevelopment/sustainable-development-goals

U.S. Commercial Service. (2009). *India: Legal and regulatory aspects of doing business in the franchising industry*. Retrieved May 21, 2018 from https://www.franchise.org/uploadedFiles/Franchise_Industry/International_Development/Country_Profiles/India_Franchise%20Sector_July%202009.pdf

U.S. Commercial Service. (2010). *Doing business in South Africa - 2010 country commercial guide for U.S. companies*. Retrieved May 21, 2018 from https://www.franchise.org/uploadedFiles/Franchise_Industry/International_Development/Country_Profiles/South%20Africa_2010.pdf

U.S. Commercial Service. (2011). *Brazil: Franchising*. Retrieved May 21, 2018 from https://www.franchise.org/uploadedFiles/FranchisingReportBrazilCCGFeb2011.doc

U.S. Commercial Service. (2012). *Doing business in Russia: 2012 country commercial guide for U.S. companies*. Retrieved May 21, 2018 from https://photos.state.gov/libraries/russia/231771/PDFs/2012-ccg-russia.pdf

USCS. (2009). *The gold key matching service*. Retrieved May 21, 2018 from http://www.export.gov/salesandmarketing/eg_main_018195.asp

USDS. (2014). *The United States and Pacific Alliance Advance Cooperation on entrepreneurship*. Retrieved May 21, 2018 from https://www.state.gov/r/pa/prs/ps/2014/12/235517.htm

WIotG. (2015). *The water institute of the Gulf*. Retrieved May 21, 2018 from https://thewaterinstitute.org/about/our-commitment

Woods, T. A. (2014). Innovation in agri-food clusters: Theory and case studies. *American Journal of Agricultural Economics*, *96*(2), aat082. doi:10.1093/ajae/aat082

Chapter 2
Student and Faculty Perceptions of Social Media Use and Relationships Inside and Outside the Higher Education Classroom

Julie A. Delello
https://orcid.org/0000-0002-4326-8096
The University of Texas at Tyler, USA

Kouider Mokhtari
The University of Texas at Tyler, USA

ABSTRACT

This study examined student and faculty perceptions of social media use inside and outside the classroom. Three hundred and ninety-six students and fifty faculty members at a regional university campus in the south central United States voluntarily completed an online survey soliciting quantitative and qualitative data about their perceptions of social media use. Results revealed important findings highlighting similarities, differences, and insights among student and faculty perceptions of social media use in the classroom, their views about whether social media use constitutes a distraction, and how each group views social media relationships in and out of the classroom. These findings are quite consistent with prior and emerging research about social media use and have implications for how institutions of higher education can explore meaningful ways of incorporating social media in the classroom with the goal of strengthening teaching and learning.

INTRODUCTION

New and emerging advances in social media and mobile technologies have allowed us to communicate with others almost anywhere and at any time. In fact, the New Media Consortium predicted that social media will be used as a platform for continuous sharing of information and collaboration in education

DOI: 10.4018/978-1-7998-3355-0.ch002

over the next five years (Adams Becker, Cummins, Davis, Freeman, Hall Giesinger, & Ananthanarayanan, 2017). According to a survey by the Pew Research Center, Americans build connections and share information with one another through the social media technologies Facebook, Youtube, and Twitter (Smith & Anderson, 2018). In 2019, there are predicted to be approximately 221 million Facebook, 200 million You Tube, and 69 million Twitter users in the United States (Clement, 2019). According to Clement (2019), there were also 203 million daily Snapchat users. These social media consumers are predicted to spend an average of 2 hours and 22 minutes per day networking socially (GlobalWebIndex, 2019).

Social media have also permeated institutions of higher education. According to the *NMC Horizon Report: 2015 Higher Education*, the use of social media by faculty members and students on campuses is expanding (Johnson, Adams Becker, Estrada, & Freeman, 2015). Research has suggested that more than half of college students are continuously connected to popular social networking sites (SNS) such as Facebook, YouTube, Snapchat, Instagram, and Twitter (Smith & Anderson, 2018). Mirroring the general population, researchers reported that 70% of faculty utilized SNS and of those, 55% used social media for managing their professional image (Seaman & Tinti-Kane, 2013). In another study, almost 85% of faculty noted having a Facebook account, two-thirds (67%) were on LinkedIn, and 50% used Twitter (Faculty Focus, 2011). However, other research has highlighted much lower percentages of faculty (50%) utilizing social media for instruction especially among those over age 55 (Kumi-Yeboah & Blankson, 2018).

Heiberger & Junco (2015) suggested that in order to improve educational outcomes, faculty should be using social media like Facebook and Twitter to engage students. But to what extent are faculty and students actually using social media in the classroom? The purpose of this study was to examine student and faculty perceptions regarding the potential benefits or detriments in the classroom. Do they view social media as beneficial for instruction? Or do they see them as a mere distraction? A related purpose of this research was to highlight how faculty and students felt about extending social media relationships outside the classroom. To date, little is known about the effects of teacher-student social media relationships in the higher education setting (Metzger, Finley, Ulbrich, McAuley, 2010; Hershkovitz & Forkosh-Baruch, 2013).

BACKGROUND

Social media technologies have the capability to change "the nature of the way we communicate, access information, connect with peers and colleagues, learn, and even socialize" (Johnson, Adams Becker, & Cummins, 2012, p. 6). The term social media is defined as "a group of Internet-based applications that build on the ideological and technological foundations of Web 2.0 and that allow the creation and exchange of user-generated content" (Kaplan & Haenlein, 2010, p.61). Web 2.0 refers to the second generation of the World Wide Web which promotes greater collaboration and sharing across the Internet through social media platforms such as Facebook, Pinterest, and Instagram which integrate visual tools with digital technologies (Delello & McWhorter, 2014). The evolution from Web 1.0 to 2.0 has allowed for more user participation and improved interaction and socialization (Eteokleous-Grigoriou & Ktoridou, 2014). Moreover, Web 3.0, combining the features of Web 1.0 and 2.0, will continue to evolve as a more intelligent web poised to allow users to co-construct and share additional information with one another.

Use of Social Media Among Young Adults

The Keiser Foundation (2013) reported that, "The continued expansion of high-speed home Internet access, the proliferation of television content available online, and the development of compelling new applications such as social networking and YouTube, have all contributed to the increase in the amount of media young people consume each day" (p. 3). Sites such as YouTube, Facebook, Instagram, Snapchat, Twitter, and Pinterest have continued to be most popular for adults 18 to 29 years of age (Perrin & Anderson, 2019). In fact, 80% of users, ages 18-29, use Snapchat and 76% of these users reported using Instagram daily with most using the SNS multiple times per day (Perrin & Anderson, 2019). More than half (51%) of the 18-24 year olds also reported that they would have a difficult time if they were unable to use social media (Smith & Anderson, 2018).

Use of Social Media among College Students

According to the New Media Consortium [NMC], the use of social media is positioned to change the face of education (Adams Becker, Cummins, Davis, Freeman, Hall Giesinger, &, Ananthanarayanan, 2017). Students will use SNS to embark on collaborative learning and "seamless sharing and communication" (Adams Becker, et al, 2017, p. 20). The Institute of Politics at Harvard Kennedy School (2017), reported that "Facebook (87%), Twitter (47%), Instagram (45%), Pinterest (37%), Snapchat (34%) and Tumblr (19%) are all more popular among college students than among young Americans who are not in, or never have attended, college" (para. 5). Quan-Haase and Young (2010) noted that majority of college students visited multiple social media sites daily to stay connected with their family and friends. This may be correlated to the upsurge in the use of Internet ready mobile devices as 97% of college students own a Smartphone (Neilson, 2016) and are connected to a digital device more than 141 hours per week (Refuel Agency, 2015).

In the United States, young adults spend close to 200 minutes per day on their mobile device and much of this time is devoted to using SNS (Clement, 2019). In fact, Wang, Chen, and Liang (2011) reported that college students post or respond to social media 6-8 hours per day and much of this takes place during school hours. Moreover, Junco & Cotton (2011) noted the typical college student sends an average of 96 text messages and receives nearly 104 text messages per day, and spends an average of 1.4 hours on Facebook.

Use of Social Media among College Faculty

A Pearson report, *How Today's Higher Education Faculty Use Social Media* suggested that almost 65% of faculty use social media for personal use while 45% of faculty used social media for their professional careers outside of teaching. Most of this use was with the social media platform Facebook with over half of faculty surveyed reporting using it monthly (Moran, Seaman, & Tinti-Kate, 2012). Additonally, some studies have shown that more than 80 percent of professors may be using social media as part of their teaching (Moran, Seaman, & Tinti-Kate, 2011). Research supports the notion that younger faculty use social media 10% more than those with more than 20 years of teaching experience; 30% of faculty have used social media to post content for students outside class, and 20% have asked students to comment or add additional posts online (Moran, et al., 2011). It is not surprising to also note that research has shown that faculty who teach online are somewhat more likely to visit social media sites than those who only teach face-to-face (F2F) courses (Tinti-Kane & Seaman, 2010). Additionally, an additional 30% of faculty do not use social media at all (Moran, et. al, 2011).

Use of Social Media in Teaching and Learning

Social media may be leveraged by faculty to create learning opportunities within the classroom (Delello, McWhorter, & Camp, 2015; Helvie-Mason, 2011; Junco, Hemberger, & Loken, 2010). For example, Irwin, Ball, Desbrow, and Leveritt (2012) reported that 78% of the students they surveyed felt that Facebook could be an effective tool for classroom instruction. More recently, another study reported that 75% of education students and 68% of business students noted that Facebook enhanced their course learning experience (Hall, McWhorter, & Delello, 2017). Additonally, Thalluri and Penman (2015) suggested that creating content specific discussions and reflections on Facebook are valuable approaches to using SNS for teaching and learning. Delello, McWhorter, & Camp (2015) conveyed that Facebook, used as part of a course, was beneficial in terms of engaging students and building community and personal meaning. Some researchers also reported that students perceived that using Pinterest supported both course engagement and motivation (Delello & McWhorter, 2014). Additionally, the use of Twitter has also been shown to promote active participation of educator communities, develop social presence, and reshape the way students learned (Delello and Consalvo, 2019; Baisley-Nodine, Ritzhaupt, & Antonenko, 2018).

Visual Social Media

Delello and McWhorter (2014) found that "A powerful social media trend is the blending of visual tools with innovative digital technologies such as Facebook, Twitter, Pinterest, and Instagram" (p. 371). This is consistent with prior reseach indicating that students spend 1-5 hours per day networking and even more on the weekends posting pictures and videos across social media sites (Knight-McCord, Cleary, Grant, Herron, Jumbo, Lacey… Emanuel, 2015). Videos, emojis, Bitmojis, GIFs, and memes are also quickly becoming part of a college students' social media landscape. Colleges are transforming their social media use to reach students. According to the New York Times, students coming to campus are from "a generation that rarely reads books or emails, breathes through social media, feels isolated and stressed but is crazy driven and wants to solve the world's problems (not just volunteer)" (Pappano, 2018, para. 4). These Generation Z students, ages 14 to 23, spend three hours or more a day using YouTube (Pearson, 2018). Moreover, Zmikly (2016) reported that the SNS Snapchat has become "impossible to ignore" as it has "a rapidly growing user base and content reaching 41 percent of all 18-to-34-year-olds in the United States" (para. 3).

In a survey conducted by the Harris Poll and commissioned by Tenor (2017), 36% of millennials ages 18-34 prefer to use visual expressions such as emojis and GIFs. In 2015, the word emoji, defined as an icon used to express an idea or emotion in electronic communication, was added to the Oxford Dictionary as word of the year (Oxford University Press, 2019). Billions of emojis have been shared across SNS and the 'face with tears of joy' emoji has been shared over two million times (Rothenberg, 2019; see Figure 1). Similar to an emoji, Bitmoji, owned by Snapchat, is an avatar that can be created, personalized, and shared across SNS (see Figure 2). A GIF (Graphical Interchange Format) is a short animated video or series of images that consistently loop. For example, Microsoft created a GIF of an angler fish to accompany their phrase "empowering every student and educator to achieve more" (see https://gph.is/2zzwhJa); it was uploaded into giphy (giphy.com) to allow users to share across SNS.

Figure 1. Face with Tears of Joy Emoji

Similar to a GIF, a meme is a virally transmitted photograph or image. Facebook, Twitter, and Tumblr have all incorporated GIFs or memes within their social media pages.

Students have reported that the use of such visuals communicate their feelings better than words (Tenor, 2017). Moreover, some educators are finding the use of visual social media tools like Bitmojis and GIFs are positively reinforcing to their students (King, 2019). Additionally, 37% of educators surveyed incorporated visual images like memes, emojis, and GIFs to help facilitate a lesson noting that the use of social media allowed them to better relate to their students (Will, 2017).

According to Educause (2006), "Any technology that is able to captivate so many students for so much time not only carries implications for how those students view the world but also offers an opportunity for educators to understand the elements of social networking that students find so compelling and to incorporate those elements into teaching and learning" (p. 2).

Figure 2. Example of a Bitmoji

CHALLENGES TO A LACK OF SOCIAL MEDIA PRESENCE

College students who do not have mobile devices such as cell phones may feel disconnected, feel anxious and/or have a fear of missing out (FoMO) if not connected (Alt, 2018). Studies have shown that students may even develop nomophobia or "the fear of being out of mobile phone contact" (SecurEnvoy, 2012, para. 1). Researchers at The University of Buffalo reported that smartphones are reinforcing to students and students would rather be deprived of food than give up their phone (O'Donnell & Epstein, 2019). Other researchers found that females who were common users of Pinterest and Instagram and males who spent time using Facebook were more likely to have cell phone addictions (Roberts, Yaya, & Manolis, 2014).

Even though faculty may be increasing their use of social media in the classroom, concerns over student privacy and the integrity of student work exist (Seaman & Tinti-Kane, 2013; Waycott, Thompson, Sheard, & Clerehan, 2017). Garrison (2017) suggested that social media has not been purposed for education but rather used as a means to share personal "selfies and titillating bits of information" (p. 153). Some researchers have suggested that "distractions, stress, anxiety, and social media plague the college classroom making it tough for faculty to teach" (Myatt & Kennette, 2017, para. 2). In fact, over 75% of faculty surveyed noted that digital media has increased communication but also added additional stress and work hours to their loads (Seaman & Tinti-Kate, 2013). And, a large proportion of faculty say Facebook (53%) and Twitter (46%) have "negative" value for use in class (Moran, et al., 2011). For example, researchers found that Facebook use was negatively correlated to both grade and study habits (Junco & Cotton, 2012; Kirschner & Karpinski, 2010). Sanger (2010) noted that "There is no reason to think that repurposing social media for education will magically make students more inspired and engaged" (p. 18). Others have also suggested that undergraduate students may have difficulties in judging the legitimacy of the validity of content shared on social media (Adams Becker, et al., 2017). According to Hall, Delello, & McWhorter (2017) students may also have personal reasons against using sites such as Facebook which must be considered by instructors before making SNSs mandatory.

Social Media Relationships Outside the Classroom

Previous studies have reported that informal or out-of-classroom communication between students and teachers results in positive outcomes, including greater academic achievements (Pascarella, 1985) and a greater sense of well-being of both teachers and students (Roorda, Koomen, Spilt, & Oort, 2011). Helvie-Mason (2011) conveyed that students may use social media for mentoring or to stay connected to their instructors after the course has ended. In one study, almost 40% of students reported being friends with faculty outside of class (Sheldon, 2015). Students suggested that befriending faculty outside of class allowed them to see their instructors as real persons (DiVerniero & Hosek, 2011). Sturgeon and Walker (2009) found that students were more willing to communicate with their instructors if they were already friends on a social media site. Another study reported that faculty members who posted personal tweets on Twitter were more credible to students than those who only posted scholarly ones (Johnson, 2011).

However, the notion of faculty-student social media relationships remains somewhat controversial. Due to the pervasive nature of social media, many times, the boundaries between the professional and personal lives of faculty and students may become blurred. Cain and Fink (2010) noted that even though faculty may want to use social media for instruction, negative results may occur when knowing too much about one's private life outside of the classroom. When it comes to connecting with students personally, research indicates that only 32% of faculty have friended undergraduate students and about half (55%) connected with some students after graduation; Many students reported that they would terminate their social media connection with their instructor at the end of the semester (DiVerniero & Hosek, 2011). Furthermore, because social media policies are non-existent or "gray" at almost 25% of institutions of higher education, the potential risk of crossing personal-professional boundaries may be high (Pomerantz, Hank, & Sugimoto, 2015).

Social Presence Theory

Researchers have suggested that learning through social media may create a sense of "social presence" between faculty-student and student-student (Delello & Consalvo, 2019). Garrison, Anderson and Archer (2000) defined *social presence* as the "the ability of participants in a community of inquiry to project themselves socially and emotionally, as 'real' people (i.e., their full personality), through the medium of communication being used" (p. 94). Gunawardena and Zittle (1997) argued that social presence is really about the degree to which a person is perceived as real within the mode of communication. Akyol, Garrison, and Ozden (2009) noted that students value social presence as a means to "share ideas, to express views, and to collaborate" (p. 76). Kawachi (2013) reported that social media can strengthen social presence by creating a sense of trust between individuals. And, according to Aragon (2003), "Social presence is one of the most significant factors in improving instructional effectiveness and building a sense of community" (Aragon, 2003, p. 57). However, research by Kaplan (2010) suggested that there are degrees of social presence within SNSs, which range from low to high. For example, the use of blogs would be considered low, social media like Facebook were ranked in the mid-range, and virtual worlds such as Second Life are considered high in terms of social presence.

Garrison (2017) questioned whether social media can move beyond a superficial role and promote an effective educational learning experience. Rather than a human-machine interaction, students and faculty must view the social media experience as a human-human relationship (Kerhwald, 2007). Researchers also noted that many times, when social media is integrated into the classroom, students may use it differently than it was assigned for such as non-academic discussions (Arnold & Paulus, 2010). It is evident that additional research is needed to better understand the challenges associated with the use of social media within the classroom (Delello & McWhorter, 2015) and the nature of the relationships of social presence and social media use in the classroom (Dunlap and Lowenthal, 2009).

FOCUS OF THE CURRENT CHAPTER

This chapter presents findings from a descriptive study, in which the authors explored faculty and student perceptions regarding the use of social media in the higher education classroom. A standard mixed-methods survey design was used to gather data pertaining to faculty and student perceptions of social media inside and outside the classroom. The online survey consisted of closed and open-ended questions allowing us to uncover as well as to more fully describe faculty and student views and opinions about social media use in the classroom and explore perceived personal and social media relationships between faculty and students outside of the classroom. The following two overarching research questions guided the study.

1. What are students and faculty perceptions of social media use in the classroom?
2. What are students and faculty views of social media relationships outside the classroom?

Study Participants

Participants in this study consisted of 396 students and 50 faculty members from a mid-size, regional four-year university in the Southwestern United States. Table 1 presents a description of student demographic characteristics. Of the 396 students participating, 120 (30%) were male and 272 (69%) were

Table 1. Student demographic profiles (N=396)

	Count	Percent
Gender		
Male	120	30%
Female	272	69%
Other	4	1%
Ethnicity		
African-American	40	10%
Asian-American	33	8%
White	245	66%
Hispanic	66	17%
Native-American	2	.5%
Other	9	2%
Year in College		
Freshmen	7	1.77%
Sophomore	36	9.09%
Junior	83	20.96%
Senior	89	22.47%
Graduate	181	45.71
College Affiliation		
Arts & Sciences	70	17.72%
Business	105	26.58%
Education & Psychology	74	18.73%
Engineering	37	9.37%
Nursing & Health Sciences	81	20.51%
Pharmacy	14	3.54%
University College	7	1.77%
Other	7	1.77%

female. Four students (1%) reported their gender as other. Two hundred-forty five (62%) were White, 66 (17%) Hispanic, 40 (10%) Black, 33 (8%) Asian, 2 (.5%) Native American, and 9 (2%) were reported as other. There were almost as many graduate students as undergraduates participating in the study. Additionally, the students represented all seven of the university's colleges: Nursing (81/21%); Education and Psychology (74/19%); Business and Technology (105/26%); Pharmacy (14/4%); Engineering (37/9%); Arts and Sciences (70/18%); and the University College (7/2%). Seven (2%) stated other.

Table 2 presents a description of faculty demographic characteristics. of the 50 faculty members participating in the study, 22 (44%) were male and 28 (56%) were female. Forty (80%) of the faculty who responded were White, 4 (8%) were Hispanic, 2 (4%) were Black, and 4 (8%) were Asian. Respondents included all faculty ranks with the majority representing Assistant Professors (38%), followed by Non-Tenure Track Faculty consisting of lecturers and other (28%), Associate Professors (20%), and Full

Table 2. Faculty demographic profiles (N=50)

	Count	Percent
Gender		
Male	22	44%
Female	28	56%
Ethnicity		
African-American	2	4%
Asian-American	4	8%
White	40	80%
Hispanic	4	8%
Academic Rank		
Lecturer	7	18%
Assistant Professor	19	38%
Associate Professor	10	20%
Full Professor	7	14%
Other	5	10%
Percent Classes Taught		
Face-to-Face	59.71%	
Online	16.0%	
Hybrid	21.6%	
Years Teaching Experience		
Mean	13.78	
Standard Deviation	9.17	

Professors (14%). Additionally, faculty members reported from one to 38 years of teaching experience (Mean=13.78; SD=36.1). The majority of faculty reported teaching face-to-face (62%) as compared to online classes (38%). It is worth noting that Non-Tenure Track (71.7%), Assistant (66.1%) and Associate Professors (68.5%) reported teaching more face-to-face than online classes. Inversely, Full Professors reported teaching more online (60.7%) than face-to face classes (39.3%).

Survey Instruments

Data was gathered using two versions of an online survey in order to examine student and faculty perceptions of social media use and relationships inside and outside the classroom. The student survey contained four demographic questions regarding gender, ethnicity, classification, and major. Additionally, four open-ended questions on the survey asked students whether they were allowed to use social media during class time, whether social media was required as a part of any course they were taking, whether using social media in the classroom was viewed as a distraction, and if they were "friends" with their instructor on any social media platform outside of school. Furthermore, each question had an open-ended comment box to allow participants the opportunity to provide additional feedback.

The faculty survey closely mirrored the student survey and asked demographic questions regarding gender, ethnicity, years of teaching experience, and rank. Additionally, faculty members were asked four questions regarding whether students were permitted to use social media in class, the types of social media they used as part of a course requirement, whether they viewed social media as a distraction, and if students had access to their personal social networks (e.g. Facebook, Twitter) outside of school. Faculty were also given the opportunity to provide additional feedback on the open-response items.

Students and faculty participants received an email invitation with a follow-up reminder at two weeks, four weeks, and six weeks to complete an anonymous online Qualtrics survey, which was approved by the university's Institutional Review Board (IRB). For purposes of survey quality design considerations as well as data triangulation and trustworthiness, the survey was piloted with five students and three faculty members and subsequently revised prior to administration.

DATA ANALYSIS

Three types of analyses were conducted when exploring faculty and student perceptions social media use and connections in and outside the classroom. Descriptive statistics such as frequencies, means, and standard deviations were used to analyze participant responses to quantitative questions. A Pearson Chi Square was utilized, where appropriate, to determine if differences exist among participants by independent variables such as gender, ethnicity, faculty rank, years of teaching experience. To complement these analyses, a constant-comparative method was used to analyze participant responses to open-ended questions. Following qualitative research guidelines, the open-ended responses, the authors individually compared and contrasted the data, creating codes and developing meaningful themes, and discussed the themes emerging from the data (Miles & Huberman, 1994).

RESULTS

Student and Faculty Perceptions of Social Media Use in The Classroom

Social Media Use in Courses

When students were asked specifically if they were allowed to use social media in class, 34% reported yes and 66% said no. When asked if their instructors used social media as part of a course, 29% reported using no social media. However, students reported that 71% of their instructors used some type of social media for instruction. Specifically, students noted that You Tube (23%), LinkedIn (8.1%); Google (6.6%), Facebook (6%), Skype (5.7%), Twitter (5.6%), Pinterest (5.4%), Instagram (2.4%), and My Space (.9%) were incorporated into their courses.

In the open-ended comment boxes provided in the survey, students reported using other social media such as Tumblr, Zoom, FaceTime, Blackboard, electronic portfolios, and the Internet. Some students also noted they used various websites related to course content.

Similarly, when faculty were asked whether they allowed students to use social media in the classroom, 28% stated yes and 72% reported no. Male (16%) faculty allowed students to use more social media than female (12%) faculty. Assistant Professors (38%) reported the most use, followed by Non-

Tenure track (28%), Associate Professors (20%), and Full Professors (14%). Of those social media sites reported, faculty members noted they used YouTube (29%), Google+ (13%), Twitter (8.4%); Facebook (6%); LinkedIn (3.6%), Skype (3.6%); Pinterest (2.4%), and other online sites such as Tumblr or Zoom (9%). However, when stratified by gender, female faculty (56%) reported higher uses of all social media in their classrooms than male faculty (44%). In fact, female faculty used all types of social media more than their male counterparts.

Regarding the role social media plays in course instruction, only the use of Google produced a statistically significant difference across gender ($\chi^2 = 3.18$, $p > .05$). Moreover, for female faculty who used social media, the majority used just one site for instruction in a course (16%) while male faculty used two or more (10%).

Social Media as a Distraction

When faculty and students were asked whether they felt distracted when students use electronic devices (e.g. iPhone, iPad, laptop, tablet) during class and why, slightly more than half (52%) of faculty stated yes but only 23% of students thought the use of devices were a distraction. For the students who freported feeling distracted, 22% of the responses were related to the use of social media. For faculty who reported feeling distracted, females (29.8%) were more distracted by their students' use of social media than males (17%).

In response to this question, students reported 167 open-ended statements regarding the use of social media as a distraction in the open-ended comment boxes. A qualitative analysis of these data yielded six themes, which were categorized as: Disengagement, sound media, light media, location, privacy, and irrelevance. Subthemes included texting, talking, chatting, surfing, in-class, and online (see Table 3).

Furthermore, among students who did not think the use of social media was a distraction, some reported that social media use was *irrelevant* and "just a part of everyday life". Other students noted a need to respect other students' privacy. For example, one student stated, "I respect others privacy, no matter if they are taking notes on their device or texting someone, it's not my place to be in their business".

Similarly, the themes emerging from faculty comments focused on policies, strategies, disrespect, and distraction. For the first theme of distraction, faculty responses ranged from "Actually it bothers me" to "YES! and I think all the other students do too" to "It does not distract my ability to teach at all, as taking notes and looking at the internet appear the same to me." When faculty were asked whether they felt distracted when students used electronic devices such as cell phones, iPads, laptops, and tablets, 52.08% reported not being distracted while 47% reported feeling distracted. One faculty stated that they were unsure whether devices in class were a distraction to students noting, "I can't speculate on student's reactions." Another theme that emerged was that of disrespect as one faculty member reported that even though they were not distracted, they felt disrespected and felt the students "were missing out by not participating."

The third theme that emerged was related to policies guiding socia media use in the classroom. Some faculty felt that it was important to have a policy in place. In fact, one faculty member even cited a particular business school that changed its policy to not allow electronic devices in class.

Finally, faculty noted they were working on strategies to integrate more social media devices in the classroom. Two faculty members reported allowing devices for class assignments and one noted, "I have been working on strategies to use the smartphone as part of class activities. This seems to diminish us-

Table 3. Themes related to using social media devices in the classroom

Themes	Subthemes	YES	NO
Disengaged		• Because somethings might be more interesting than the class. • They are not on task • I am easily distracted by just about anything, to be honest - but, if they are doing something other than course work, I am curious/nosy, and wish to see what it is. • Concentration is affected • Usually it's for something social and not school related. • Yes, because it demonstrates the student is not fully present with the class.	• Not as much as the professors get distracted or annoyed by it. It's not a bad distraction. Sometimes it keeps me awake. • I don't feel distracted all of the time, but there are some moments when it can get that way. As a student, when you see someone doing something other than listening to the lecture, it can be distracting to know what they're doing.
Sound Media	• Texting • Chatting • Talking • Surfing	• Because seeing someone texting or messing around on social media on their phone can take my attention away from what the professor is saying. • If they are taking notes it is fine, but if they are on social websites or chatting it is distracting and aggravating. • I get distracted when people talk about what they see on Facebook or other social media sites. • Yes, at times because students will be surfing the web, using social media and other things.	
Light Media		• I tend to watch their screen or get distracted with the noise from typing.	
Location	• In Class • Online	• If they're right in front of me, it can be distracting if they're not using it for the class. • Sometimes yes. If someone is sitting in front of me and they're playing a video game or on Pinterest it distracts me.	• I sit close up front or am writing notes. • I don't pay attention to other students. I sit in the front and don't see what is on their computers. • I'm doing everything online, so I am not distracted by other students.
Privacy			• ...It's not my place to be in their business.
Irrelevance			• It doesn't really matter if they do. • It's part of everyday life.

age during regular lecture, which I view as a distraction." Another member wrote, "But it depends on what they are doing. I let them use devices as writing or research tools when we are writing in class".

Student and Faculty Perceptions of Social Media Relationships Outside The Classroom

When students were asked whether they were connected to any of their instructors' personal social networks outside of school (e.g. "friend or tweet" them?)", 16% reported doing so while an overwhelming majority reported not doing so. For those stating yes, three specific social media were mentioned which included LinkedIn, Facebook, and Twitter. For example, one student commented, "Yup, I'm friends with several of my professors on Facebook. We're cool." Another student noted, "Friends of FB and following on Twitter."

Similarly, when faculty were asked whether they allowed students to access their personal social networks. 29% stated yes while 71% reported no. In terms of gender, female (16.3%) faculty friended students slightly more than male faculty (12.2%). Of those faculty who reported yes, a few stated that they only allow students to "friend" them personally after they have "finished the course" or "graduated". One faculty noted, "Only on LinkedIn" while another stated, "I have a professional Facebook page for students". For those faculty who conveyed they did not allow students personal access to their social media, some stated it was "unprofessional" or that departmental policies did not allow for it. Faculty who taught F2F courses (52.5%) stated they would allow personal friendships as compared to just 47.5% of those teaching in online environments.

While social media relationships are occurring inside and outside the classroom between faculty and students, it is unclear whether these relationships truly create a sense of social presence and if so, to what degree. Thus, the role that social media play in the academic performance of students should be further investigated.

IMPLICATIONS FOR PRACTICE

This study revealed important findings. One key findings pertains to a significant discrepancy among faculty (52%) and students (23%) with respect to whether social media use is or is not a distraction inside the classroom. A second key finding relates to how the participants view social media relationships both inside and outside the classroom. Current research (e.g., Moran et al. 2012), suggest that a little over one-third of faculty use social media in the higher education classroom. The findings of our survey reveled a slightly lower percentage of faculty (28%) using social media for instruction. Moran et al. (2011) found that less experienced faculty use social media 10% more than faculty with more than 20 years of experience. In this study, more experienced faculty (M=13.77; SD=9.26) or those who were full professors were less likely to use social media than any other faculty ranking (non-tenure track, assistant professors, associate professors). It is interesting to note that those faculty who used the least social media and who had been employed in higher education the longest taught the most online courses. This is somewhat inconsistent with earlier research noting that faculty who teach online tend to visit social media sites more often (Tinti-Kane, et al., 2010).

Social media use was reported as the primary distraction in college classrooms. Papacharissi and Mendelson (2011) noted that "online media serve as functional alternatives to interpersonal and mediated communication, providing options or complements for aspects of an individuals' environment that are not as fulfilling" (p. 214). It is important to educate students on how to be responsible with their social media use (e.g. silencing devices, limiting use). However, it is just as important to create engaging classrooms which hold a student's attention. Delello & McWhorter (2015) noted that we must find ways to use social media in class as a tool for engagement as well as instruction. Hall, Delello, & McWhorter (2017) and Leder (2012) suggested that social media use over other learning management systems (LMS) such as discussion boards can enhance student communication and engagement leading to greater social presence. However, the lack of social presence may lead to lower levels of student satisfaction and perceived learning (Richardson, Maeda, Lv, Caskurlu, 2017) as well as higher levels of frustration (Wei, Chen, and Kinshuk, 2012). In this study, most of the social media (e.g. Facebook, Twitter, YouTube) reported by students to be utilized for instruction would be classified as having a medium degree of social presence (Kaplan, 2010). It would be valuable to instructors to make sure they are matching the appropriate social media with the task in order to creates higher degrees of satisfaction.

Contrary to prior research indicating that interpersonal boundaries between faculty and student are shifting as we become more connected via social media; only 17% of students and 29% of faculty noted forming social media friendships outside of the classroom. In 2012, Wakefield found that 32% of faculty connect with undergraduates and almost 55% connect with graduate students. However, research has shown that interpersonal boundaries between faculty and student are shifting as we become more connected to one another and the world at large (Schwartz, 2012). Faculty need to set clear expectations for use in class but also boundaries—including their policies on friending students on personal social media pages. These should correlate to the college/department policies which should be created if they are not already in place.

CONCLUSION AND RECOMMENDATIONS

Because technological devices and the use of social media have become an integral part of college students' everyday life, there is value in finding new ways to incorporate information communication technologies into the classroom, which may encourage students to collaborate and communicate better with fellow students and faculty. On the other hand, faculty need to find ways to limit distractions these media can potentially create while promoting student learning and engagement. Additionally, while social media relationships are occurring both inside and outside the classroom between faculty and students, it is unclear whether these relationships truly create a sense of social presence and if so, to what degree. Thus, the role that social media play in the academic performance of students should be further investigated.

The findings of this study have important implications for faculty professional development opportunities where faculty work together to discuss and explore new and innovative ways of incorporating social media in the classroom. Questions examined in this study could form a basis for faculty professional development. For instance, should socia media be systematically incroprated in teaching and inside the higher education classroom? What is the role of students and faculty in determining which social media to inrporate in the classroom? In what ways can research, policy, and best practices help in developing a campus-wide vision for incorporating social media inside and outside the classroom? What curricular changes need to be made so that social media can be meaqnigfully incorporated in teaching and learning? What data should be collected to help ensure the impact of social media use in and outside of the classroom is systematically documented? Exploring these questions, and others, can be used to develop a research-informed strategy for socia media use that could potentially strengthen instruction and increase student and faculty engagement in teaching and learning. Suggestions for a start of such a professional development program could begin with having junior faculty team with more experienced faculty members to collaborate on how they use technology and further foster opportunities for teaching and scholarship between the group members. Groups of faculty could offer workshops or share examples of using technology at faculty meetings. Additionally, incentives could be offered such as academic innovation or teaching awards for those faculty members that use technology in their classrooms in new ways to promote student engagement and learning.

Limitations

Although this study has achieved its aims, it has unavoidable limitations, which must be taken into consideration when interpreting its findings. First, because of resource and access constraints, the study was conducted in only one, relatively small, public university setting. Thus, it is conceivable that different findings could be attained if the study were replicated in another university setting, at a different time, or with a different target group of students and faculty.

Second, the survey is a self-report instrument, which generates perceptual data about respondents use of social media inside and outside the classroom. These data reflect what respondents think they do and not what they actually do. For purposes of this study, student and faculty responses were viewed at face value. In addition, it is likely that some respondents may have had difficulty understanding some of the questions despite efforts to field-test the survey to evaluate administration procedures and survey items to identify potential issues prior to fullscale data collection. Some students may have alternative views and arguably different definitions of social media. A few noted that social media included the Internet, electronic portfolios, and video conferencing tools such as Zoom. In addition, students and faculty may not have been totally forthright regarding the personal nature of their online relationships with one another. Thus, there is a need in future research to take these liniting factors into consideration when exploring student and faculty perceptions aboit social media use inside and outside the classroom.

Finally, while the student sample size was adequate, the faculty sample size was much smaller. This makes it difficult to examine potential associations across variables within the faculty dataset, as statistical tests generally require larger sample sizes to ensure a representative distribution of the population. On the other hand, it is worth noting that this faculty sample represented approximately 20% of all faculty surveyed at the university and closely reflected the total faculty population in terms of age, gender, ethnicity, and other related characteristics. All faculty who responded to the surveys completed 100% of the survey questions. However, the response rate for female faculty members was 69%, compared to a response rate for male faculty members of 31%. Additionally, the majority of the faculty participants were White and thus the experiences of other faculty respondents may be underrepresented. Faculty respondents reflect the demographics of the institution as there are more females than men and only 17% of the full-time faculty are members of minority groups. These aspects of the study, and perhaps others, can potentially limit the generalizability of the study and should be interpreted with caution.

REFERENCES

Adams Becker, S., Cummins, M., Davis, A., Freeman, A., Hall Giesinger, C., & Ananthanarayanan, V. (2017). *NMC Horizon Report: 2017 Higher Education Edition*. Austin, TX: The New Media Consortium.

Akyol, Z., Garrison, D. R., & Ozden, M. Y. (2009). Online and blended communities of inquiry: Exploring the developmental and perceptional differences. *International Review of Research in Open and Distance Learning*, *10*(6), 65–83. doi:10.19173/irrodl.v10i6.765

Alt, D. (2018). Students' wellbeing, fear of missing out, and social media engagement for leisure in higher education learning environments: Research and reviews. *Current Psychology, 37*(1), 128-138. http://dx.doi.org.ezproxy.uttyler.edu:2048/10.1007/s12144-016-9496-1

Aragon, S. R. (2003). Creating social presence in online environments. *New Directions for Adult and Continuing Education*, *100*(100), 57–68. doi:10.1002/ace.119

Arnold, N., & Paulus, T. (2010). Using a social networking site for experiential learning: Appropriating, lurking, modeling and community building. *Internet and Higher Education*, *13*(4), 188–196. doi:10.1016/j.iheduc.2010.04.002

Baisley-Nodine, E., Ritzhaupt, A., & Antonenko, P. (2018). Exploring social presence within an online course using Twitter. *E-Learning and Digital Media*, *15*(5), 235–253. doi:10.1177/2042753018786004

Cain, J., & Fink, J. L. III. (2010). Legal and ethical issues regarding social media and pharmacy education. *American Journal of Pharmaceutical Education*, *74*(10), 1–8. doi:10.5688/aj7410184 PMID:21436925

Clement, J. (2019, July). *Reach of leading social media and networking sites used by teenagers and young adults in the United States as of February 2017*. Retrieved from https://www.statista.com/statistics/199242/social-media-and-networking-sites-used-by-us-teenagers/

Delello, J. A., & Consalvo, A. L. (2019). "I Found Myself Retweeting": Using Twitter Chats to Build Professional Learning Networks. In J. Yoon & P. Semingson (Eds.), *Educational Technology and Resources for Synchronous Learning in Higher Education* (pp. 88–108). Hershey, PA: IGI Global. doi:10.4018/978-1-5225-7567-2.ch005

Delello, J. A., & McWhorter, R. R. (2014). New visual social media for the higher education classroom. In G. Mallia (Ed.), *The social classroom: Integrating social network use in education* (pp. 368–393). Hershey, PA: IGI Global. doi:10.4018/978-1-4666-4904-0.ch019

Delello, J. A., & McWhorter, R. R. (2014). Creating virtual communities of practice with the visual social media platform Pinterest. *International Journal of Social Media and Interactive Learning Environments*, *2*(3), 216–236. doi:10.1504/IJSMILE.2014.064205

Delello, J. A., McWhorter, R. R., & Camp, K. M. (2015). Using social media as a tool for learning: A multi-disciplinary study. *International Journal on E-Learning*, *14*(2), 163–180.

DiVerniero, R. A., & Hosek, A. M. (2011). Students' perceptions and communicative management of instructors' online self-disclosure. *Communication Quarterly*, *59*(4), 428–449. doi:10.1080/01463373.2011.597275

Dunlap, J. C., & Lowenthal, P. R. (2009). Tweeting the night away: Using Twitter to enhance social presence. *Journal of Information Systems Education*, *20*(2).

Educause. (2006). *7 things you should know about Facebook*. Retrieved from https://net.educause.edu/ir/library/pdf/ELI7017.pdf

Eteokleous-Grigoriou, N., & Ktoridou, D. (2014). Social networking for educational purposes: The development of social-cultural skills through special interest groups. In G. Mallia (Ed.), *The social classroom: Integrating social network use in education* (pp. 394–416). Hershey, PA: IGI Global. doi:10.4018/978-1-4666-4904-0.ch020

Faculty Focus. (2011). *Social media usage trends among higher education faculty*. Faculty Focus Special Report. Retrieved from http://cdn.facultyfocus.com/wp-content/uploads/2011/09/2011-social-media-report.pdf

Garrison, D. R., Anderson, T., & Archer, W. (2000). Critical inquiry in a text based environment: Computer conferencing in higher education. *The Internet and Higher Education*, *2*(2/3), 87–105.

Garrison, R. D. (2017). *21st century e-learning*. Routledge.

Globalwebindex. (2019). *The latest social media trends to know in 2019*. Retrieved from https://www.globalwebindex.com/reports/social

Gunawardena, C. N., & Zittle, F. J. (1997). Social presence as a predictor of satisfaction within a computer-mediated conferencing environment. *American Journal of Distance Education*, *11*(3), 8–26. doi:10.1080/08923649709526970

Hall, A. A., Delello, J. A., & McWhorter, R. R. (2017). Using Facebook to supplement instruction in online and hybrid courses. *Journal of Innovation and Learning*, *22*(1), 87–104. doi:10.1504/IJIL.2017.085250

Heiberger, G., & Junco, R. (2015). *Meet your students where they are: Social media*. National Education Association. Retrieved from http://www.nea.org/assets/docs/HE/HigherEdSocialMediaGuide.pdf

Helvie-Mason, L. (2011). Facebook, friending, and faculty–student communication. In C. Wankel (Ed.), Teaching Arts and Science with the New Social Media. Emerald Group Publishing Limited. doi:10.1108/S2044-9968(2011)0000003007

Hershkovitz, A., & Forkosh-Baruch, A. (2013). Student-teacher relationship in the Facebook era: The student perspective. *International Journal of Continuing Engineering Education and Lifelong Learning*, *23*(1), 33–52. doi:10.1504/IJCEELL.2013.051765

Irwin, C., Ball, L., Desbrow, B., & Leveritt, M. (2012). Students' perceptions of using Facebook as an interactive learning resource at university. *Australasian Journal of Educational Technology*, *28*(7), 1221–1232. doi:10.14742/ajet.798

Johnson, K. (2011). The effect of Twitter posts on students' perceptions of instructor credibility. *Learning, Media and Technology*, *36*(1), 21–38. doi:10.1080/17439884.2010.534798

Johnson, L., Adams Becker, S., & Cummins, M. (2012). *The NMC horizon report: 2012 higher education edition*. Austin, TX: The New Media Consortium.

Johnson, L., Adams Becker, S., Estrada, V., & Freeman, A. (2015). *NMC horizon report: 2015 K-12 edition*. Retrieved from http://cdn.nmc.org/media/2015-nmc-horizon-report-HE-EN.pdf

Junco, R. (2012). Too much face and not enough books: The relationship between multiple indices of Facebook use and academic performance. *Computers in Human Behavior*, *28*(1), 187–198. doi:10.1016/j.chb.2011.08.026

Junco, R., & Cotton, S. (2011). *A decade of distraction? How multitasking affects student outcomes*. Paper presented at A Decade in Internet Time Symposium on the Dynamics of the Internet and Society, Oxford Internet Institute, University of Oxford, Oxford, UK. 10.2139srn.1927049

Junco, R., & Cotton, S. (2012). The relationship between multitasking and academic performance. *Computers & Education*, *59*(4), 1–10. doi:10.1016/j.compedu.2011.12.023

Junco, R., Heiberger, G., & Loken, E. (2010). The effect of Twitter on college student engagement and grades. *Journal of Computer Assisted Learning*, *27*(2), 119–132. doi:10.1111/j.1365-2729.2010.00387.x

Kaplan, A. M., & Haenlein, M. (2010). Users of the world, unite! The challenges and opportunities of Social Media. *Business Horizons*, *53*(1), 59–68. doi:10.1016/j.bushor.2009.09.003

Kawachi, P. (2013). Online social presence and its correlation with learning. *International Journal of Social Media and Interactive Learning Environments*, *1*(1), 19–31. doi:10.1504/IJSMILE.2013.051653

Keiser Foundation. (2013). *Generation M²: Media in the lives of 8-to-18-year-olds*. Retrieved from https://kaiserfamilyfoundation.files.wordpress.com/2013/04/8010.pdf

King, T. (2019, July). Bitmojis, GIFs, and Snaps in the Classroom? Oh My! *Faculty Focus*. Retrieved from https://www.facultyfocus.com/articles/effective-classroom-management/bitmojis-GIFs-and-snaps-in-the-classroom/

Kirschner, P. A., & Karpinski, A. C. (2010). Facebook and academic performance. *Computers in Human Behavior*, *26*(6), 1237–1245. doi:10.1016/j.chb.2010.03.024

Knight-McCord, J., Cleary, D., Grant, N., Herron, A., Jumbo, S., Lacey, T., ... Smith, R. (2016). What social media sites do college students use most? *Journal of Undergraduate Ethnic Minority Psychology*, *2*(21), 21–26.

Kumi-Yeboah, A., & Blankson, H. (2018). Social media and use of technology in higher education. In M. Khosrow-Pour (Ed.), *Teacher Training and Professional Development: Concepts, Methodologies, Tools, and Applications, 4* (Vol. 41, pp. 932–952). Information Resources Management Association. doi:10.4018/978-1-5225-5631-2.ch041

Metzger, A. H., Finley, K. N., Ulbrich, T. R., & McAuley, J. W. (2010). Pharmacy faculty members' perspectives on the student/faculty relationship in online social networks. *American Journal of Pharmaceutical Education*, *74*(10), 188. doi:10.5688/aj7410188 PMID:21436929

Miles, M. B., & Huberman, A. M. (1984). *Qualitative data analysis* (Vol. 16). Newbury Park, CA: Sage.

Moran, M., Seaman, J., & Tinti-Kane, H. (2011). *Teaching, learning, and sharing: How Today's Higher Education Faculty Use Social Media*. Pearson Learning Solutions and Babson Survey Research Group.

Moran, M., Seaman, J., & Tinti-Kane, H. (2012). *Social Media in Higher Education*. Pearson Learning Solutions and Babson Survey Research Group.

Myatt, B. M., & Kennette, L. N. (2017, Jan.). *Towards a positive u*. Faculty Focus. Retrieved from http://www.facultyfocus.com/articles/effective-classroom-management/towards-positive-u/

Nielson. (2016). *2016 Nielson social media report. Social studies: A look at the social landscape*. Retrieved from https://www.nielsen.com/content/dam/corporate/us/en/reports-downloads/2017-reports/2016-nielsen-social-media-report.pdf

O'Donnell, S., & Epstein, L. H. (2019). Smartphones are more reinforcing than food for students. *Addictive Behaviors*, *90*, 124–133. doi:10.1016/j.addbeh.2018.10.018 PMID:30390436

Oxford University Press. (2015). *Word of the year 2015*. Retrieved from https://languages.oup.com/word-of-the-year/word-of-the-year-2015

Pappano, L. (2018). The iGen shift: Colleges must change to reach the next generation. *New York Times*. Retrieved from https://www.nytimes.com/2018/08/02/education/learning/generationz-igen-students-colleges.html

Pascarella, E. (1985). College environmental influences on learning and cognitive development: A 32 2g critical review and synthesis. In J. Smart (Ed.), *Higher education: Handbook of theory and research* (Vol. I). New York: Agathon.

Pearson. (2018). *Beyond millennials: The next generation of learners*. Retrieved from https://www.pearson.com/content/dam/one-dot-com/one-dot-com/global/Files/news/news-annoucements/2018/The-Next-Generation-of-Learners_final.pdf

Perrin, A., & Anderson, M. (2019). *Share of U.S. adults using social media, including Facebook, is mostly unchanged since 2018*. Retrieved from https://www.pewresearch.org/fact-tank/2019/04/10/share-of-u-s-adults-using-social-media-including-facebook-is-mostly-unchanged-since-2018/

Pomerantz, J., Hank, C., & Sugimoto, C. R. (2015). The state of social media policies in higher education. *Plos*. Retrieved from http://journals.plos.org/plosone/article?id=10.1371/journal.pone.0127485

Quan-Haase, A., & Young, A. L. (2010). Uses and gratifications of social media: A comparison of facebook and instant messaging. *Bulletin of Science, Technology & Society*, *30*(5), 350–361. doi:10.1177/0270467610380009

Refuel Agency. (2015). *College explorer '15*. Retrieved from http://research.refuelagency.com/wp-content/uploads/2016/01/College%20Explorer%202015%20Final.pdf

Richardson, J. C., Maeda, Y., Lv, J., & Caskurlu, S. (2017). Social presence in relation to students' satisfaction and learning in the online environment. *Computers in Human Behavior*, *71*(C), 402–417. doi:10.1016/j.chb.2017.02.001

Roberts, J. A., Yaya, L. H., & Manolis, C. (2014). The invisible addiction: Cell-phone activities and addiction among male and female college students. *Journal of Behavioral Addictions*, *3*(4), 254–265. doi:10.1556/JBA.3.2014.015 PMID:25595966

Roorda, D. L., Koomen, H. M. Y., Spilt, J. L., & Oort, F. J. (2011). The influence of affective teacher–student relationships on students' school engagement and achievement. *The Journal of Educational Research*, *81*(4), 493–529.

Rothenberg, M. (2019). *Emoji tracker*. Retrieved from http://emojitracker.com/

Sanger, L. (2010). Individual knowledge in the Internet age. *EDUCAUSE Review*, *45*(2), 14–24.

Schwartz, H. L. (2012). From the classroom to the coffee shop: Graduate students and professors effectively navigate interpersonal boundaries. *International Journal on Teaching and Learning in Higher Education*, *23*(3), 363–372.

Seaman, J., & Tinti-Kane, H. (2013). *Social media for teaching and learning*. Pearson Learning Solutions and Babson Survey Research Group. Retrieved from http://www.pearsonlearningsolutions.com/assets/downloads/reports/social-media-for-teaching-and-learning-2013-report.pdf#view=FitH,0

SecurEnvoy. (2012). *66% of the population suffer from Nomophobia the fear of being without their phone*. Retrieved from http://www.securenvoy.com/blog/2012/02/16/66-of-the-population-suffer-fromnomophobia-the-fear-of-being-without-their-phone/

Sheldon, P. (2015). Understanding students' reasons and gender differences in adding faculty as Facebook friends. *Computers in Human Behavior*, *53*, 58–62. doi:10.1016/j.chb.2015.06.043

Smith & Anderson. (2018). *Social Media Use in 2018*. Retrieved from https://www.pewinternet.org/2018/03/01/social-media-use-in-2018/

Sturgeon, C. M., & Walker, C. (2009). *Faculty on Facebook: Confirm or deny*. Paper presented at the Annual Instructional Technology Conference, Murfreesboro, TN.

Tenor. (2017). *10 Years with the iPhone: Communication is now visual*. Retrieved from https://blog.tenor.com/10-years-with-the-iphone-communication-is-now-visual-c2b7dae6cf07

Thalluri, J., & Penman, J. (2015). Social media for learning and teaching undergraduate sciences: Good practices guidelines from intervention. *Electronic Journal of E-Learning*, *13*(6), 455–465.

The Institute of Politics at Harvard Kennedy School. (2017). *Use of social networking technology*. Retrieved from https://iop.harvard.edu/use-social-networking-technology

Wakefield, K. (2012). Should professors use facebook to communicate with students? *Faculty Focus*. Retrieved from http://www.facultyfocus.com/articles/edtech-news-and-trends/should-professors-use-facebook-to-communicate-with-students/

Wang, Q., Chen, W., & Liang, Y. (2011). The effects of social media on college student (Paper 5). *MBA Student Scholarship*. Retrieved from https://scholarsarchive.jwu.edu/mba_student/5

Waycott, J., Thompson, C., Sheard, J., & Clerehan, R. (2017). A virtual panopticon in the community of practice: Students' experiences of being visible on social media. *The Internet and Higher Education*, *35*, 12–20. doi:10.1016/j.iheduc.2017.07.001

Wei, C., Chen, N., & Kinshuk. (2012). A model for social presence in online classrooms. *Educational Technology Research and Development*, *60*(3), 529–545. doi:10.100711423-012-9234-9

Will, M. (2017). Memes, emojis, and GIFs, oh my! Teachers tell how they use social media. *Education Week*. Retrieved from http://blogs.edweek.org/teachers/teaching_now/2017/08/most_teachers_think_social_media_are_bad_for_grammar_and_spelling_but_still_use_it_in_class.html

Zmikly, J. (2016). Remix: How to use Snapchat in the classroom. *Mediashift*. Retrieved from http://mediashift.org/2016/09/remix-snapchat-classroom/

ADDITIONAL READING

Barnes, N. G., & Lescault, A. M. (2012). *Social media adoption soars as higher-ed experiments and reevaluates its use of new communications Tools.* University of Massachusetts Dartmouth. Retrieved from https://www.umassd.edu/cmr/studiesandresearch/socialmediaadoptionsoars/

Greenwood, S., Perrin, A., & Duggan, M. (2016). *Social media update 2016: Facebook usage and engagement is on the rise, while adoption of other platforms holds steady.* Pew Research Center. Retrieved from https://www.pewinternet.org/2016/11/11/social-media-update-2016/

Kehrwald, B. (2007). The ties that bind: Social presence, relations and productive collaboration in online learning environments. In ICT: *Providing choices for learners and learning.* Proceedings Ascilite Singapore 2007. Retrieved from http://www.ascilite.org.au/conferences/singapore07/procs/kehrwald.pdf

Lenhart, A., Purcell, K., Smith, A., & Zickuhr, K. (2010). *Social media and young adults.* Washington, DC: Pew Internet and American Life Project.

Pew Research Center. (2017). *Social media fact sheet.* Retrieved from https://www.pewinternet.org/fact-sheet/social-media/

Rifkind, L. J. (1992). Immediacy as a predictor of teacher effectiveness in the instructional television. *Journal of Interactive Television, 1*(1), 31–38.

KEY TERMS AND DEFINITIONS

Distraction: Something that divides one's attention.
Emerging Technology: Technologies that are currently being developed or refined.
Social Media: Interactive Web 2.0 application that allows for sharing of information across the Web.
Social Presence: Ones ability to feel connected or present in their environment.
Visual Social Media: The use and sharing of visual images across social media.
Web 2.0: Second generation of the world-wide web that allows for more collaboration and interaction.

Chapter 3
Security Issues Detected in the Computer Slot Machine Systems

Marius Popescu
Vasile Goldis Western University of Arad, Romania

Antoanela Luciana Naaji
Vasile Goldis Western University of Arad, Romania

ABSTRACT

The chapter analyzes three types of computer systems from the hardware and software point of view, slot machines, respectively. The owners of these devices, by different methods of illegal manipulation, can have total control of the chances of winning and losing, and on the other hand, they can determine at will the amounts to be recorded in accounting. The investigation techniques used in the analysis of electronic gaming machines with winnings show that they present particular problems of cyber security. As a result, solutions have been proposed to solve these problems, such as those concerning the verifications that the legal metrology offices must carry out.

INTRODUCTION

Slot-machines are means of gambling represented by the specific machines, equipment and specific installations assembled from a constructive point of view in a unit through which random elements are generated independently based exclusively on chance, but also on the dexterity or ability of the participants in the gamble (Aragay et al., 2018), without involving in any form or by any means the organizer, his staff or other persons in the selection or the generation of chance and in determining the results of the game (Kelly et al., 2018; Lister et al., 2016; Auer&Griffiths, 2015).

The gambling machine is provided with electronic control keys which in fact represent an assembly of two dependent computer systems, in which one of the systems controls the operation of the other, with two electromechanical meters, a counter for the amounts of money or the points entered in the device (IN), respectively a counter for the points or sums of money that leave the device, respectively the counter (OUT). From the analysis of the norms mentioned in the specialized literature (Simion&Baciu,

DOI: 10.4018/978-1-7998-3355-0.ch003

2015) it results that, the electronic meter aims to keep track of the amounts of money introduced in the device,respectively paid to the potential winners, and the mechanical meter has the same purpose, which is why they are interconnected, constituting in reality a double record of the activity of the respective devices,more precisely, a fair exposure for a period of time of the obtained amount of the devices, the granted profits and the payments made to the winners. As standard for a slot machine type of gambling machine, the manufacturer of the device offers a number of 3 keys, a key for entering or canceling points, a key for reading, respectively for canceling electronic counters, distributed to the operator from the game room as well as a «master» key for the owner of the gambling machine, with which the following operations can be performed: reading, respectively, canceling the electronic counters, configuring the game card, initializing the game card (Brito, 2016). In fact, by using a special key, the owners of the devices can intervene on the slot machine gambling machines, respectively on the mechanical meter, having the "out" function, through the same access point to the electronic counters, unjustifiably increasing the winnings (exits), which were to be given to potential players. In this way you can practically hide the amount of money collected for the respective devices from the players and represented at the points indicated by their counters. On the same occasion, the electronic counters could be set to "zero" in all their pages under the pretext of a better highlighting of their activity, contrary to the legal norms prohibiting such actions.

The modification of the electronic counters is done for the same purpose, of the concealment of the real receipts that were obtained, and these illegal ways of manipulating the gaming machines have the following major consequences.First of all, by this way the amounts of money collected from the games on these devices are being withdrawn as "under the counter".Secondly, by the interventions on the electronic and mechanical meters, the chances of winning set in the machine are eliminated. According to the legal provisions, the respective devices are set to give certain percentage of winnings. Obviously this win can only be given after the game on the machines and the receipts made. Or, in the conditions in which through the false key winnings can be awarded, logically the machine does not give winnings, until new receipts are made or if gives very small winnings.The machine offers a win to the player when, on the display panel, or in a delimited area of the video screen, certain preset types of image configurations appear, resulting from the independent, "rotation" with a (pseudo) random speed of the drums and respectively by being set by the player, type of action transmitted also to the modern non-mechanical equivalents publicly exploited or to the computer simulators installed locally or online (Mujal, 2010).

METROLOGICAL NORMS

The metrological verification norm establishes the technical conditions that slot machines have to fulfill in order to be placed on the market, respectively, to approve their further operation, at the annual verification carried out by the *Legal Metrology Bureau* (LMB). According to this, requirements regarding the monitoring, recording and highlighting of all the operations of use refer to: total receipts (total *in*), total payments granted (total *out*) and connection/disconnection means of play from the power supply. The values of the indexes of the meters will increase all the time, the meters being "designed" to add the money that "enters-IN" or "exit-OUT" from the gaming machine, similar to the meters of electricity, or water from homes, or the mileage on the cars. The game organizers periodically re-authorize their means of play, through a functioning authorization. The verification activity begins when the economic operator requests this in writing - a verification request for initial, periodic technical verifications or reparations. When a means of play is checked, the game program is measured, it is checked if it works according to the earnings table, then the program is measured and compared with the standard. The owners of gaming means are obliged to allow access, not to the program code, but to the file itself that the LMB employees measure by calculation algorithms for the sum of control ((Manogaran&Lopez, 2018; Mars et al. 2018). In principle, two programs are identical if these amounts coincide. In case of suspicion, there is also a bit-by-bit comparison of the standard program with the program in the respective game medium. The standard is made by LMB or is made available by the software manufacturer. The control amount is calculated either by a classic algorithm with EPROM (Erasable Programmable Read Only Memory) readers, or by some calculation algorithms of the control amount, through an application made available by the software manufacturer.In principle, two programs are identical if these amounts coincide. In case of suspicion, there is also a bit-by-bit comparison of the standard program with the program in the respective game medium. The standard is made by LMB or is made available by the software manufacturer. The control amount is calculated either by a classic algorithm with EPROM (*Erasable Programmable Read Only Memory*) readers or by some calculation algorithms of the control amount, through an application made available by the software manufacturer.

Hereinafter we will describe the hardware and software investigation techniques for 3 types of slot machines, in which intruded computer systems were installed. Also, some security solutions are proposed.

HARDWARE ANALYSIS

From the diagrams of the key reading module (Figure 1) we can see the connection mode with the other blocks of the gaming device, having the following results: the key reading circuit is connected to the electromechanical impulse meter which displays the outputs; the game mode is connected to the electronic key reading circuit through a different number of connections, which depend on the number of keys required (cancellation and entry points, small and large statistics); the connection of the central unit to the electromechanical output meter is made for some devices by means of a diode and the others have a common connection with that of the electronic key reading circuit.

The diagrams for connecting the electronic key reading circuits with the central units and with the output meters of the analyzed game devices are shown in Figure 2.

The numbering of the pins of the electronic circuit was made in accordance with the pins of the connector in which it is mounted. Thus, pins 1 to 10 belonging to the connector correspond to the face of the printed circuit, and the pins 11 to 20 are behind it. The electronic key reading assembly (Figure 3) is performed using a microcontroller (μC) of type PIC16F84A (Erol et al., 2007). The *iButton* proximity reader reads the key code, called digital access key. The abovementioned circuit is provided with a *Central Processing Unit* (CPU), a 68-bytes RAM (*Random Access Memory*) data memory, a 1024x14-bits*flash* program memory, a non-volatile memory type EEPROM (*Electrically Erasable Programmable ReadOnly Memory*) with a capacity of 64 bytes, input/output interfaces with a total of 13 pins and a timer. This device performs the automatic processing of the input data using the computer program/ software stored in the program memory.

The processing results are transposed into commands on the output interface. In addition to the PIC16F84A μC on the module, there is also a ULN2003AG type circuit (used in other computer applications (Huang et al, 2011), which is used to adapt the μC's outputs with the inputs of the game module and the electromechanical impulse meter which it controls (Sari et al., 2017), a 4 position switch used

Figure 1. Electrical diagram of the electronic key reading module type A

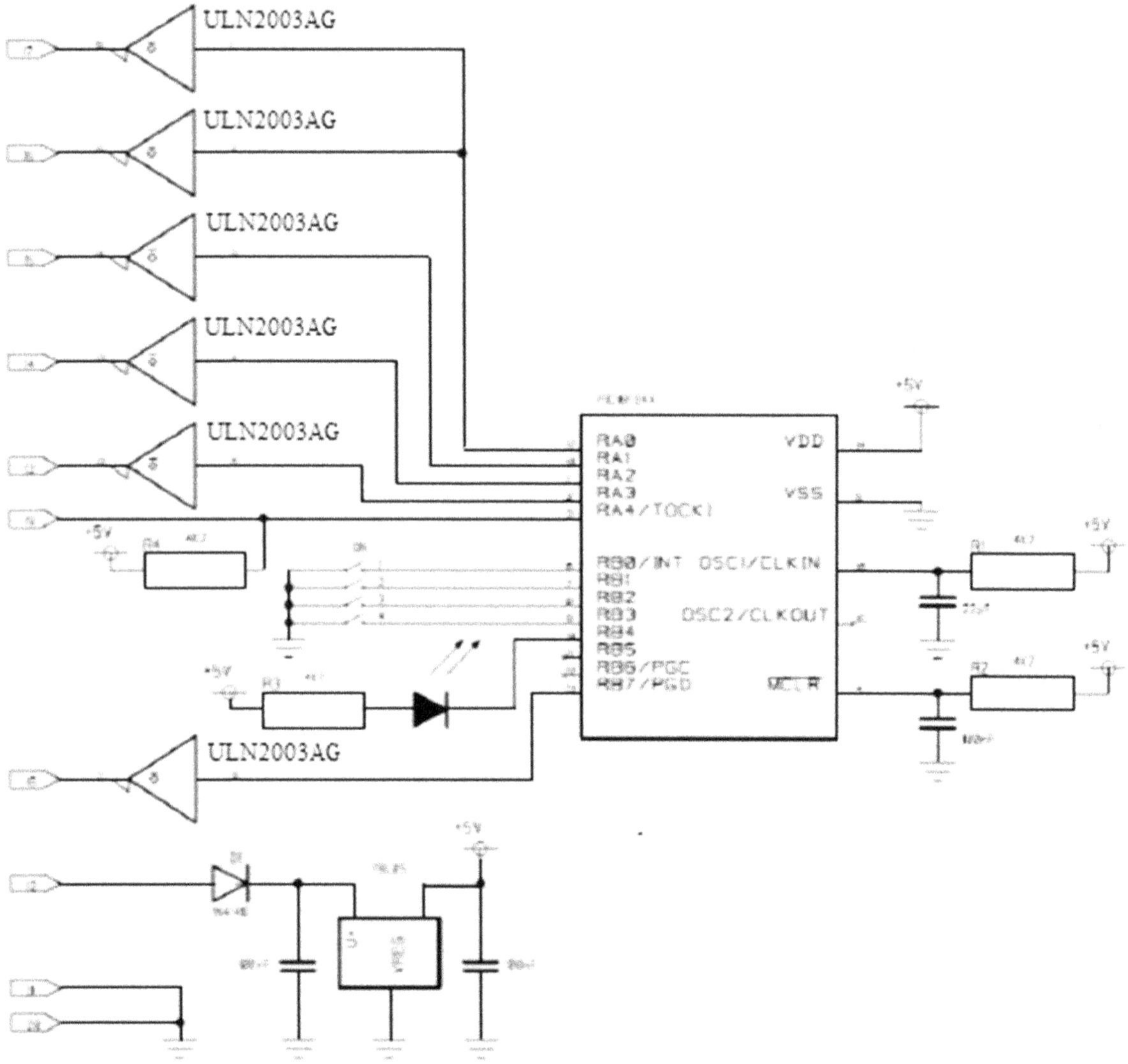

in the learning phase of a key - for the selection of the function - as well as a 5V source - made with a 78L05 type circuit (YiJie et al.,, 2010). The inputs to the PIC16F84A circuit represent the output of the *iButon* electronic key connector and the four-position switch. The outputs from the PIC16F84A μC are passed through the ULN2003AG integrated circuit and are available on pins 13, 14, 15, 16, 17 and 18 of the electronic assembly. The outputs on pins 16 and 17 are controlled from the same output pin of the PIC16F84A circuit. The interconnection of the electronic key reading circuit with the game module and the electromechanical pulse meter is realized by a cable form.

This type of electronic circuit is not mentioned in the *Technical Verification Norms* of the LMB, where the composition of the *slot-machine* type of devices is described. Also, there is no mention of the form of cable that connects all the modules of a gaming device and, which, not being protected in any way, can be modified with ease - this fact being able to cause the operation of the gaming device in a different way from the approved one. In these technical norms, no requirement is specified regarding

Figure 2. Schemes for connecting the electronic key reading circuit to the analyzed gaming devices

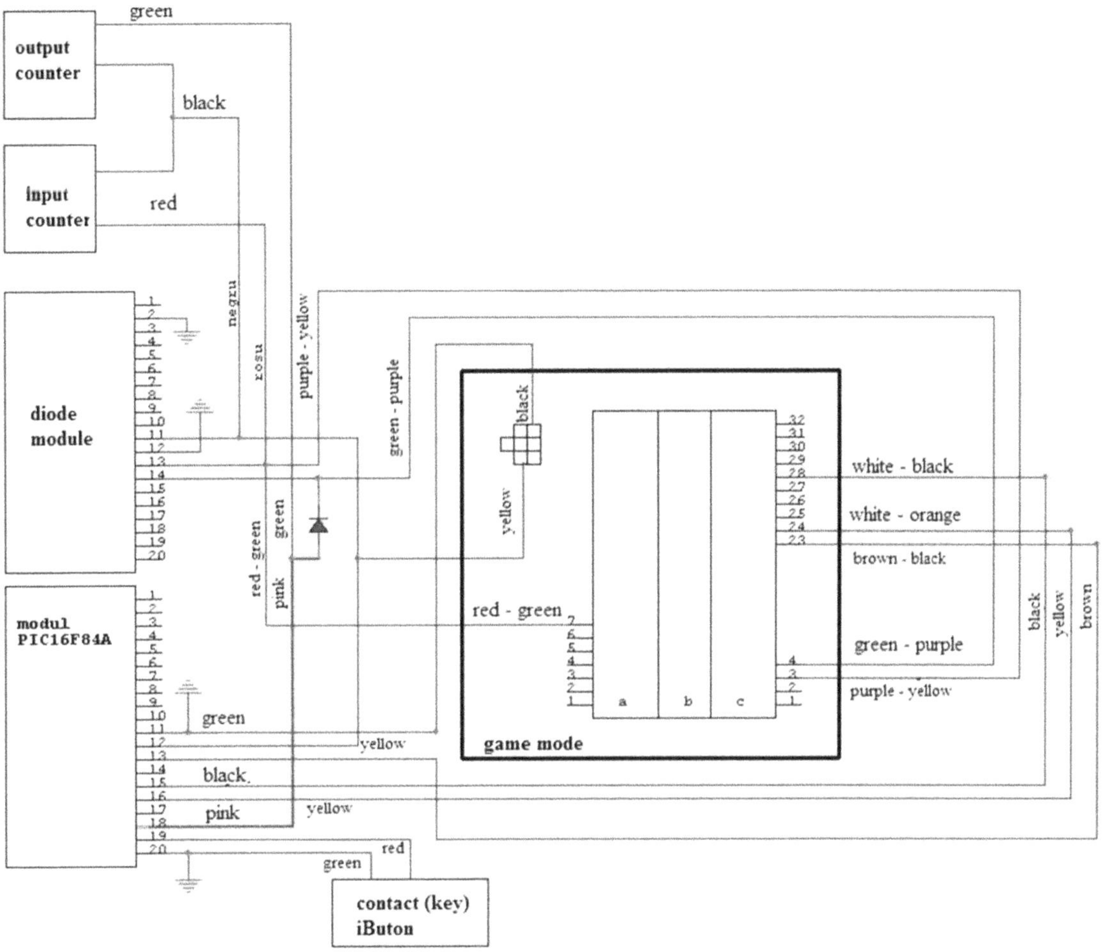

the verification of the electronic key reading module and the cable form of the gaming machine. In these technical norms, no requirement is specified regarding the verification of the electronic key reading module and of the cable form of the gaming machine.

For analyzing the data content of each of the three μC type PICI6F84A from the analyzed game devices, the microchips can be extracted from the printed circuit boards (because the microchips are not bonded to the printed circuit as are the other electronic components, but are mounted on a socket – fact which facilitates the replacement or reprogramming of the integrated circuits) and connects to a programming device (*ChipProg-40 type*) (Ahmad et al, 2017). As a rule, μC circuits have read protection enabled, which does not allow access to the data content (Trivedi&Awad, 2015; Lin et al., 2005). Therefore, the functional analysis of the electronic key reading circuit is necessary in order to determine the learning method of a key, the number of different keys that can be learned and the reaction of the electronic mode, for all possible combinations of the four positions of the switch.

Consequently, the following tests were retained (Figure 4):

Figure 3. Electronic key reading circuit: a) front view - the 4 positions of the switch on the key reading circuit; b) back view.

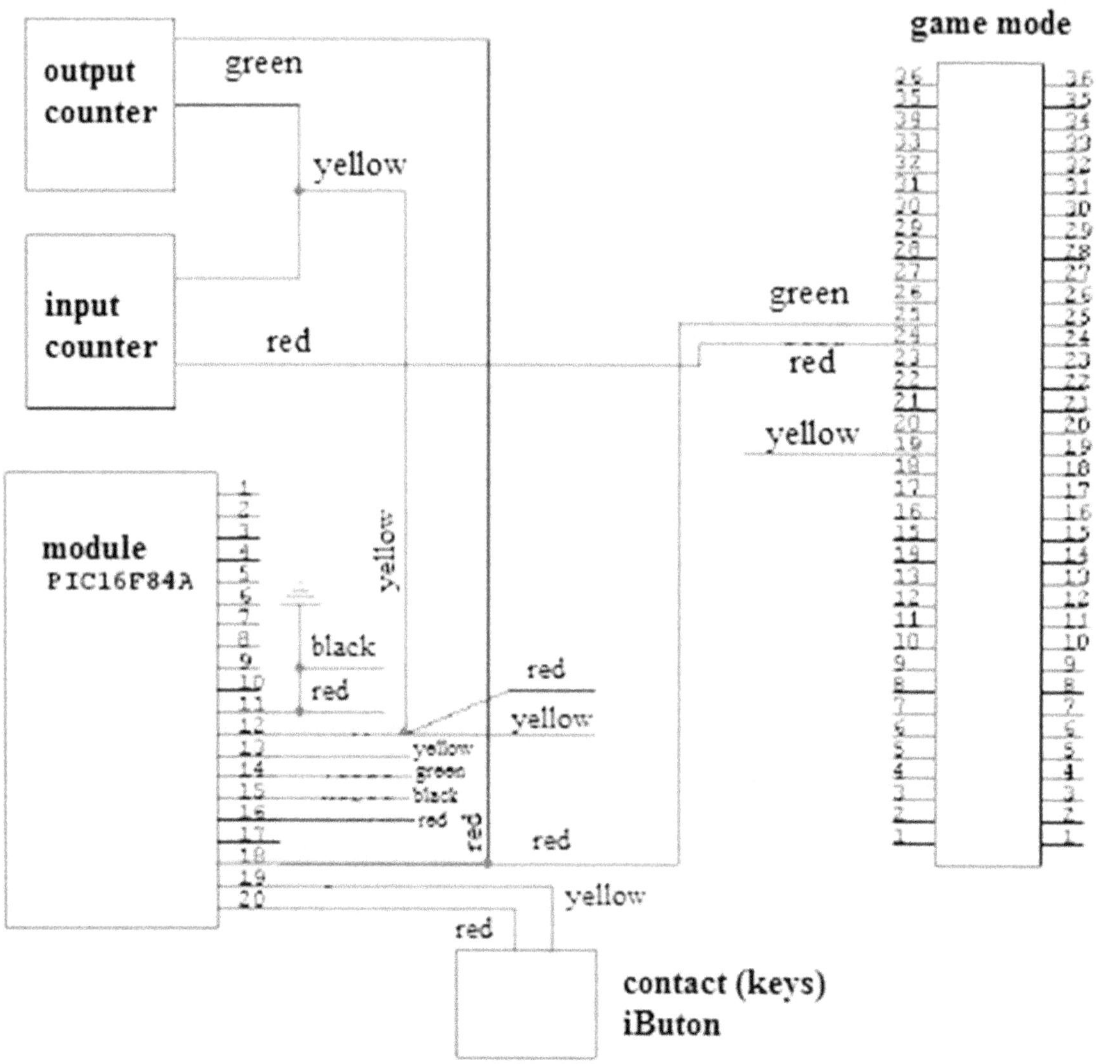

1. Key no.1 has a special regime, in the sense that once it has been contacted and read, all other previously learned keys will be deleted; the code of that key was entered in the program in μC, being the first key read, when the module was new;
2. We enter in the learning mode, in the sense that the key that will be placed on the reader (after all the keys have been deleted in advance) it will become the programming key through which the keys for each function can be defined;
3. For learning the function keys, the programming key is placed on the reader, when the LED will flash, the four positions of the switch corresponding to the function will be switched, putting the key that will perform this function, and after the reading, the LED power goes out indicating the storage of its value; key no.4 could not be used to become a programming key or function key;

Figure 4. Keys of the analyzed gaming devices

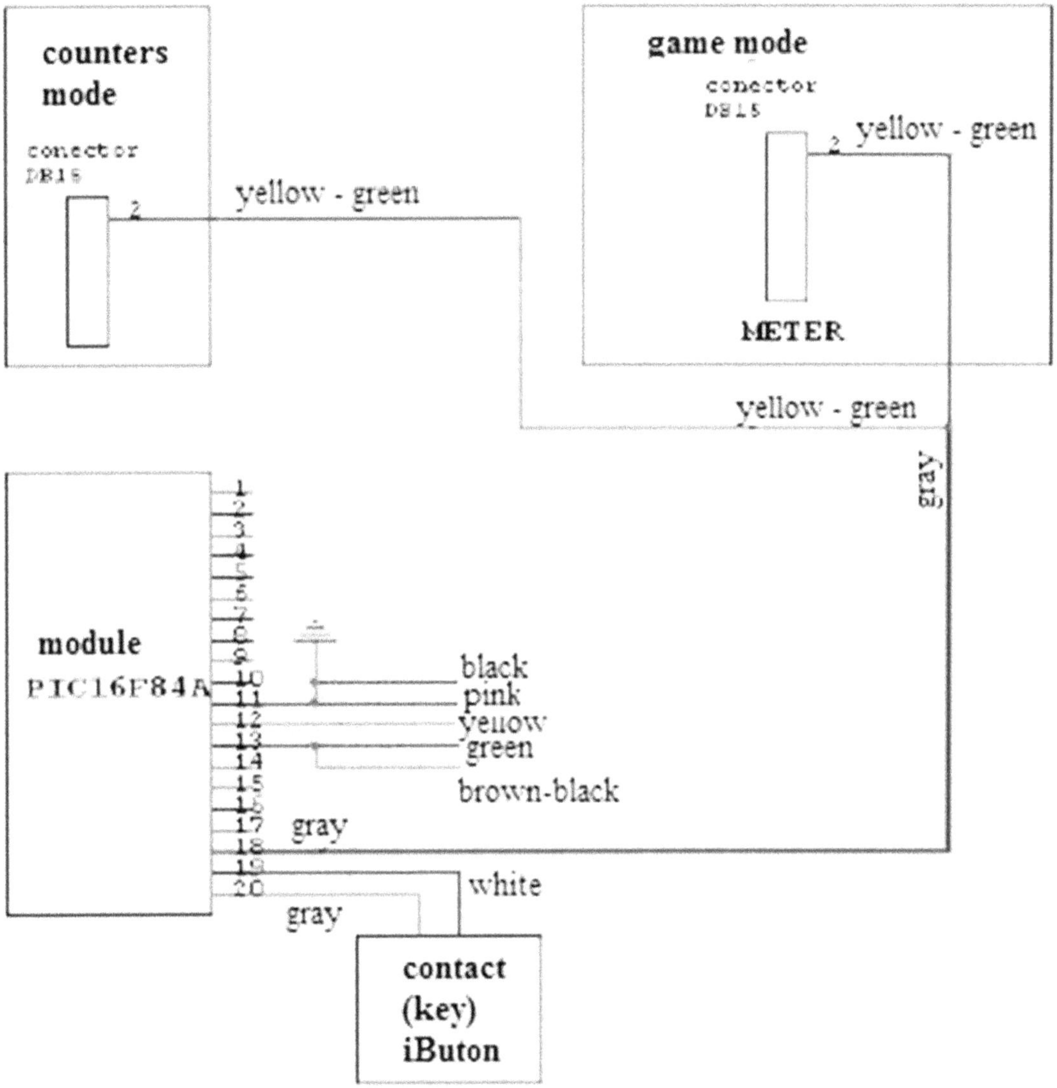

4. Once the key serial number of the key is read, the μC will execute the function for which it was programmed;
5. In addition to the two special keys (respectively, deletion and programming) the PIC16F84A circuit can learn another six keys for another six functions;
6. The electronic circuit allows the programming of several keys for the same function, but does not allow the programming of a key for several functions - when reading the key the circuit will execute the last function related to the key.

Using a single key and programming one at a time all the combinations/positions of the switch (SW), changes were observed at the output pins of the electronic key reading mode, data presented in Table 1, where the ON status (open) it means “1” logic, and the OFF state represents “0” logic.Because the key

Table 1. Possible selectable functions with switch positions SW1-SW4 and the status of the output pins of the key decoding circuit

Function	SW 1	SW 2	SW 3	SW 4	Pin 13	Pin 14	Pin 15	Pin 17	Pin 18	Observations
1	1	1	1	0	1 V	1 V	1 V	12 V	12 V	
2	1	1	0	1	1 V	1 V	12 V	1 V	12 V	
3	1	1	0	0	1 V	1 V	12 V	12 V	12 V	
4	1	0	1	1	12 V	12 V	12 V	12 V	100 Pulse	Total time 11,8 s
5	1	0	1	0	1 V	12 V	1 V	12 V	12 V	
6	1	0	0	1	1 V	12 V	12 V	1 V	12 V	
7	1	0	0	0	1 V	12 V	12 V	12 V	12 V	Annulation points
8	0	1	1	1	12 V	12 V	12 V	12 V	1000 Pulse	Total time 118 s
9	0	1	1	0	12 V	1 V	1 V	12 V	12 V	
10	0	1	0	1	12 V	1 V	12 V	1 V	12 V	
11	0	1	1	0	12 V	1 V	12 V	12 V	12 V	Introduction points
12	0	0	0	1	12 V	12 V	12 V	12 V	100 Pulse	Total time 11,8 s
13	0	0	1	0	12 V	12 V	1 V	12 V	12 V	Small statistics
14	0	0	0	1	12 V	12 V	12 V	12 V	12 V	Large statistics

reading circuit is electrically connected to the electromechanical output pulse meter on pin 18, the presence of the pulses will cause the value displayed by it to change. For all measurements, image captures were taken from the used oscilloscope screen (Figure 5). At the first measurement (the left side of each Fig5a-o) the oscilloscope probes were connected to pins 13, 14, 15 and 16, and at the second measurement (the right side of each Fig5a-o) the probes were connected to pins 13, 14, 15 and 18. Since the signal on the pin 16 is identical to the one on the pin 17 it was not necessary to measure, thing which results from the wiring diagram of the electronic module.

The duration of the low signal level on pins 13, 14, 15 and 16 is caused by the time the key was held on the reader. The length of the time period the pulse for generating impulses on pin 18 is independent of the key holding time on the reader. It has been observed that even if the number of impulses corresponding to the function (100 or 1000) has not been completed and the key is put back on the reader, the number of impulses will increase by 100, for functions 4 and 12, and by 1000, for function 8. The impulses generated for functions 4, 8 and 12 have a period of 120 ms and the filling factor ½, and their amplitude is 11 V.

After the verifications that were carried out, besides the type A machine, there were identified two other types of gambling devices meaning the type of electronic machine with winnings, which use other two types of electronic circuits for reading electronic keys - hereinafter referred to as type B circuits (Figure 6a) and type C (Figure 6b). These circuits, compared to the one marked type A, use another µC, type PIC16F628A, which has the same number of pins and performs the same functions, respectively, reading *iButton* electronic keys, key recognition and the execution of the associated function.

Figure 5. Screenshots of the oscilloscope illustrated in and explanations: a) function 1; b) function 2; c) function 3; d) function 4; e) function 5; f) function 6; g) function 7; h) function 8; i) function 9; j) function 10; k) function 11; l) function 12; m) function 13; n) function 14.

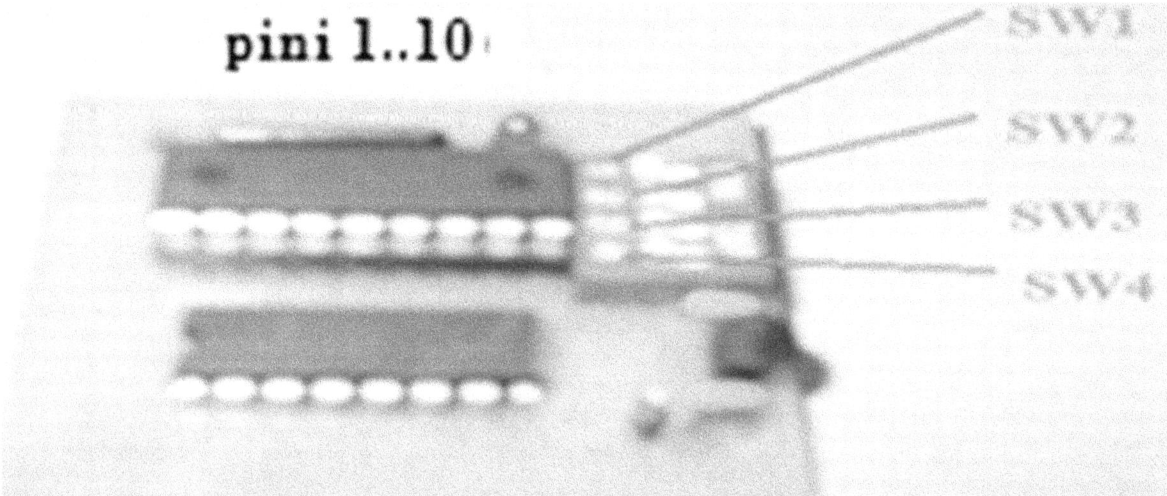

For this type of device there is no information on how the electronic keys are learned and how their association with the function is performed. However, it has been observed that in these cases, the electromechanical pulse meters which show the total of the outputs are connected to the central unit of the game apparatus as well as to the key reading module.

The circuits were developed around the µC PIC16F628A Microchip supply (Bekiroglu, 2008). The microcontroller is equipped with a central processing unit, a 224-byte data memory, a 2048x14-bit *flash* program memory, a 128-byte EEPROM non-volatile memory, an input/output interface with a total of 16 pins, two 8-bit timers, two comparators, a UART (*Universal Asynchronous Receiver / Transmitter*) interface and a CCP/PWM (*Capture, Compare and Pulse Width Modulation*) module - all interconnected by integrated circuit specific technologies.

The device automatically processes the input data, using the software implemented in the program memory, and the processing results are commands transmitted to the output interface.

In addition to the µC PIC16F628A on the respective modules are mounted also: a circuit of the type ULN2003AG, 2 selection blocks with *jumpers* (with 2 and 3 pins), a quartz of 10MHz, 5 LEDs and various passive components (resistors and capacitors).

On the type B circuit there is also mounted a SN74LS07N type circuit and the 3-pin selector which is used to choose which of the µC outputs controls the electromechanical output meter. As the type B circuit has no selection made, the black jumper between the middle pin and one end must be connected, while at type C the pin 18 of the µC, passed through interface circuit ULN2003AN (Aranjo et al., 2012).

On the type B circuit there is also mounted a SN74LS07N type circuit and the 3-pin selector which is used to choose which of the µC outputs controls the electromechanical output meter.

As the type B circuit has no selection made, the black jumper between the middle pin and one of the ends must be connected, while at the type C circuit the pin 18 of the µC is selected, passed through the interface circuit ULN2003AN (Aranjo et al., 2012).. In the type B electronic circuit, the interface of the µC PIC16F628A with the central unit of the gaming device is realized by means of the integrated

Figure 6. Electronic circuits used for reading electronic keys: a) circuit type B; b) circuit type C.

pini 11 ... 20

circuit SN74LS07N (used in other computer applications (Pico Martin, 2011), and the interface with the electromechanical output meter is realized by means of the circuit ULN2003A (also used in other computer systems (Shah et al., 2014). The previous solution was chosen as the electromechanical meters are controlled with 12V pulses, and the 5V supply voltage for μC and the SN74LS07N circuit is not performed locally as in the case of the electronic key reading circuit of the first device.

In the type C electronic circuit, the interface of the μC PIC16F628A with the central unit of the gaming device and the electromechanical output meter is realized by means of the integrated circuit ULN2003A for which a 2 volt power supply mode was used: 12V for the ULN2003A circuit that makes the connection with the central unit and the electromechanical output meter, respectively, 5V for μC.

The inputs to PIC16F628A represent the output of the *iButton* electronic wrench connector and the 2-pin jumper selector, and the outputs in μC are passed through the interface circuits and are available on pins 12, 13, 14 and 15 of the electronic assembly.

The signal collected from the pin 11 of the command module controls the electromechanical pulse meters and is selected by means of a jumper, between those available on pins 17 and 18 of the μC.From a practical point of view, it is observed that the electronic key reading circuit controls the operation of the central unit of the gaming machine. If no electronic key is used, then the central unit is in the way a game can be played, and if an electronic key is used, it will exit the game mode and enter in a manner appropriate to the function of the key used, respectively, displaying partial or integrated accounting data, changing the parameters of the gaming machine, displaying and/or deleting events, etc.

The diagrams of the 2 types of electronic circuits made with µC PIC16F628A are shown in Figure 7, specifying that the electronic circuit corresponding to the type C electronic key reader is provided with a series of non-implanted electronic components that are not represented in the diagram, because they do not modify the µC connection scheme.

Figure 7. Electrical diagram of the SC Baum SRL electronic key reading module: a) type B; b) type C.

In all types of key reading circuits, the µC is located on a socket, not being attached to the printed circuit, which allows it to be easily replaced or reprogrammed.The keys are type *Read Only Memory* (ROM) containing a unique 64-bit series written on the date of manufacture. Data transfer between this memory and another circuit is performed on a single wire at a rate of 16.3 kbps, the transmission protocol being developed by Dallas Semiconductor.

The memory housing is laser-engraved and contains the following information: device type, date of manufacture, and a unique 64-bit serial number. With a DS9490R converter - a USB adapter with built-in support for *iButton* - and the *OneWireViewer* application, all the aforementioned electronic keys can be identified (Haight, 2009; Linke, 2009), the data obtained are presented in Table 2. The information (series, CRC, family code and type) on key no. 7 was easily read, resulting in the series (0000139111E1), the CRC (52), the family code (01) and the key type (DS1990R # F5). It is noted that the data identified coincides with those displayed by the *OneWireViewer* program.

The key number 4 is of another EEPROM type, causing a different behavior of the electronic key reading circuit, its content (Figure 8) or the data contained in it can be modified by a programmer.

Table 2. Types and addresses of electronic keys

No.	Name	Name / alternative	Device address	Description
1	DS1990A	DS2401/ DS2411	5D00000CEC566D01 01 6D 56 EC 0C 00 00 5D	64 unique serial number bits
2	DS1990A	DS2401/ DS2411	A1000010F8D24E01 01 4E D2 F8 10 00 00 A1	64 unique serial number bits
3	DS1990A	DS2401/ DS2411	48000010F927C501 01 C5 27 F9 10 00 00 48	64 unique serial number bits
4	DS1971	DS2430A	B50000022CEA4714 14 47 EA 2C 02 00 00 B5	EEPROM organized as a 256-bit and 64-bit page - a single programmable application registry time
5	DS1990A	DS2401/ DS2411	2C000010AD19DF01 01 DF 19 AD 10 00 00 2C	64 unique serial number bits
6	DS1990A	DS2401/ DS2411	FF00001392BAAF01 01 AF BA 92 13 00 00 FF	64 unique serial number bits
7	DS1990A	DS2401/ DS2411	520000139111E101 01 E1 11 91 13 00 00 52	64 unique serial number bits
8	DS1990A	DS2401/ DS2411	D0000010F923D101 01 D1 23 F9 10 00 00 D0	64 unique serial number bits
9	DS1990A	DS2401/ DS2411	D9000010F99C3101 01 31 9C F9 10 00 00 D9	64 unique serial number bits
10	DS1990A	DS2401/ DS2411	120000157AA40A01 01 0A A4 7A 15 00 00 12	64 unique serial number bits
11	DS1990A	DS2401/ DS2411	56000015956C7401 01 74 6C 95 15 00 00 56	64 unique serial number bits

Figure 8. Content of the key number 4: a) main memory; b) application registry

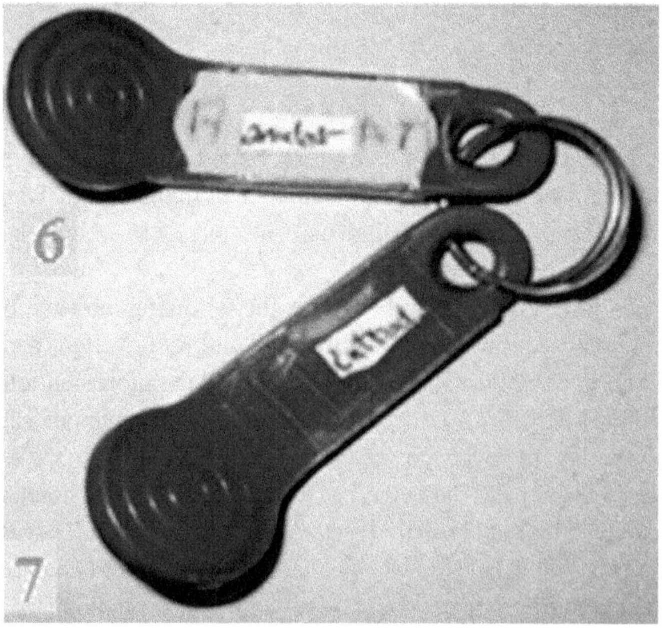

Following the verifications carried out on the analyzed gaming devices the following were found:

1. in all gaming devices the electronic key reading circuit also controls the electromechanical pulse meter which displays the total outputs;
2. the µC with which this circuit is performed is not subject to the metrological checks required to attest the identity with the model of the checks performed when obtaining the model approval;
3. based on the same model approvals, game devices with different electrical connections between the electronic key reading module and the central unit were identified, so a different number of electronic keys.

In Figure 9 are illustrated details of the first gaming device analyzed, from which a pink wire is observed that starts from the electronic key reading module (Figure 10a) and connects with a blue wire from the cable (Figure 10b) leading to the electromechanical meters (Figure 10c). On the respective wire the control flows to the electromechanical output meter. At the same time, the function selector switches when learning a key have been set on the code of the 1000 impulse generation function (see Table 2).

Figure 9. Modification made at the first gaming device analyzed: a) the pink wire connected to the key reading circuit; b, c) the path of the pink thread.

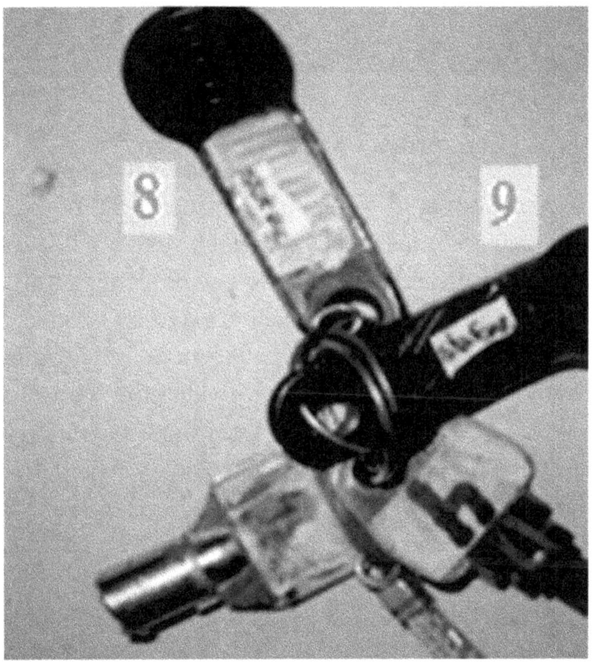

In Figure 10 the connection between the electronic key reading circuit and the output counter (Figure 10a) of the second gaming device analyzed (Figure 10b) was marked with a red arrow. You can see the connection between the brown color wire (Figure 10c), which comes from the key reading circuit, and the blue wire that comes from the central unit and through which the output meter is controlled. Then,

Figure 10. Modification made to the second gaming device: a) the brown wire connected to the key reading circuit; b, c,) the route of the 3 wires leading to the mechanical meter: indication IN 0484845 and indication OUT 0412816.

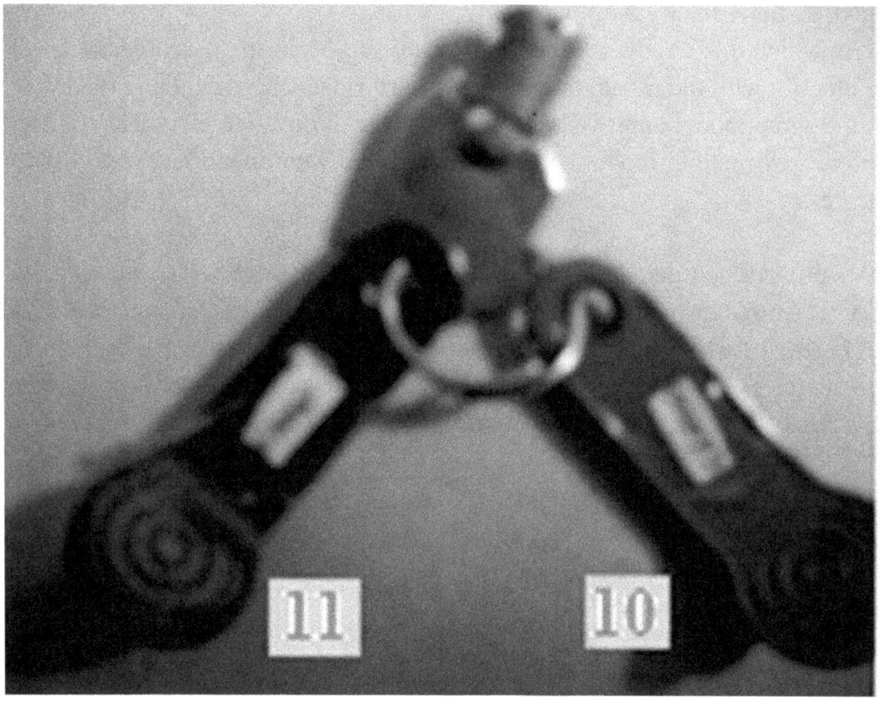

through the three wires - red, blue and yellow - the controls of the input and output electromechanical meters are transferred to a second set of meters on the front panel of the game machine. The set of counters placed inside the unit are not protected by a metal box.

In Figure 11a the connection between the electronic key reading circuit and the electromechanical output meter of the third gaming device analyzed was indicated by a red arrow (Figure 11b). You can see the connection between the brown color wire coming from the key reading circuit (Figure 11c), and the blue wire coming from the central unit and through which the electromechanical output meter is controlled.Then, by means of the three tricolor wires (red, yellow and blue), the controls of the electromechanical meters of input and output are sent to a second set of meters located on the front panel of the electronic gaming device.

From the electrical diagrams it turns out that the electronic key reading module (called "*Electrical key system*") does not control the electromechanical pulse meter which displays the total outputs, which means that the game devices have been modified and no longer correspond with the model approval

SOFTWARE ANALYSIS

Another stage of the analysis consisted in verifying the *logs* of the gaming devices and determining their behavior when the electromechanical output meter is operated by using an electronic key. The *logs* of the devices were read, using the electronic keys to access the respective pages. Gaming devices were

Figure 11. Photographs inside the third gaming device: a) the brown wire connected to the key reading circuit; b, c, d, e) route of the 3 wires to the electromechanical counter (indication IN counter 0499761 and indication OUT counter 0398516).

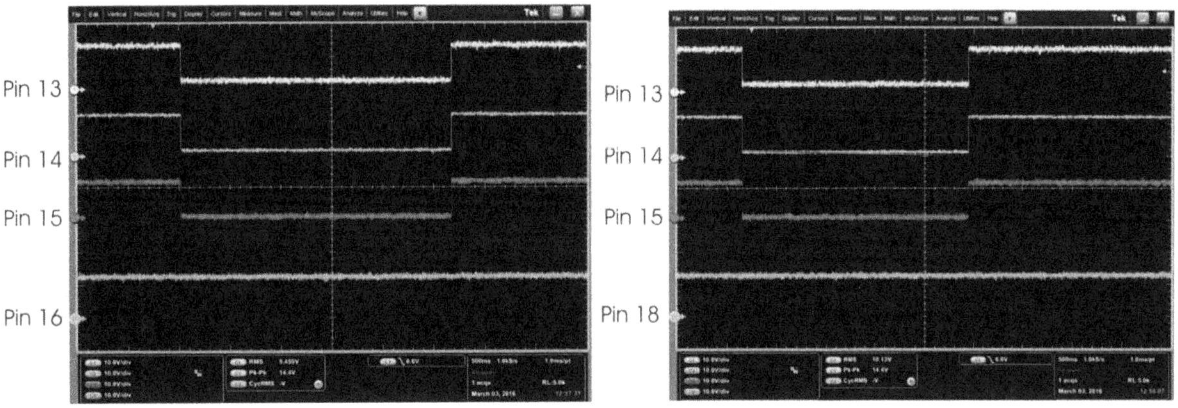

started and photographs were taken with *statistics* and *status* pages, using the keys that determine the generation of impulses.

In the first device analyzed, the electronic key reading module is directly connected to the central unit and the electromechanical output meter, but when operating the meter using an electronic key on the device display, no error message will be displayed. The model has a *log* page available where the errors from the electromechanical meter (Total IN and Total OUT) are counted, the number of starts and the number of errors of the game device, being possible to delete this information.

The machine has several statistics/accounting pages and an event page (Figure 12) - some of these pages can be deleted.

Figure 12. Print screens with the statistics/accounting pages of the first device analyzed

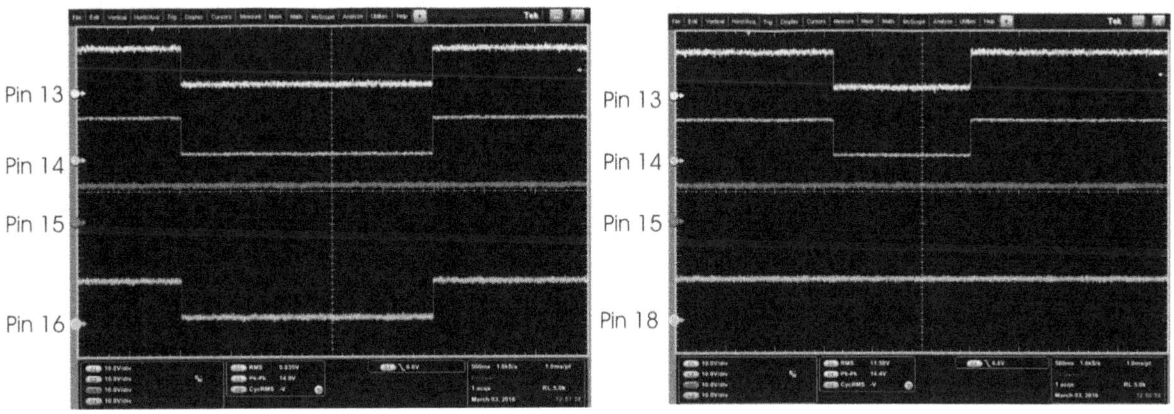

Since the electronic key reading module of this device is connected directly to the output electromechanical meter, and the central unit is connected to the meter by means of a diode, the influence of the diode present in the circuit was determined, proceeding as follows: the key that determines the generation of a number of 1000 pulses by the electronic key reading module has been put and it was observed that when on the meter output of the key reading module the voltage is about 2V on the meter output of the central unit of the gaming device is about 4V as a result of the separation made by the diode 1N4148 (Ocaya, 2006) - voltage level sufficiently high so that the central unit of the device does not detect the actuation of the electromechanical output meter, or its interruption.

Then, the diode was eliminated from the circuit, resuming the operation of the electromechanical meter and it was observed that when on the meter output of the key reading module the voltage is about 2V on the meter output of the central unit of the gaming device is about 1V, moment when the central unit notices the actuation/interruption of the electromechanical output meter.

Just as the first one, the second game device allows the deletion of information using the "*clear PAGE*" command, an action which results in zero for all indications/displays.

On the third gaming device, the pages were accessed and photographed "ACCOUNTING--->MASTER ACCOUNTING" (Figure 13a), "ACCOUNTING--->HARD METER COPY", "ACCOUNTING--->SECURITY ACCOUNTING" (Figure 13b,c) and "EVENT LOG--->MACHINES EVENTS" (Figure 13d-j). In the *log* pages accessed, the last 100 events are retained / stored - entering / exiting the service mode, interrupting the electromechanical meters, actuation of the *cash* button (credit is introduced by the operator), etc. Compared to the other devices analyzed, these *logs* cannot be removed / deleted, and the page "EVENT LOG ---> MACHINE EVENTS" 5/5 shows the error message "E7 1" (Figure 13f) generated as a result for the use of the meter key and the actuation of the *cash* button.

Figure 13. The status and accounting pages of the third gaming machine analyzed

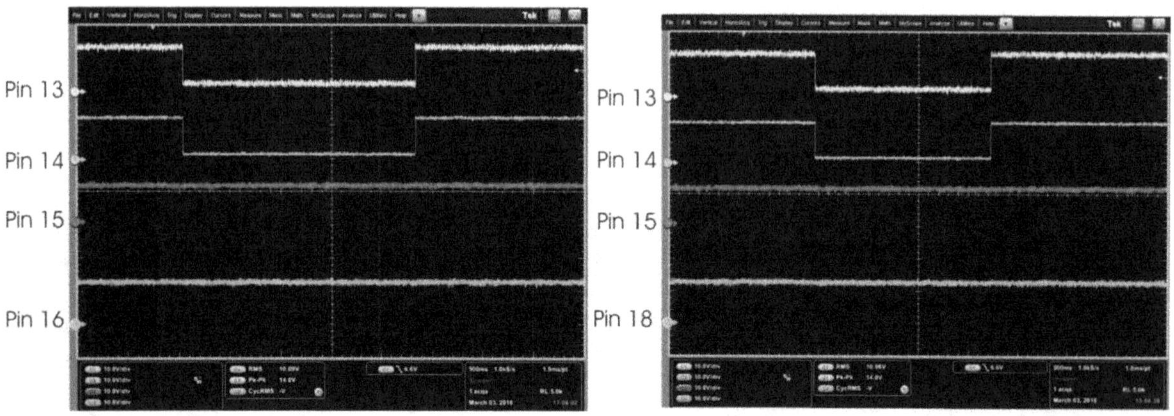

If the electronic key is placed, which causes 100 pulses to be generated, an error message "*E7 1: Meters disconnected*" will appear on the display, and the value of the electromechanical output meter will be increased by 100, because the central unit is connected to this device directly with the electromechanical output meter and the key reading module. The page "ACCOUNTING ---> HARD METER COPY" does not change its content, and on the page "EVENT LOG ---> MACHINE EVENTS" 1/5 a new error message "E7 1" is generated as a result of using the meter key.

The menu with the page with the copy of the mechanical meter values is made by accessing the path "ACCOUNTING ---> HARD METER COPY". Similarly, the events are performed if the electronic key is placed which determines the generation of 1000 pulses, in which case the electromechanical output meter will be increased by 1000 and following the operation of the output meter the error message "E7 1" will be displayed. The page "ACCOUNTING ---> HARD METER COPY" will not change its content, and the page "EVENT LOG ---> MACHINE EVENTS" 1/5 will receive the error message "E7 1" generated as a result of using the meter key.

For this device, a test was also performed which consisted of disconnecting the cable leading to the electromechanical meters, in which case the central unit displayed the error message "E7 1" on the device display. As a result of the key control module present in the analyzed gaming device, it was observed that if a key was associated with functions 4 or 12 (according to Table 1), then when reading this key the circuit will generate a train of 100 pulses with a total duration of 11.8 s, and if it is associated with function 8 (according to Table 1), then at its reading a train of 1000 pulses with a total duration of 118 s will be generated.

FINAL RESULTS AND PROPOSED SOLUTIONS

It was found that µC for the analyzed devices activated the reading protection, which did not allow access to the data content, although extreme methods were used, which in other situations worked (keeping µC for a period at a low temperature - 5 hours) - or power supply µC at an overvoltage, +15 V).

The only possibility to read the *logs* of the electronic gambling machines with winnings was to read the electronic keys, in order to access the respective pages and to photograph them.

The keys used for electronic gambling machines with winnings can be programmed to perform any function, but the effect of placing them on other gaming machines cannot be determined. The key 4 is of type EEPROM, and the data contained can be modified with the help of a programmer. Key no.1 is the master key for all three machines analyzed and with which all the other previously learned keys can be deleted. Keys no.2, 3, 5, 6, 7, 8, 9, 10 and 11 can be used for running any function - the functions used depend on the type of game device - after they are learned by the electronic key reading circuit - keys can be learned to perform up to six different functions. The following pins are connected to the electronic key reading circuit of the first gaming device analyzed: 13, 14, 15 and 16 at the game mode and 18 at the electromechanical output pulse meter. According to Table 1 the keys for functions 4, 7, 8, 11, 12, 13 and 14 have been defined. When placing the key associated with function 7, the points available in the player's account are canceled, and when placing the key associated with function 11, points can be entered in the player's account. By putting the keys associated with functions 13 and 14 in a row, the pages corresponding to the small and large statistics will be displayed on the game device display. When placing the keys associated with functions 4 and 12 the output counter will be increased by 100 points - for each key, and when putting the key associated with the function 8 the output counter will be increased by 1000 points. The following pins are connected to the electronic key reading circuit, of the second gaming device analyzed: 13 to the game mode and 18 to the electromechanical pulse output meter.

According to Table 2 the keys for functions 4, 7, 8 and 12 have been defined. When placing the key associated with function 7, the points available in the player's account are canceled, and by placing the keys associated with functions 13 and 14 in turn, the corresponding pages will be displayed for the small and large statistics. When placing the keys associated with functions 4 and 12 the output counter will be increased by 100 points - for each key, and when putting the key associated with the function 8 the output counter will be increased by 1000 points.

The following pins are connected to the electronic key reading circuit, to the third gaming device: 13, 15 and 16 to the game mode and pin 18 to the electromechanical output pulse meter. The keys for functions 4, 7, 8, 12, 13 and 14 have been defined, according to Table 1. When placing the key associated with the function 7, the points available in the player's account are canceled, and by placing the keys associated with functions 13 and 14 in turn, the pages corresponding to the small and large statistics will be displayed on the gaming machine display. When placing the keys associated with functions 4 and 12 the output counter will be increased by 100 points - for each key, and when the key associated with function 8 is placed, the output counter will be increased by 1000 points.

The key control circuit, related to the electronic gambling machine with winnings, reads the code entered in an *iButton* capsule (electronic key), and if it is identified it will run the function associated with the respective key, if it has been programmed. The pins 13, 14, 15, 16 and 17 of the circuit are connected to the numerical inputs - intended for the keys - of the game module, specifying that depending on the game module more pins or only a single pin can be connected. When reading the key to which the function 4, 7, 13 or 14 has been associated (according to Table 1), the gaming machine will cancel or enter points in the player's account, or it will display pages with statistics or settings. If the key has been assigned the function 4 or 12 (according to Table 1), then when reading this key the control circuit will generate a train of 100 pulses, and if the key is associated with function 8 (according to Table 1), then once with its reading, a train of 1000 pulses will be generated

One of the proposed solutions for checking gaming devices of the type analyzed is to follow the connection of the pin 18 of the electronic key reading circuit to the game module and to the electromechanical contour of output pulses. The specification in the LMB technical norms of the requirements regarding the verification of the electronic key reading module and the cable form of the gaming machine is another proposed requirement. Also, the mention of the cable shape that connects all the modules of a gaming device

The microcontroller with which the reading circuit is made must be subjected to the metrological checks required for attesting the identity with the model of the checks carried out when obtaining the model approval. The current technical verification rules require for the verification to be performed by comparing bit by bit with a standard program owned by LMB, or by verifying the amount of control displayed by the game device and, therefore, establishing the authenticity of the EPROM memory content can only be performed by representatives of legal metrology by comparing bit by bit with a standard program. At the first electronic gambling machine with winnings, it can be established whether the connection with the electromechanical pulse meters is interrupted. Thus, it follows from the contents of the service manual that two error codes are allocated for the meters (Figure 14).

It was found that using an electronic key the game machine reacted in the sense that an error was displayed "E7 1: Meters disconnected" - recorded in the event *log* - when operating the electromechanical output meter through the key reading module. As the log of the machine is organized as a circular stack with 100 events, it turns out that to clear the error message it is sufficient to use the service key 50 times. At each use of the service key - the only key for this type of central unit - the gaming device will store two events: "EE06: *Service mode entered*" and "EE07: *Service mode left*".

The second analyzed gaming device did not respond to the actuation of the electromechanical output meter by the key reading module. The microprocessor of the central unit of this gaming machine

Figure 14. Error codes

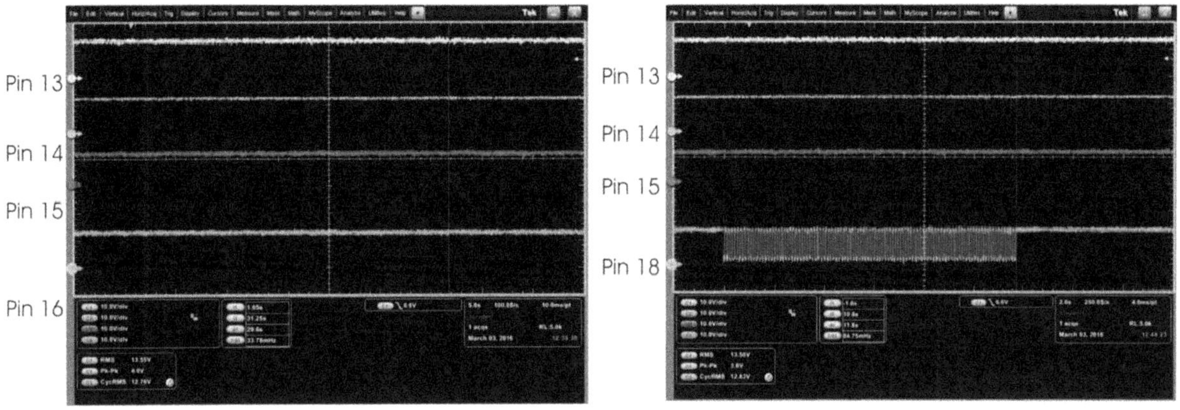

identifies the disconnection of the electromechanical meters, counters these errors in an event page, but does not display any error message on the screen and allows the content of the event page to be deleted.

At the third gaming device analyzed, because between the game module and the electromechanical output meter a diode is mounted, the central microprocessor will not notice when, using an electronic key, the output meter is actuated by the key reading module - the respective gaming machine not displaying any error message. If the diode is removed, the gaming device will display an "E7: METER Errors" error message, when the electromechanical output meter is actuated by means of the electronic key. In this model of central unit, although the errors of the electromechanical meters are counted, nevertheless the program allows their deletion.

It was not possible to determine the operation mode for the electronic key reading circuits of type B and C, although they fulfill the conditions of a special program in the sense that when reading an electronic key a number of pulses is generated for increasing the electromechanical output meter.

CONCLUSION

On the analyzed gambling devices of slot machine type, it has been found that methods and techniques can be used that can lead to illegal winnings, which is a serious security problem and must be remedied in the future.

Thus, it was observed the devices allow the implementation of two computer programs, one in the memory of the central unit of the gaming device (usually, subject to control by the authorities), and the other implemented in the memory of the microcontroller on the decoding/reading circuit of the electronic keys (which it is not subject to the control of the metrological authorities and whose authenticity cannot be established). Moreover, the microcontroller of the reading circuit, in the memory of which the computer program (the software) is written, can be easily changed, as it is mounted on a socket, and the software implemented in this circuit, together with a minor modification of the cable shape of the gambling device, can lead to the modification of the indications of the electromechanical output meter by using an electronic key

On the other hand because the memories in the central unit of the electronic gambling machines with winnings are subjected periodically to the verifications carried out by the legal metrology representatives and since the gambling machines were checked from a metrological point of view, it follows that extracting and reading the contents of the EPROM memories would not bring any additional information in order to influence the performed analysis. Furthermore, comparing the amount of control displayed by a gambling machine with the reference one is not sufficient, because even if the program is modified it can be done in such a way that the amount of control displayed is the one entered in the approval documentation.

REFERENCES

Ahmad, M. A., Leo, K. W., Mohamad, G. H. P., Ahmad, A., Hashim, S. A., Chulan, R. M., & Baijan, A. H. (2017, September). *Development of the automated bunker door by using a microcontroller-system*. Paper presented at theIOP Conference Series: Materials Science and Engineering, Selangor, Malaysia.

Aragay, N., Barrios, M., Ramirez-Gendrau, I., Caballero, A.G., Garrido, G., Ramos-Grille, I., … Vallès, V. (2018). Impulsivity profiles in pathological slot machine gamblers. In Comprehensive Psychiatry (vol. 83, pp. 79-83). Elsevier Inc.

Aranjo, B., Soori, P. K., & Talukder, P. (2012, July). *Stepper motor drives for robotic applications*. Paper presented at thePower Engineering and Optimization Conference, Melaka, Malaysia.

Auer, M. M., & Griffiths, M. D. (2015). Testing normative and self-appraisal feedback in an online slot-machine pop-up in a real-world setting. In S. Kopp (Ed.), *Frontiers in psychology* (pp. 6–339). Bielefeld University. doi:10.3389/fpsyg.2015.00339

Bekiroglu, E. (2008). Microcontroller-based full control of ultrasonic motor with frequency and voltage adjusting. *Sensors and Actuators. A, Physical*, *141*(1), 151–159. doi:10.1016/j.sna.2007.07.018

Brito, M. (2016). *Slot machine game with symbol lock-in*. US Patent 9,520,031.

Erol, Y., Balik, H. H., Inal, S., & Karabulut, D. (2007). Safe and Secure PIC Based Remote Control Application for Intelligent Home. *International Journal of Computer Science and Network Security*, *7*, 179–182.

Haight, C. M. (2009). *One Wire Viewer and an iButton Quick Start Guide*. Retrieved March 11, 2009, from https://www.maximintegrated.com/en/design/technical-documents/tutorials/4/4373.html

Huang, H. C., Tsai, F. C., & Mao, T. F. (2011). Intelligent Embedded Control Warning System For Car Reversing. In *Measuring Technology and Mechatronics Automation* (pp.849-852). Third International Conference Shangshai. 10.1109/ICMTMA.2011.785

Kelly, K., Schwarz, C. J., Gomez, R., & Marsh, K. (2018). Time required to move cash through slot machines. *Journal of Money Laundering Control*, *21*(3), 376–384. doi:10.1108/JMLC-09-2017-0049

Lin, E. I., Eskicioglu, A. M., Lagendijk, R. L., & Delp, E. J. (2005). Advances in Digital Video Content Protection. *Proceedings of the IEEE*, *93*(1), 171–183. doi:10.1109/JPROC.2004.839623

Linke, B. (2009). *One Wire Viewer User's Guide*. Retrieved September 25, 2009, from https://pdfserv.maximintegrated.com/en/an/AN3358.pdf

Lister, J. J., Nower, L., & Wohl, M. J. A. (2016). Gambling goals predict chasing behavior during slot machine play. *Addictive Behaviors*, *62*, 129–134. doi:10.1016/j.addbeh.2016.06.018 PMID:27347654

Manogaran, G., & Lopez, D. (2018). Spatial cumulative sum algorithm with big data analytics for climate change detection. *Computers & Electrical Engineering*, *65*, 207–221. doi:10.1016/j.compeleceng.2017.04.006

Mars, P., Chen, J. R., & Nambiar, R. (2018). Learning Algorithms: Theory and Applications in Signal Processing, Control and Communications. In Stochastic Learning Automata (pp.6-9). CRC Press Taylor & Francis.

Mujal, I., & Colilles, E. (2010). *Estudi d'un amplificadorclasse D*. Retrieved May 10, 2010 fromhttps://upcommons.upc.edu/bitstream/handle/2099.1/9546/memoria.pdf

Ocaya, R. O. (2006). An experiment to profile the voltage, current and temperature behaviour of a P–N diode. In *European Journal of Physics* (Vol. 27). IOP Publishing Ltd. doi:10.1088/0143-0807/27/3/015

Picó Martín, I. (2011). *Sistema Microcomputador per a l'Enregistrament de Dades GPS d'una Bicicleta Elèctrica*. Projecte/Treball Final de Carrera. UniversitatPolitècnica de Catalunya.

Sari, M. B., Sanjaya, Y., & Djamal, M. (2017). Pengembangan Spektrometer Cahaya Tampak Menggunakan LED RGB untuk Menentukan Konsentrasi Glukosa. In Risalah Fisika (vol. 1, pp.21-27). doi:10.35895/rf.v1i1.40

Shah, J., Modi, B., & Singh, R. (2014, February). *Wireless home appliances controlling system*. Paper presented at theInternational Conference on Electronics and Communication System, Coimbatore, India.

Simion, C., & Baciu, A. M. (2018). *Legislation related to the field of gambling*. Retrieved November 20, 2018from https://iclg.com/practice-areas/gambling-laws-and-regulations/romania

Trivedi, A., & Awad, M. (2015). ARM Based Handy and Portable Oscilloscope Using Graphical Display. *Journal of Engineering Research and Applications*, *5*, 106–109.

YiJie, Z., Xiangjun, Z., & Weibo, L. (2010, December). *Power supply's design and research for high voltage circuit of active ECT*. Paper presented at the Conference on Power System Technology, Hangzhou, China.

Chapter 4
How to Integrate Emotions in Dialogues With Pedagogic Conversational Agents to Teach Programming to Children

Elizabeth Katalina Morales-Urrutia
Universidad Técnica de Ambato, Ecuador

Jose Miguel Ocaña
Ejército Ecuatoriano, Spain

Diana Pérez-Marín
Universidad Rey Juan Carlos, Spain

ABSTRACT

Pedagogic conversational agents are interactive systems that allow students to dialogue with them about a certain domain to learn. PCAs have been used in multiple domains from pre-primary education to university, in roles such as teacher, student, or companion. In this chapter, Alcody, a PCA to teach programming to children, is enhanced with a new proposal to manage emotions in the dialogue with students. The goal is that when children are learning to program, Alcody can help them with the emotions associated to the learning. Six emotions have been integrated into Alcody: happiness, anger, sadness, fear, surprise, and disgust. A description of how a PCA to teach programming can modify its face and verbal expressions according to the emotion detected in the student. This is given for any other researcher that would like to incorporate emotions in dialogues between PCAs and students.

INTRODUCTION

A Pedagogic Conversational Agent (PCA) can be viewed as a computer system that interacts with a student in a natural language to learn about a certain domain (Johnson, Rickel, & Lester, 2000). Such agents can assume the role of a lecturer, instructor, tutor, or even of a peer student. One of the first sample PCAs in the role of teacher was Herman the Bug (Lester, Converse, Kahler, Barlow, Stone & Bhogal, 1997), in

DOI: 10.4018/978-1-7998-3355-0.ch004

the role of student was Betty (Biswas, Roscoe, Jeong & Sulcer, 2009), and in the role of companion was Jake & Jane (Arroyo, Woolf, Royer, & Tai, 2009), which incorporated some basic affective interaction.

The use of PCAs has proved to have benefits for education. One reported benefit is "the Persona Effect," which is described as the presence of an interactive agent in an educational computer environment that has a positive influence in the students' perception of their learning experience (Lester et al. 1997). Another reported benefit is "the Proteus Effect," according to which students are motivated to achieve the features of the agents and become more like them (Bailenson, Yee, Blascovich, & Guadagno, 2008; Yee & Bailenson, 2007). Finally, another reported benefit is "the Protégé Effect," in which students can make greater efforts to teach their conversation agents than to study on their own (Chase, Chin, Oppezzo & Schwartz, 2009).

Previous work with agents such as JARO (Pérez-Marín & Caballero, 2013) show the viability of designing agents that dialogue with the students to learn about a certain domain. JARO was friendly and good evaluated in a proof-of-concept experiment with adults. However, in the last decades, advances in natural language dialogue, affective computing, machine learning, virtual environments, and robotics are making possible to design enhanced versions of PCAs with a higher potential impact on education (Johnson & Lester, 2018).

As can be seen in Figure 1, JARO is a static figure. It does not have any possibility of affective interaction, answering the same to all students irrespectively of their emotional state. The need of integrating some kind of affective interaction between the students and the agent was reinforced in the experiments performed with Dr. Roland (Tamayo-Moreno, 2017). Dr. Roland was designed to be used with children from Pre-Primary Education to Primary Education. Children regard PCAs as their friends. Friends have emotions, and learning can be enhanced when emotions are associated (Arroyo et al. 2009).

Figure 1. Snapshot of JARO (source: Pérez-Marín & Caballero, 2013)

Moreover, it has also been detected an increasing interest in teaching programming to children (Pérez-Marín, Hijón, Bacelo, Pizarro, 2018). Alcody (Morales-Urrutia, Ocaña, Pérez-Marín, Tamayo, 2017) is a PCA designed to be used by children to teach them programming. In a first version, it was as Dr. Roland or JARO and it did not integrate affective interaction. In this paper, it is presented, the next step that is to enhance the agent with animated facial expressions and gestures, and able to manage the students' emotions in the dialogue with them. In particular, six emotions have been integrated into Alcody: happiness, anger, sadness, fear, surprise, and disgust. A description of how a PCA to teach programming can modify its face and verbal expressions according to the emotion detected in the student is given for any other researcher that would like to incorporate emotions in dialogues between PCAs and students.

The paper is organised into five sections: Section "Background" describes the related work in Pedagogic Conversational Agents, and teaching programming to children; Section "Alcody" focuses on the description of the PCA enhanced with affective interacting; Section "Learning Environment" describes how to integrate the management of emotions in the dialogue with the children to teach them programming; and, finally Section "Conclusion and future work" ends the paper with the main conclusions and lines of future work.

BACKGROUND

Pedagogic Conversational Agents

Human-Computer Interaction (HCI) studies how to design, evaluate and implement interactive computer systems to be used by human beings (Dix, Finlay, Abowd, & Beale, 2004). Among the frequently discussed usability guidelines for interface design, two can be highlighted: "speak the user's language" and "minimize the user memory load". Both of them can be achieved by using pedagogical agents, that is, static or animated anthropomorphic interfaces that talk to the students on a certain topic of study from University down to Primary education (Johnson et al. 2000; Bada & Meneses, 2012).

Pedagogic Conversational Agents can use textual and/or multimedia resources to keep the dialogue with the student (Konzen, Oliveira, Kist, Anjos, Moraes, Freitas, Müller, Mega Vitta, 2011; Herpich, Voss, Nunes, Medina, 2016). In the 90s, agents needed that students typed their answers to their questions (shown on the screen or verbally reproduced as they have been previously recorded) or they have an interaction based on the use of mouse. This is the case of agents such as Herman the Bug (Lester et al. 1997). Figure 2 shows a sample of one of the modes of the agent. No affective interaction was integrated in Herman the Bug, the modes of the agent change depending on the verbal or non verbal answer.

In 2003, there was a novelty in the interaction by introducing the possibility of projecting the agent on a wall. It was the case of Sam (Ryokai et al. 2003), a personalized agent to help children develop storytelling skills. Figure 3 shows a sample snapshot of Sam. As can be seen, the agent was behind a toy castle, which serves as sample scenery for the story. However, Sam was rather static as it was not designed neither to interact in the dialogue with the student nor with affective interaction, just to be there to encourage children to tell their story.

AutoTutor can be referenced as one of the agents able to simulate empathy with the student through facial impressions and gestures (Graesser, Mello, Craig, Witherspoon, Sullins, McDaniel, & Gholson, 2008). According to their authors, Autotutor is able to increase the average score in an exam up to 1 point (0.8) with respect of no using the agent in the study of the exam.

Figure 2. Snapshot of Herman the Bug (source: Lester et al. 1997)

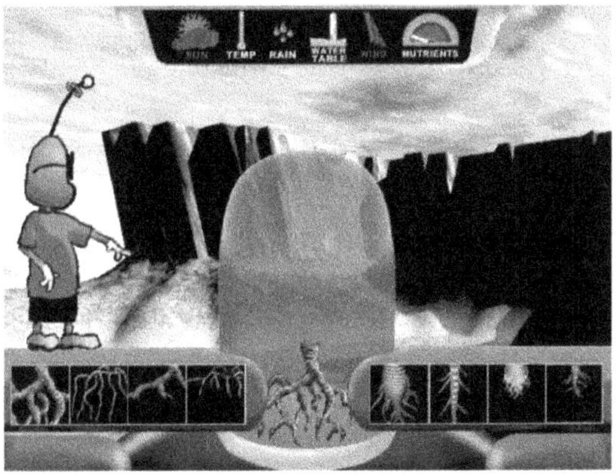

Figure 3. Pedagogic agent Sam (source: Ryokai et al., 2003)

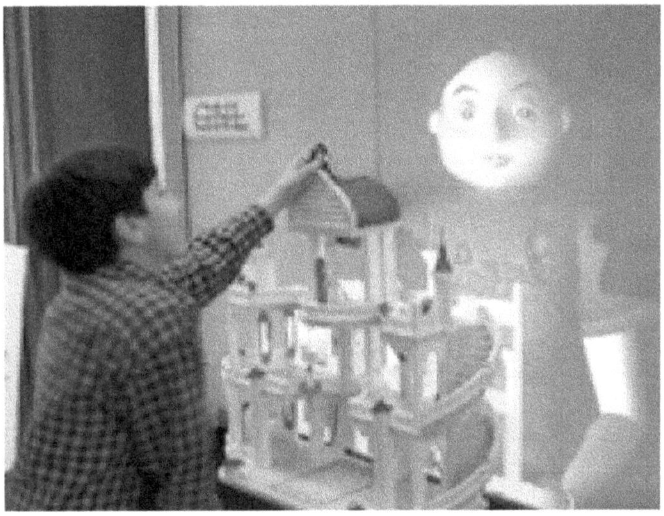

Herman the Bug, Sam and Autotutor have the role of teacher. In 2009, Betty was created as a PCA to interact with the student in the role of student. The core idea was that students could make a greater effort to help Betty to pass the exam, instead of just studying for themselves with positive results (Biswas et al. 2009). Lately, Mr. Davis was introduced as an agent in the role of teacher to supervise that students did not introduce misconceptions in their answers to Betty. Figure 5 shows a snapshot of Betty and Mr. Davis.

Neither Betty nor Mr. Davis show any affective interaction to the student. The agents Jake & Jane were pioneers in showing emotional facial expressions according to the taxonomy developed by Ekman (1999). To find out how the student feel, they ask the student and show a different expression (see Figure 6). Jake and Jane behaved just the same with the only difference of being a boy or a girl to find out whether boy/girl students would choose a boy/girl agent, and to find out more about gender differences in the difference between the boy/girls and their agents (Arroyo et al. 2009).

Since 2013, more agents started to integrate affective interaction such as the Geranium system with the Gera agent (Griol & Callejas, 2013). As can be seen in Figure 7a, Gera could be happy, ashamed,

Figure 4. Snapshot of AutoTutor (source: Graesser et al., 2008)

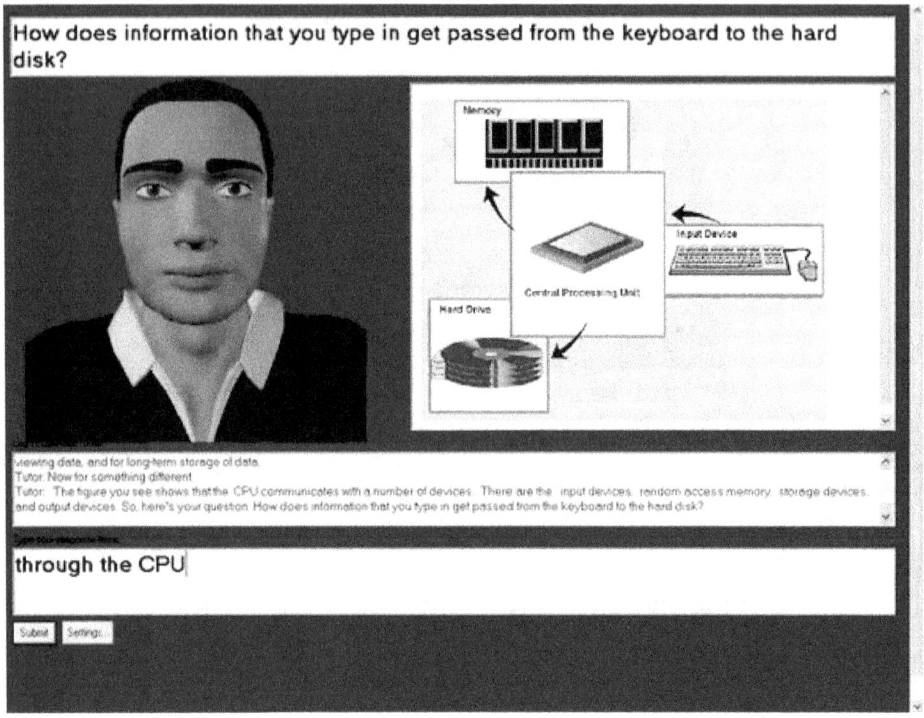

Figure 5. Snapshot of Betty and Mr. Davis (source: Biswas et al., 2009)

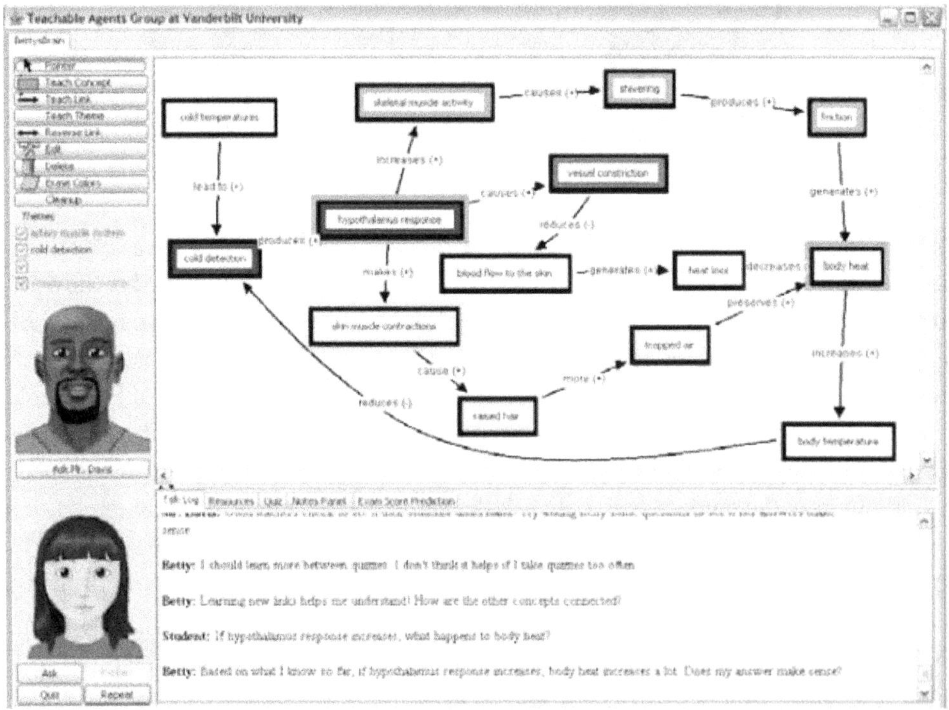

Figure 6. Snapshot of Jake and Jane showing different emotions (source: Arroyo et al., 2009)

sad or surprised. For instance, if the answer of the student was correct, Gera was happy. On the other hand, if the answer of the student was incorrect, Gera was sad (see Figure 7b). Gera could talk, wait or listen depending on the emotion.

The most recent agents can be more realistic in appearance because they can keep the expectatives of the students that the dialogue is going to be quite as natural as with a human tutor (if limited to the domain of the knowledge under the study). Figure 8 shows a snapshot of a realistic agent. Although it does not have affective interaction.

A meta-analysis of 20 experiments using PCAs with gestures revealed that the effect of the gestures of the agent in the multimedia environment on the transfer of knowledge was small-to-medium (David, 2018). It indicates that gestures of the PCA are beneficial for student learning. An important factor to introduce affective interaction is to identify the reason of the emotion and to learn how to manage it (Gil´Adi, 1998)

Learning how to manage emotions should be learnt (Goleman, 1999). This learning starts in the childhood. Each child is unique with his or her interests, abilities and needs. There are many different modes to remember, think and learning. Moreover, emotion is associated to how the brain learns (Barbieri Delgado, Castro Arenas, Gonzales Remache, Lopez Torres, & Ratti Scudellari, 2016).

Figure 7. a) Gera showing different emotions; b) Snapshot of the Geranium system showing a sad Gera (source: Griol & Callejas, 2013)

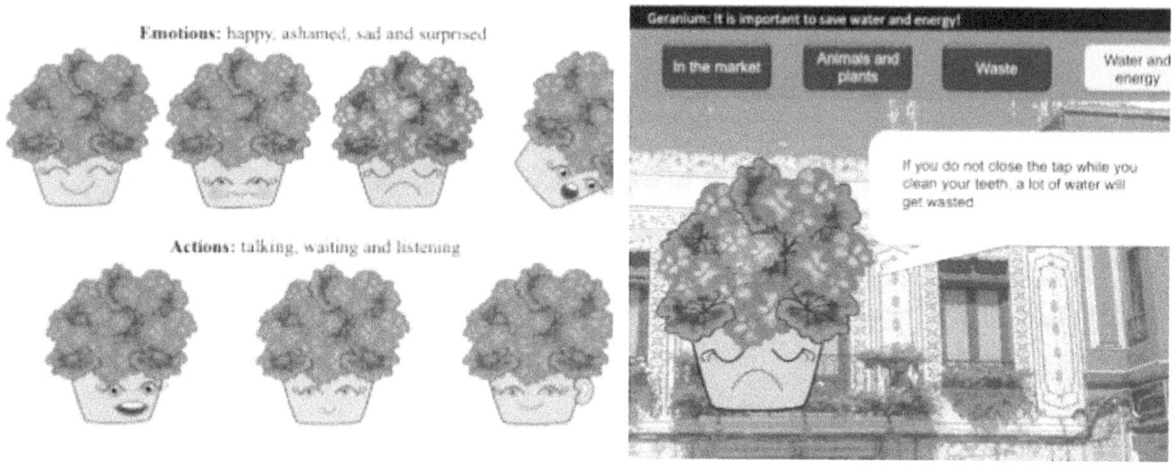

Figure 8. Snapshot of a quite realistic agent (source: Griol & Callejas, 2013)

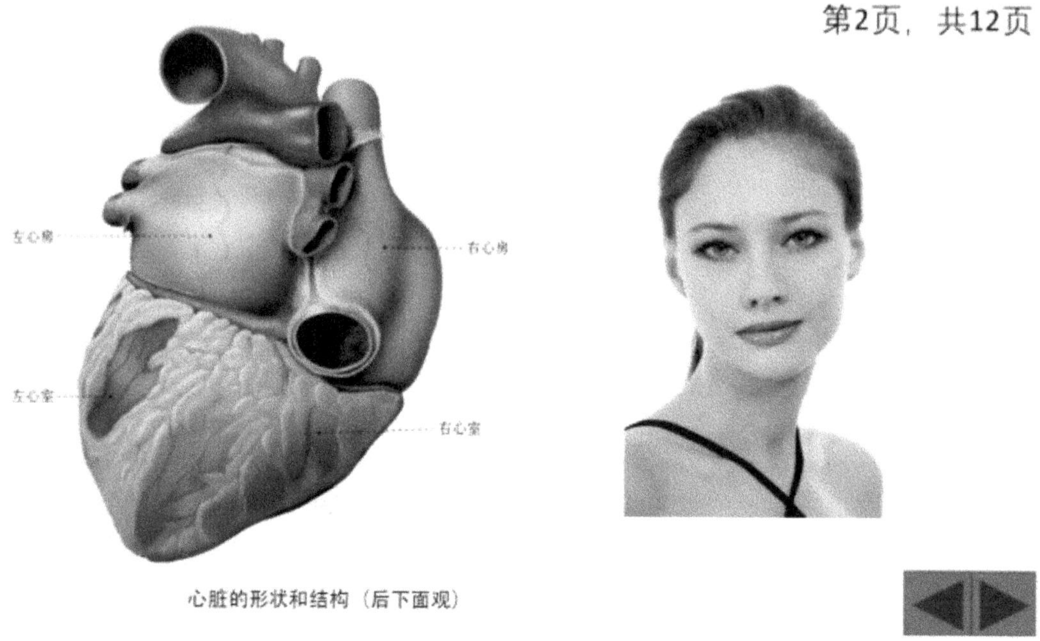

Teaching Programming to Children

Learning how to program can induce changes in the way that people think (Papert, 1980; Resnick, 1996). It is useful not only to develop the ability of solving problems for Computer Engineers, but in general for any person. This could be one of the reasons why teaching Computer Science programming in Primary Education has recently attracted a great deal of interest all over the world (Heintz, Mannila, & Färnqvist, 2016, see Table 1).

A commonly-used approach to teach Computer Science to children is using Scratch, which can be described as an authoring environment, developed by the Lifelong Kindergarten research group at the MIT Media Lab, to design interactive media by snapping together programming-instruction blocks (Resnick, Maloney, Monroy-Hernandez, Rusk, Eastmond, Brennan, et al. 2009).

Other approaches include making your own program (Campe & Denner, 2015), using Lego WeDo or Mindstorms EV3 robots (Sović, Jagušt & Seršić, 2014). With respect to unplugged approaches, they are commonly used in countries with limited resources, but also in developed countries where Computer Science is considered an interesting option, but which lack enough trained teachers and/or computers connected to Internet (Brackmann, Barone, Casali, Boucinha, & Muñoz-Hernández, 2016). In unplugged approaches, the concepts of Computer Science are transmitted by means of storytelling, or free exercises available in Code.org.

ALCODY

From the review of the literature, and up to our knowledge, no PCA has previously been used to teach programming to children. In Morales-Urrutia et al. (2017) a first proposal of how a PCA could be used to teach programming to children was presented. As there was no precedent into how to design such type of agent, the MEDIE methodology to design PCAs was followed (Tamayo-Moreno, 2017).

According to MEDIE, the first step is to communicate with the teaching team. This is the reason why the first step to create the agent has been to talk to programming teachers. They asked for a colourful agent with robot shape for 6-12 years old children. A first prototype in Spanish can be seen in Figure 9.

Table 1. Worldwide interest in teaching programming (based on Heinz et al., 2016)

Country	Content	Form	Primary	Secondary
Australia	Digital Technologies	Own subject and integrated	Compulsory	Compulsory
England	Computing	Replaces existing subject	Compulsory	
Estonia	Programming	Integrated	Compulsory	Compulsory
Finland	Programming	Integrated	Compulsory	
New Zealand	Programming and Computer Science	Own subject		Elective
Norway	Programming	Own subject		Elective
Sweden	Programming and Digital Competence	Integrated	Compulsory	Elective
South Korea	Informatics	Own subject	Compulsory	Elective
Finland	Computer Science	Own subject	Compulsory	Compulsory
USA	Computer Science	Own subject		Elective
Macedonia	Computers and basics of programing	Own subject	Compulsory	

Figure 9. A first Spanish prototype of iProg (old name of Alcody, source: Morales-Urrutia et al. 2017)

iProg (old name for Alcody) had two scenarios as requested by the teachers: storytelling and chef. The storytelling option allows the children to create scenes, choose characters, objects and personalize the actions of the characters by using the iProg language programming. The goal is that children while interacting with iProg can create a story at the time that they are playing and programming.

iProg chef allows to apply the cooking-programming metaphor (Pérez-Marín, Hijón-Neira, Martín-Lope, 2017) that compares cooking with programming, and algorithms with recipes that should be followed step by step. The goal is that children while interacting with iProg can cook a recipe at the time that they are playing and programming.

To make how the PCA would be used clear the following learning paths are provided. In the case of iProg storytelling, the agent will first ask for a goal of the story, and later the agent will register the objects that the child chooses, and how they are used to fulfil the story goal. In the case of iProg chef, the agent will ask for a recipe to cook, and the child will execute step by step as in an algorithm the steps together with iProg to get the meal ready. In both scenarios, it is taken into account how children have an initial mental structure based on before (initial analysis of the elements and actions to perform), during (the programming), and after (the result of the programming and how it has affected the elements of the story in iProg storytelling or the meal in iProg chef).

However, iProg does not have affective interaction. The next prototype is called Alcody, as it has been designed to have facial expressions and gestures. Alcody has human shape with technologic colour (although it can be changed by the student) and according to the needs of the students should talk in a friendly voice (Morales-Urrutia, Ocaña, Pérez-Marín, Tamayo-Moreno, in press). Figure 10 shows the interface of Alcody and how it changes in different colours.

The eyes of Alcody has been designed to have expression, and the idea is to change the agent according to the emotion or the dialogue with each student. Regarding the general form of ALCODY, it is an ovoid with round details and complements, basic and simplified figures that define the character's shape are used; the strokes are fine and do not have a great impact on the global structure, the structure is more simplified and represents technology, it is related to the experience that children have in handling media and devices. In addition, the glasses represent an identity element of the character's name

Figure 10. Alcody (source: Morales-Urrutia et al. in press)

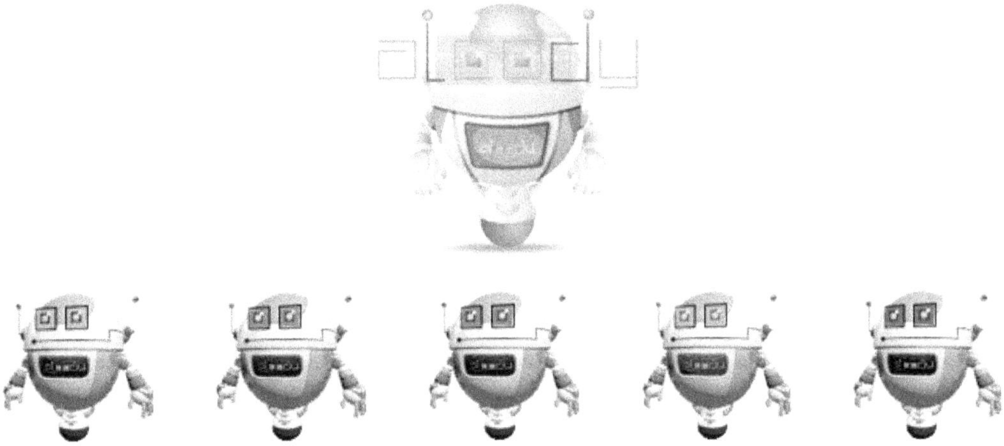

ALCODY, even though it is not 100% visible, this is planned as a differentiation strategy that deepens the character's identity.

Alcody pretends to be a quite realistic state-of-the-art intelligent robot that guides children in their programming learning. The assigned name presents qualities of a multifaceted being with personality, cheerful, friendly, active, with immediacy in communication and guidance of activities. Alcody has a compact form, where the most representative elements stand out like the eyes and the screen of the body, while the complementary elements that do not interact in the interface have been simplified to facilitate the interaction of the character.

Learning in children requires immediate and adequate responses to what they are experiencing within the educational process. Children can feel easily invaded by their emotions and they need a guide to accompany and empathize with them. Emotions have had several contextualizations and various approaches both in social, private or educational life, so expressions come hand in hand with feelings and emotions.

For Darwin, facial expressions were learned socially, and therefore they would have to vary in different parts of the world. Against all odds he saw that facial expressions of certain emotions were repeated wherever he went. Ekman (1972) claimed that all basic emotions are universal, primitive, independent of culture, has its own facial expression, activates organism and brain in a specific way and prepares the body for an action. Ekman proposed a classification of universal emotions: happiness, sadness, fury, fear, disgust and surprise.

An analysis of how to integrate each emotion in Alcody has been carried out to empathize with the children according to the mood they may be. Alcody is able to move arms and hands to make gestures as a communication strategy with children. The result is six designs for Alcody (see Figure 11): Happiness (see Figure 11a), Fury (Figure 11b), Sadness (Figure 11c), Fear (Figure 11d), Surprise (Figure 11e), and Disgust (Figure 11f). In the following paragraphs, it is explained how the face and body of Alcody was modified according to the terms (Robb, 2015) identified in the dialogue with the student.

In the case of happiness, the eyes of Alcody are bright, cheeks are raised and lips form a semiarch. The terms shown in Table 2 serve as trigger to find out when happy Alcody should be used.

A sample dialogue between happy Alcody and a child could be the following:

Figure 11. Emotions in Alcody (source: prepared by the authors). A) Happiness: 1. bright eyes, 2. raid cheeks, 3. lips forming a semiarch. B) Fury: 1. the eyebrows go down and come together, 2. the look becomes enraged, 3. narrow lips. C) Sadness: 1. raised eyebrows inside, 2. depressed corners, 3. raised chin. D) Fear: 1. raised upper eyelids, 2. lips slightly stretched horizontally in the direction of the ears, 3. raised eyebrows and joints, 4. tense lower eyelids. E) Surprise: 1. raised eyebrows outward, 2. raised eyelid, 3. chin down. F) Disgust: 1. uneven eyebrows, 2. lip curves up, 3. narrow eyes, 4. cheeks rise

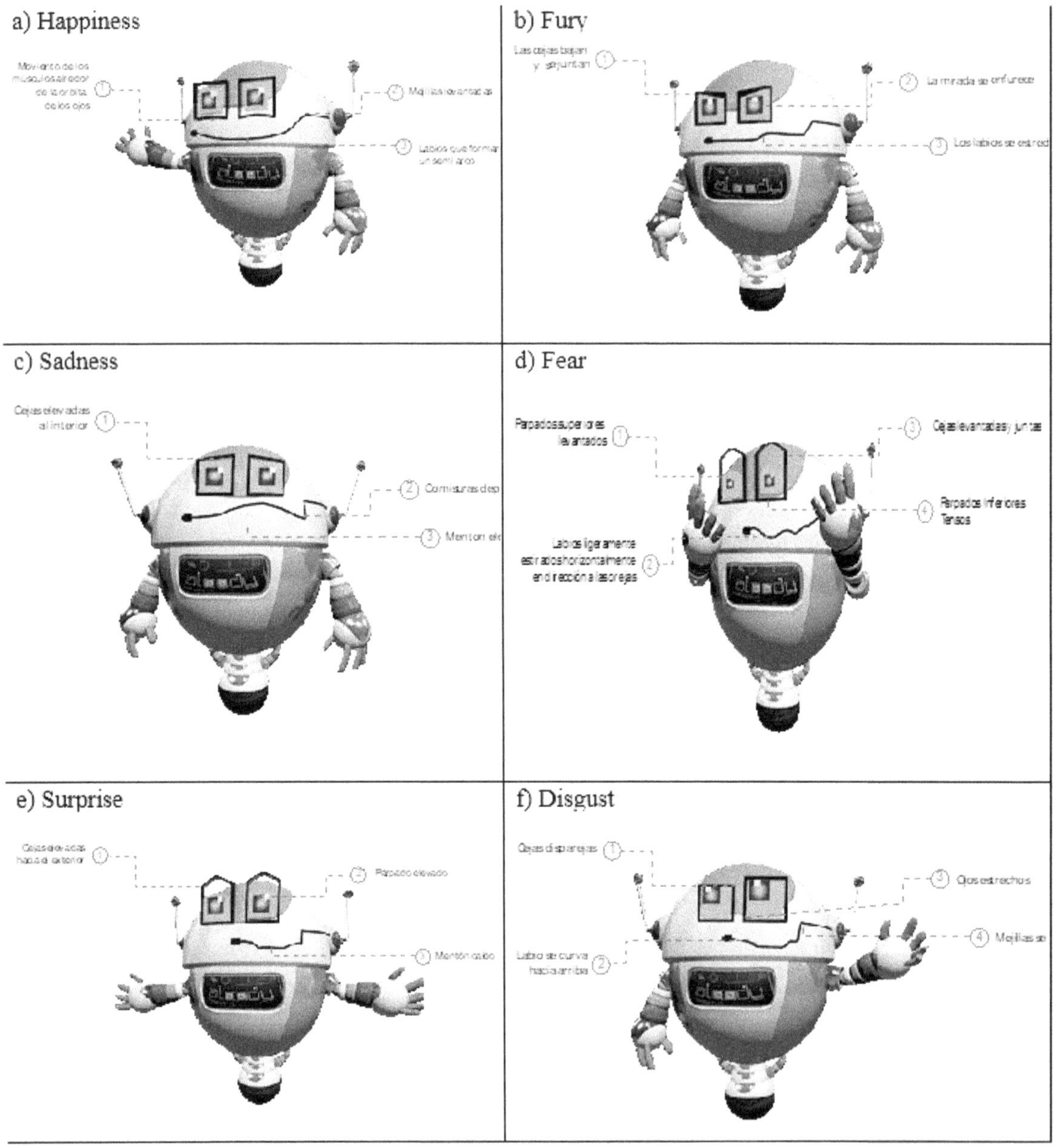

Table 2. Happiness terms for Alcody

Released,	Cheerful	Happiness
euphoric		
entertaining	Interested	
curious		
important	Proud	
sure		
respected	Accepted	
satisfied		
brave	Powerful	
provocative		
affectionate	peaceful	
hopeful		
sensitive	Intimate	
joker		
open	Optimistic	
inspired		

Table 3. Fury terms for Alcody

Sorry	Injured	Fury
Devastated		
Attacked	Threatened	
Jealous		
Resentful	Full of hate	
Outraged		
Rabid	Crazy	
Furious		
Provocative	Aggressive	
Hostile		
Enraged	Frustrated	
Irritated		
Introvert	Distant	
Suspicious		
Skeptical	Critical	
Sarcastic		

Figure 12. Sample messages provided by: a) Fury Alcody, b) Sad Alcody (source: prepared by the authors). The dialogue is translated in the text of the paper.

Child: *I feel* **brave** *today!*

Alcody: *Every day you have an opportunity to achieve your goals.*

In the case of fury, the eyebrows go down and come together, the look becomes enraged and the lips narrow. The terms shown in Table 3 serve as trigger to find out when angry Alcody should be used. A sample dialogue between angry Alcody and a child could be the following (see Figure 12a):

Child: *I am* **furious***! He has hit me!*

Alcody: *When we are irritated we can hurt others. Helping others can make us feel better.*

In the case of sadness, Alcody has raised eyebrows inside, depressed corners and raised chin. The terms shown in Table 4 serve as trigger to find out when sad Alcody should be used. A sample dialogue between sad Alcody and a child could be the following (see Figure 12b):

Child: *I feel so* **helpless***... (crying)*

Alcody: *Sometimes we think that everything is wrong, and we feel bad. Being sad is not bad, but remember that you are in charge of your emotions. You can choose to breath to try to calm yourself and become happier.*

In the case of fear, Alcody has upper eyelids raised, lips slightly stretched horizontally in the direction of the ears, raised eyebrows and joints, tense lower eyelids. The terms shown in Table 5 serve as trigger to find out when scared Alcody should appear on screen.

A sample dialogue between scared Alcody and a child could be the following (see Figure 13a):

Child: *I am so* **worried***...*

Alcody: *May be you had a nightmare. That could make us feel scared. It is not bad to have a nightmare. Remember that you are in charge of your emotions. Try to breath to calm yourself and become happier.*

In the case of surprise, Alcody has eyebrows raised outward, eyelid raised, and chin down. The terms shown in Table 6 serve as trigger to find out when surprised Alcody should appear on screen. A sample dialogue between surprised Alcody and a child could be the following (see Figure 13b):

Child: *I am* **shocked***! It is too much homework!*

Alcody: *I understand that you are surprised that you have so much to do. It seems that it is really complicated and that it is impossible to finish all. I am sure that if you try to solve them one by one, you will be able to finish them and go back to playing.*

Finally, in the case of disgust, Alcody has uneven eyebrows, lip curves up, narrow eyes, and cheeks rise. The terms shown in Table 7 serve as trigger to find out when disgusted Alcody should appear on screen. A sample dialogue between disgusted Alcody and a child could be the following:

Figure 13. Sample messages provided by: a) Scared Alcody, b) Surprised Alcody (source: prepared by the authors). The dialogue is translated in the text of the paper.

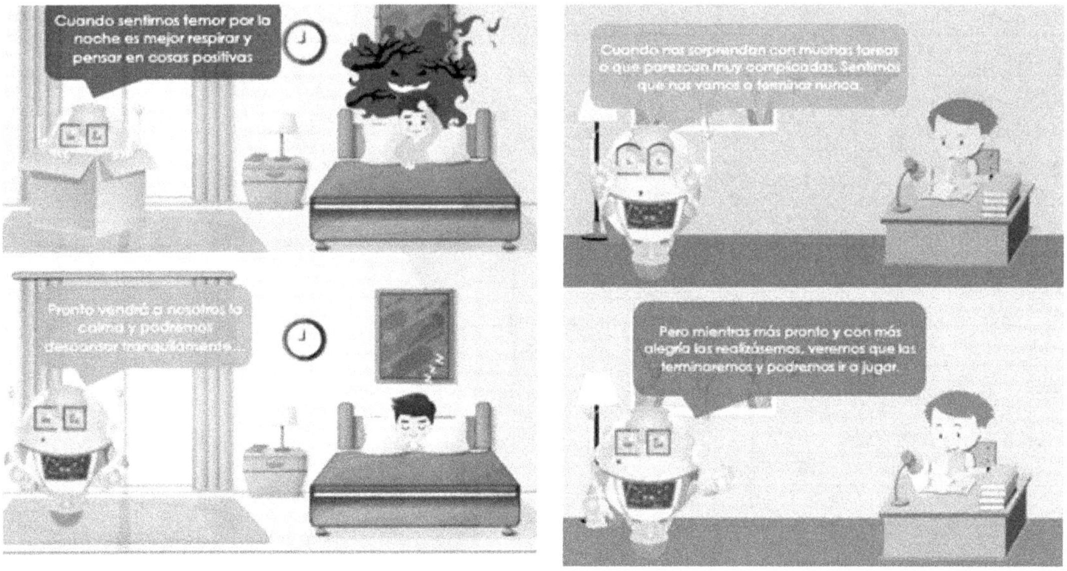

Table 4. Sadness terms for Alcody

Regretful	Guilty	Sadness
Ashamed		
Ignored	Abandoned	
Victimized		
Helpless	Desperate	
Vulnerable		
Melancholic	Depressed	
Empty		
Helpless	Alone	
Isolated		
Apathetic	Bored	
Indifferent		

Table 5. Fear terms for Alcody

Ridiculous	Humiliated	Fear
Disrespected		
Alienated	Rejected	
Outcast		
Insignificant	Submissive	
Useless		
lower	Insecure	
Insufficient		
Worried	Anxious	
Overwhelmed		
Frightened	Scared	

Table 6. Surprise terms for Alcody

Shocked	Surprised	Surprise
Dejected		
Disappointed	Confused	
Perplexed		
Thunderstruck	Amazed	
Impressed		
Enthusiastic	Excited	
Energetic		

Table 7. Disgust terms for Alcody

<table>
<tr><td>Moralist</td><td rowspan="2">Dissatisfied</td><td rowspan="8">Disgust</td></tr>
<tr><td>Reluctant</td></tr>
<tr><td>Disgusting</td><td rowspan="2">Disappointed</td></tr>
<tr><td>Unruly</td></tr>
<tr><td>Disgust</td><td rowspan="2">Horrible</td></tr>
<tr><td>Hateful</td></tr>
<tr><td>Aversion</td><td rowspan="2">Abstinence</td></tr>
<tr><td>Hesitant</td></tr>
</table>

Child: *I am* **reluctant** *to finsih this program. I hate to program!*

Alcody: *Sometimes there are tasks that we do not enjoy. However, we have to learn to make an effort to complete them. Remember that after programming, you can go back to other tasks that you prefer. Even, may be, with time you start enjoying programming!*

It should also be noted that the expression of emotions in the agent is momentary. Alcody personifies the emotion and returns to a neutral state (see Figure 10). The core idea is that emotions can change.

LEARNING ENVIRONMENT

For the integration of the conversational agent Alcody and the emotions in a learning environment, a friendly interface has also been designed. First of all, children are able to personalize their profile. They can modify their name, gender, colour for the interface and for Alcody. The goal is to provide children with a feeling of relevance to their preferences in the digital learning environment. Figure 14 shows the personalization interface of the learning environment for the children.

In order to identify the user's mood at the start of activities or questionnaires, the user is asked to select on the scale of emotions (happiness, disgust, anger, fear, surprise, or sadness) the most appropriate emotion with their current mood. The answer is recorded in the database of the application. Similarly, at the end of the activity or questionnaire, children are asked to identify their emotion. This information is recorded to identify the behavior of the mood of the children during the use of the application and its correlation with the learning results in the final exam. Figure 15 shows the emotion scale.

The six emotions of Alcody are related with recommendations through messages that are given according to the sentences typed by each child. Whenever a term shown in Tables 2-7 is in a sentence typed by a child, Alcody answers with the correlated message. See Figure 16 for sample dialogues with affective interaction integrated in the chat with Alcody. The text translated to English would be the following:

Alcody: *Hello Elizabeth. How do you feel now? (the scale emotion is shown)*

Elizabeth: *I am upset.*

Figure 14. Snapshot of the personalization interface (source: prepared by the authors). The text translated to English is: To edit the profile, Select the colour for Alcody, Update and Cancel.

Alcody: *I hope that you feel better after working with me. Please, look at this dialogue.*

Message: *When things are not done in the way we like, we can react badly with people we love. However, it is better to explain ourselves with tranquility and affection, so they will treat us in the same way.*

Moreover, when children click on the thumbnail image that accompanies each of the recommendations, a pop-up window with the image of the recommendation is launched, allowing the user to see it properly, and each of these messages is intended to induce a change in the children. The core idea is to encourage them and support them in their programming learning process.

Each of the messages of positive recommendations has related intentions, which are similar sentences that are constantly updated, allowing to continue nourishing the database. Whenever the agent finds out sentences that cannot classify automatically, a supervision mode is provided to teach Alcody as shown in Figure 17.

Figure 15. Snapshot of the scale of emotions for Alcody (source: prepared by the authors). The text translated to English is the following: Hello Elizabeth. How do you feel now? …Write your message.

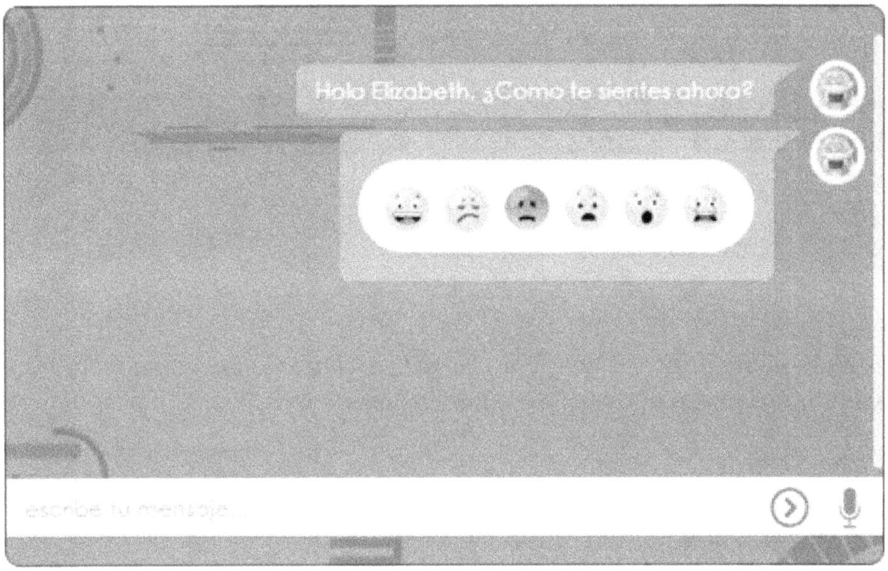

Figure 16. Sample dialogues with affective interaction with Alcody (source: prepared by the authors). The dialogue is translated in the text of the paper.

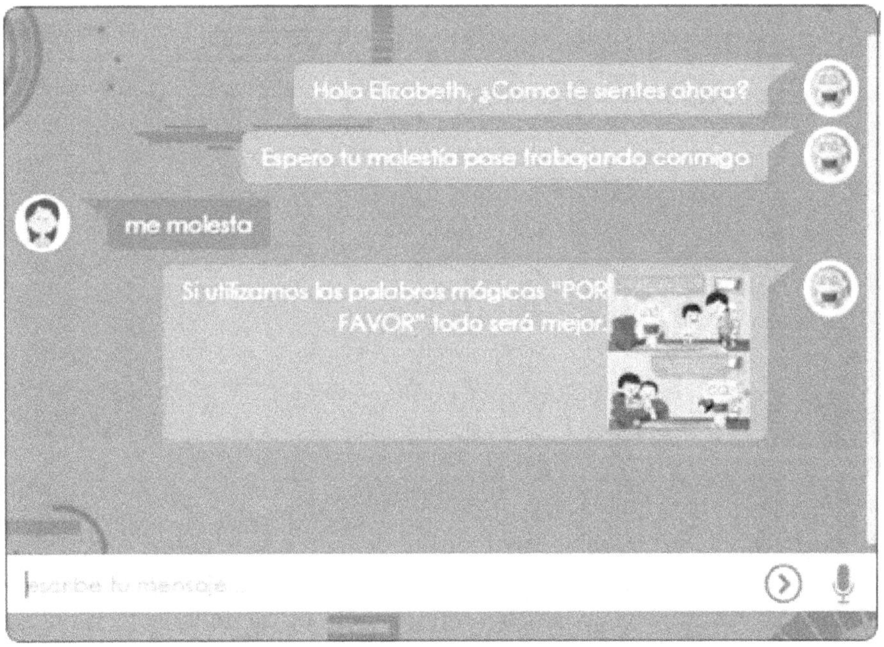

Figure 17. Supervised learning mode of Alcody (source: prepared by the authors). The text translated to English is the following: To assign messages. Sentence. Repetitions. Intention. Action. "It makes me feel angry".

Figure 18. Learning environment of Alcody, the chat is in AIchat (source: prepared by the authors). The sample program is translated in the text of the paper. The button is "Run".

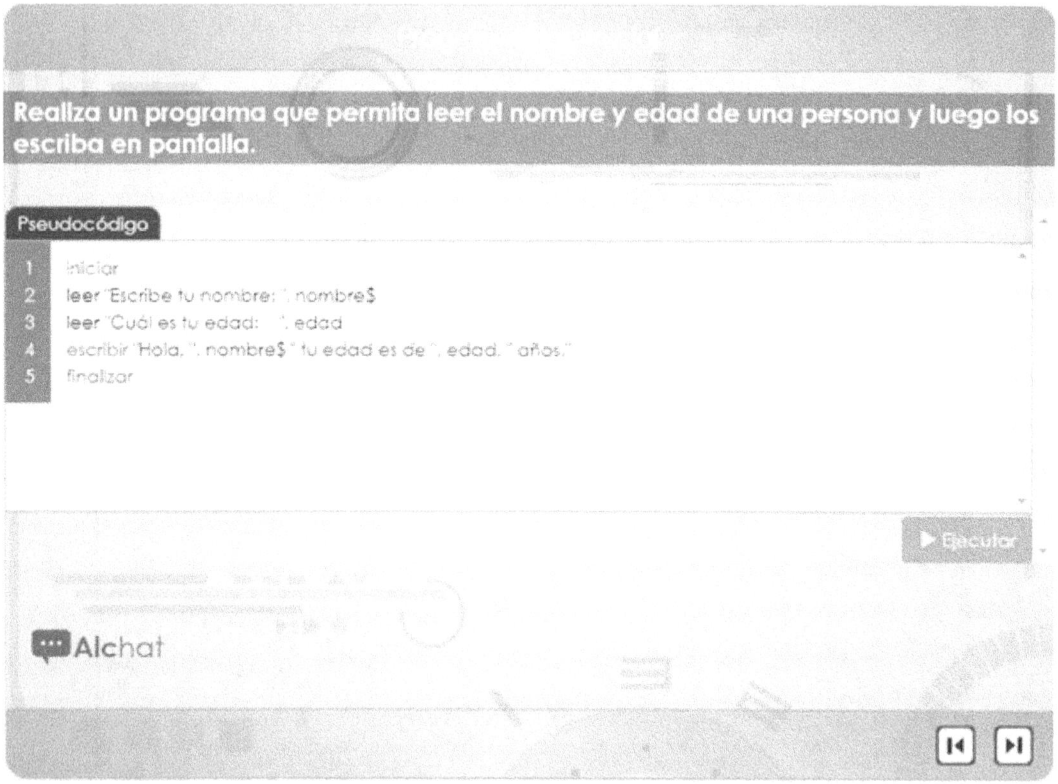

The chat is integrated into the learning environment of Alcody (see Figure 18) while the student is typing the program requested. Students have also a "Run" button to test whether their program works, or if they should modify something. A sample exercise could be the following: Write a program that reads the name and age of someone and writes that information on screen. P-Code:

```
start
read "Write your name:", name$
read "Write your age:", age
write "Hello ", name$ "you are", age, "years old."
finish
When the child press the "Run" button, the result is shown on the screen:
Result:
Write your name: María Sol
Write your age: 8
Hello, Maria Sol, your age is 8 years old.
```

Alcody can also provide some debugging support, indicating if there is something missing such as "=" or "start". Help messages in the sentences written by the students are also identified. Whenever a help message is detected because the student has written something similar to "I do not know how to program" a tutorial is triggered as shown in Figure 19. The text of the dialogue translated to English would be the following:

Alcody: *Hello María Sol. How do you feel now? (the scale emotion is shown)*

Elizabeth: *I do not know how to program.*

Alcody: *I can teach you with a tutorial.*

Figure 19. Tutorial offered by Alcody (source: prepared by the authors). The translation to English would be: Alcody teachers you p-code! The names of the slides are translated below in the text of the paper.

There content of the tutorial available in Alcody (the text of the slides have been recorded with a child voice so that students can have the information both written in the slide and in audio if they have speakers in their device) is the following (it is reproduced here, in case other researcher would like to replicate it):

1. **Overview:** Alcody tells the child "Hello friend, i want to tell you that to start programming, you must write the word "start" and when you finish your program, you have to write "finish".
2. **Variable:** What is a variable? a variable is a space in memory that assigns values to a data. in alcody we have two types of variables: numerical and text. to differentiate them when writing a text variable it should finish with $. it is not necessary, unlike other programming languages, to declare all variables at the beginning of the program. you just name it and alcody will know that you have created a variable. remember that when you create a variable it does not have any value, if you want that the variable has a value you need to give it with the "=" symbol. for instance, to give the variable "age" a "1" value, then you have to type: "age = 1".
3. **Read and Write:** If you want Alcody to read a value of a variable, you have to use the "read" command. If you want Alcody to write something on screen, you have to use the "write" command. A sample program to ask for your age, save it in a variable called "age" and write it on screen would be the following:

```
START
WRITE "WHAT IS YOUR AGE?"
READ AGE
WRITE "YOU ARE ", AGE, " YEARS OLD."
FINISH
```

6. **CONDITIONALS:** If you want Alcody to make a choice, it is called in programming "a conditional" and you have to use the "if" command. a sample program to ask Alcody whether the value is greater than 10, and if so, to write "it is greater than 10" on screen would be the following:

```
START
WRITE "WHAT IS YOUR AGE?"
READ AGE
IF AGE > 10 WRITE "YOU ARE ", AGE, " IS GREATER THAN 10."
FINISH
```

7. **Loops:** If you want Alcody to repeat the same action several times, you have to use the "for" command. a sample program to make Alcody count from 1 to 5 would be the following:

```
START
FOR I=1 TO 5
 WRITE I
 NEXT I
FINISH
```

8. **Exercises**: Some sample exercises are provided to the student.

Finally, there is also a tutorial to assist students in their emotional dialogue with Alcody. It could be frustrating for the children if Alcody is unable to provide them with the correct assistance for their emotion. The tutorial of how to emotionally communicate with Alcody remember the students the terms for each emotion. Children have to read the tutorials at the beginning of the first session with Alcody. It is also advisable that they learn words to express their emotions. Learn to put a name to their emotions is the first step to be able to manage them (Siegel & Bryson, 2012). In the following sessions, they can look at the tutorial at any time during the time that they are interacting with Alcody.

CONCLUSION AND FUTURE RESEARCH

The Pedagogic Conversational Agent Alcody to teach programming to children has been integrated the possibility of managing six basic emotions in the dialogue with the student according to the sentences typed by them. This makes possible that children learn programming while they are accompanied by a PCA that it is not only a teacher but also a companion. From the state-of-the-art review, it is clear that learning is associated to emotion, and that introducing PCAs into learning environments can improve the learning scores of the students.

Alcody's facial development and emotions entailed an analysis of the basic expression characteristics of a face, in order to establish certain facial features of empathy between the user of the application, its mood and the agent. The integration of Alcody and the web interface for the interaction process through chat has been carried out under an analysis of possible textual phrases related to the emotions embodied by Alcody.

The emotions managed by Alcody are: happiness, sadness, fear, surprise, disgust and fury. These emotions are personified by Alcody. The emotion is activated in Alcody when the child writes through the keyboard or talks using the microphone. The key terms are identified and messages are given as recorded in the database. The goal is to help students to face various situations that can happen in the classroom or outside it.

In order for Alcody to continue growing in its vocabulary and to cover more terms related to emotions, the agent accepts any term expressed by the children. Later, the teacher can review the new terms. For each term, the teacher can indicate with which emotion is related. That way, when a student uses in the future those new terms, Alcody will have learnt which messages should provide.

To sum up, Alcody is a proposal that aims to guide the learning process of programming taking into account the emotions of the students. Alcody can also be used as a tool that gives the teacher the possibility of taking into account the mood of the students by reading the dialogue of the students with Alcody and the terms they use. This work contributes to two research fields: Pedagogic Conversational Agents with affective interaction and teaching programming for children.

As future research, it is scheduled to go to schools both in Spain and Ecuador to test the proposal with children. The hypothesis is that by using Alcody with emotions, the learning results will also be improved given the relationship between learning and emotion. If the hypothesis is proven, more emotions could be considered for Alcody such as frustration, boredom or tiredness. Moreover, more sophisticated techniques to identify the emotion could be used, such as eye tracking or sensors.

Other line of future research is to animate Alcody so that it can move on the screen. Currently it appears like in a messenger interface. Children indicated that they did not only like to have a friendly voice and face, but an agent that do sports like football. It could be possible to create scenarios in which

Alcody is dressed as a football player. Children could create programs to manage the football match, while Alcody keeps track of their emotions to answer with the most adequate assistance for each case.

REFERENCES

Arroyo, I., Woolf, B., Royer, J. M., & Tai, M. (2009). Affective gendered learning companions. In *Proceedings of the International Conference on Artificial Intelligence in Education* (pp. 41-48). Brighton, UK: IOS Press.

Bada, E., & Meneses, C. (2012). Una propuesta para la extracción de preguntas y respuetas. *XVII Congreso Internacional de Informática Educativa (TISE).*

Bailenson, J., Yee, N., Blascovich, J., & Guadagno, R. E. (2008). Transformed social interaction in mediated interpersonal communication. In E. A. Konijn, S. Utz, M. Tanis, & S. B. Barnes (Eds.), *Mediated interpersonal communication* (pp. 78–99). New York, NY: Routledge.

Barbieri Delgado, A. F., Castro Arenas, R. Y., Gonzales Remache, T. M., Raffo Lopez Torres, M. A., & Ratti Scudellari, M. C. (2016). *Mejorar el proceso de enseñanza aprendizaje para el desarrollo de las habilidades emocionales de los niños y niñas de 5 años del aula Amarilla de la Institución Educativa cuna jardín Armatambo del distrito de Chorrillos perteneciente a la UGEL 07.* Academic Press.

Biswas, G., Roscoe, R., Jeong, H., & Sulcer, B. (2009). Promoting self-regulated learning skills in agent-based learning environments. In *Proceedings of the 17th International Conference on Computers in Education* (pp. 67-74). Hong Kong: Asia-Pacific Society for Computers in Education.

Brackmann, C., Barone, D., Casali, A., Boucinha, R., & Muñoz-Hernández, S. (2016). Computational thinking: Panorama of the Americas. *International Symposium on Computers in Education (SIIE)*, 1-6.

Campe, S., & Denner, J. (2015). Programming Games for Learning: A Research Synthesis. *Annual meeting of the American Educational Research Association*, Chicago, IL.

Chase, C., Chin, D., Oppezzo, M., & Schwartz, D. (2009). Teachable agents and the protégé effect: Increasing the effort towards learning. *Journal of Science Education and Technology*, *18*(4), 334–352. doi:10.100710956-009-9180-4

Davis, R. O. (2018). The impact of pedagogical agent gesturing in multimedia learning environments: A meta-analysis. *Educational Research Review*, *24*, 193–209. doi:10.1016/j.edurev.2018.05.002

Dix, A., Finlay, J., Abowd, G., & Beale, R. (2004). *Human-Computer Interaction*. Prentice Hall.

Ekman, P. (1972). Universals and cultural differences in facil expression of emotion. In *Nebraska Symposium on Motivation* (vol. 19). Lincoln, NE: University of Nebraska Press.

Ekman, P. (1999). Facial expressions. In T. Dalgleish & M. Power (Eds.), *Handbook of cognition and emotion* (pp. 301–320). New York, NY: John Wiley & Sons.

Gil'Adi, D. (1998). *La Inteligencia Emocional en Práctica*. Carácas: Interamericana de Venezuela.

Goleman, D. (1999). *La práctica de la inteligencia emocional.* Barcelona: Kairos.

Graesser, A., D'Mello, S., Craig, S., Witherspoon, A., Sullins, J., McDaniel, B., & Gholson, B. (2008). The relationship between affective states and dialogue patterns during interactions with AutoTutor. *Journal of Interactive Learning Research, 19*(2), 293–302.

Graesser, A. C., Chipman, P., Haynes, B. C., & Olney, A. (2005). AutoTutor: An intelligent tutoring system with mixed-initiative dialogue. *IEEE Transactions on Education, 48*(4), 612–618. doi:10.1109/TE.2005.856149

Griol, D., & Callejas, Z. (2013). An Architecture to Develop Multimodal Educative Applications with Chatbots. *International Journal of Advanced Robotic Systems, 10*(3), 175. doi:10.5772/55791

Heintz, F., Mannila, L., & Färnqvist, T. (2016). A review of models for introducing computational thinking, computer science and computing in K-12 education. In *Frontiers in Education Conference (FIE), 2016 IEEE* (pp. 1-9). IEEE.

Herpich, F., Voss, G. B., Nunes, F. B., & Medina, R. D. (2016). Three-Dimensional Virtual Environment and NPC: A Perspective about Intelligent Agents Ubiquitous. In Handbook of Research on 3-D Virtual Environments and Hypermedia for Ubiquitous Learning (pp. 510-536). IGI Global.

Johnson, W. L., & Lester, J. C. (2018). Pedagogical Agents: Back to the Future. *AI Magazine, 39*(2), 33–44. doi:10.1609/aimag.v39i2.2793

Johnson, W. L., Rickel, J. W., & Lester, J. C. (2000). Animated pedagogical agents: Face-to-face interaction in interactive learning environments. *International Journal of Artificial Intelligence in Education, 11*(1), 47–78.

Konzen, A., Oliveira, O. L., Kist, L. S., Anjos, A. R., Moraes, L., Freitas, C. I. L., ... Mega Vitta, A. (2011). *Agente conversacional aplicado al juego de la educación.* XXII Simposio Brasileño de Informática en Educacion.

Lester, J., Converse, S., Kahler, S., Barlow, S., Stone, B., & Bhogal, R. (1997). The persona effect: Affective impact of animated pedagogical agents. In *Proceedings of the SIGCHI Conference on Human Factors in Computing Systems* (pp. 359-366). New York, NY: ACM. 10.1145/258549.258797

Lin, L., Ginns, P., Wang, T., & Zhang, P. (2020). Using a pedagogical agent to deliver conversational style instruction: What beneðts can you obtain? *Computers & Education, 143*, 103658. doi:10.1016/j.compedu.2019.103658

Morales-Urrutia, E. K., Ocaña, J. M., Pérez-Marín, D., & Tamayo, S. (2017). *A first proposal of Pedagogic Conversational Agents to develop Computational Thinking in children.* TEEM. doi:10.1145/3144826.3145350

Morales-Urrutia, E. K., Ocaña, J. M., Pérez-Marín, D. P., & Tamayo-Moreno, S. (to appear). Interfaz de usuario enfocado en el co-diseño con niños. *Revista Ibérica de Sistemas y Tecnologías de la Información.*

Papert, S. (1980). *Mindstorms: Children, computers, and powerful ideas.* New York, NY: Basic Books.

Pérez-Marín, D., & Caballero, C. (2013). Using Conceptual Models to Implement Natural Language Pedagogic Agent Student Conversations. *International Journal of Interactive Communication Systems and Technologies, 3*(2), 29–47. doi:10.4018/ijicst.2013070103

Pérez-Marín, D., Hijón, R., Bacelo, A., & Pizarro, C. (2018). *Can computational thinking be improved by using a methodology based on metaphors and Scratch to teach computer programming to children?* Computers in Human Behaviour. doi:10.1016/j.chb.2018.12.027

Pérez-Marín, D., Hijón-Neira, R., & Martín-Lope, M. (2017). Propuesta de Metodología basada en Metáforas para la enseñanza de la Programación a Niños. *IEEE Revista Iberoamericana de Tecnologias del Aprendizaje*, *6*(1). doi:10.1109/RITA.2018.2809944

Resnick, M. (1996). New paradigms for computing, new paradigms for thinking. In Constructionism in practice: Designing, thinking, and learning in a digital world. Mahwah, NJ: Erlbaum.

Resnick, M., Maloney, J., Monroy-Hernandez, A., Rusk, N., Eastmond, E., Brennan, K., ... Kafai, Y. (2009). Scratch: Programming for all. *Communications of the ACM*, *52*(11), 60–67. doi:10.1145/1592761.1592779

RobbK. (2015). Available at http://friedtechnology.blogspot.com/2015/11/use-this-feelings-wheel-by-kaitlin-robb.html

Ryokai, K., Vaucelle, C., & Cassell, J. (2003). Virtual peers as partners in storytelling and literacy learning. *Journal of Computer Assisted Learning*, *19*(2), 195–208. doi:10.1046/j.0266-4909.2003.00020.x

Siegel, D. J., & Bryson, T. P. (2012). *The whole-brain child: 12 revolutionary strategies to nurture your child's developing mind.* Bantam.

Sović, A., Jagušt, T., & Seršić, D. (2014). How to teach basic university-level programming concepts to first graders? *In Integrated STEM Education Conference (ISEC),* 2014 *IEEE* (pp. 1-6). IEEE. 10.1109/ISECon.2014.6891050

Tamayo-Moreno, S. (2017). *Propuesta de Metodología para el Diseño e Integración en el Aula de un Agente Conversacional Pedagógico desde Educación Secundaria hasta Educación Infantil* (Doctoral Dissertation). Universidad Rey Juan Carlos. Available on-line at http://hdl.handle.net/10115/14691

Yee, N., & Bailenson, J. (2007). The Proteus effect: The effect of transformed self-representation on behaviour. *Human Communication Research*, *33*(3), 271–290. doi:10.1111/j.1468-2958.2007.00299.x

KEY TERMS AND DEFINITIONS

Alcody: A pedagogic conversational agent co-designed with children to teach them how to program.

Anger: A strong feeling of annoyance, displeasure, or hostility. The person may have the impulse of hurting something or someone.

Disgust: A feeling of revulsion or strong disapproval aroused by something unpleasant or offensive.

Fear: An unpleasant emotion caused by the threat of danger, pain, or harm. The person may hide from others or the object that causes the feeling. S/he can be still and do not move or run from it.

Happiness: The state of being happy. It is different from joy as happiness is momentary, the person feels good and excited and smiles for a time. Joy is a more permanent state.

Primary Education: It is the education phase for children from 6 to 12 years old. Each year corresponds to a grade from first grade (6 years old) to sixth grade (12 years old). Primary education provides children with an elementary understanding of maths, language, science, as well as skills for their lives.

Programming: Creating programs as a sequence of instructions in a certain computer language.

Sadness: Sadness is the opposite state of happiness. The person feels bad and tends to like to be alone. S/he may cry and has feelings of disadvantage, loss, despair, grief, helplessness, disappointment, and sorrow.

Surprise: The emotion that someone feels when an unexpected or astonishing event happens.

Chapter 5
OAuth 2.0:
A Framework to Secure the OAuth-Based Service for Packaged Web Application

Shawon S. M. Rahman
University of Hawaii at Hilo, USA

Md Alam Hossain
Jashore University of Science and Technology, Bangladesh

Nazmul Hossain
https://orcid.org/0000-0002-3743-5193
Jashore University of Science and Technology, Bangladesh

Md Zobayer Hossain
https://orcid.org/0000-0002-5687-3121
Jashore University of Science and Technology, Bangladesh

Md Hassan Imam Sohag
Jashore University of Science and Technology, Bangladesh

ABSTRACT

OAuth is an open security standard that enables users to provide specific and time-bound rights to an application to access protected user resources. It stored on some external resource servers without needing them to share their credentials with the application. Unlike websites, for locally installed packaged web applications, the main security challenge is to handle the redirect response. The OAuth flow initiated from packaged web apps is similar to the OAuth flows explained in the current literature. However, for packaged web apps, it is difficult to define an HTTP endpoint as redirection endpoint since these apps are locally installed. The authors have proposed a novel method to execute OAuth flow from such applications with the help of a web runtime framework that manages the life cycle of these applications. They have compared their approach with another two existing approaches. After conducting experiments, they have found their approach blocking all illegal OAuth flow executions. The approach also delivers better OAuth response handling time and power consumption performance.

DOI: 10.4018/978-1-7998-3355-0.ch005

INTRODUCTION

OAuth 2.0 is an authorization framework that enables applications to gain restricted access to user accounts on an HTTP service, like as GitHub, Facebook, Twitter, and Digital Ocean. OAuth 2.0 is working by delegating user authentication to the service which is hosting the user account and authorizing 3rd party applications to access user account. OAuth 2.0 gives authorization flows (Hardt Ed., 2012) for desktop and web applications, and mobile devices.

A packaged app is a zip file that contains all the resources and it enables a Browser Operating System app to function, with an app explicit in the zip's primary directory. The app explicit gives niceties about the app like as its description, icons are used to recognize the installed app and like this. The package is then used to install the app to Firefox Operating System devices. Once installed the app runs on the device but it is still able to access resources on the Web, like as a database on a web server.

The principle aim of this dissertation is to handle the illegal OAuth flows in a packaged web application system to secure the OAuth protocol. Considering the major security problems of OAuth flow, we designed and develop a packaged web application with a Web Runtime feature that gives more security facilities than other existing approaches. We developed the packaged web application with the help of app packaging which is a secure app development process. As the app package contains web resources like HTML files, JavaScript files, CSS files, media assets (icon image, audio files, etc.) along with an installation descriptor file (or referred to as configuration document). The JavaScript files inside these packages are special, in the sense; those can make calls to standardized web APIs and to custom JavaScript APIs (Charlie, C., Ben, L., Benjamin, G. Z. & Christian, S., 2011) available on the underlying platform. These custom APIs provide access to various platform services from web content enabling these kinds of web apps to implement various use cases on that platform, like the native apps running on it. We created a special web framework or runtime, W, of the platform manages the lifecycle of such packaged web apps. Here managing lifecycle refers to the cycle of installation, instantiation, and uninstallation of an app. When a packaged web app is instantiated, W creates a web view: a window where the web contents of the app are rendered on to using a web rendering engine. Within the context of the web view the app invokes various standard and custom JavaScript APIs. The runtime framework W also takes care of ensuring secure access to platform services using JavaScript API calls. During execution W may create more than one web view as per the JavaScript API call from the app code. This approach will give both the security and better performance factor.

The problem can be stated in short as from existing approaches of OAuth protocol, we selected better twos to compare with our new developed architecture OAuth for packaged web application with respect to different security and performance scenarios. The steps of the methodology are depicted as follows:

Each OAuth based systems have their own mechanism for OAuth flow. The existing approaches have their own architecture, OAuth flow, security mechanism, etc. So, we found the similarities and dissimilarities of those existing approaches to find out the problems and vulnerabilities.

To find out the architectural requirements of an OAuth based system for a client-side web application, we have analyzed the architectures of two existing systems. We gathered some useful information by studying and browsing the internet about those approaches. For more information, we tested them on the internet and analyzed the various sections of those systems. From all that information, we have tried to draw the general architectures of those approaches mentioned above. As, in this research our task is to propose, design and develop a new OAuth architecture, we must derive some architectural requirements.

Figure 1. Steps of methodology

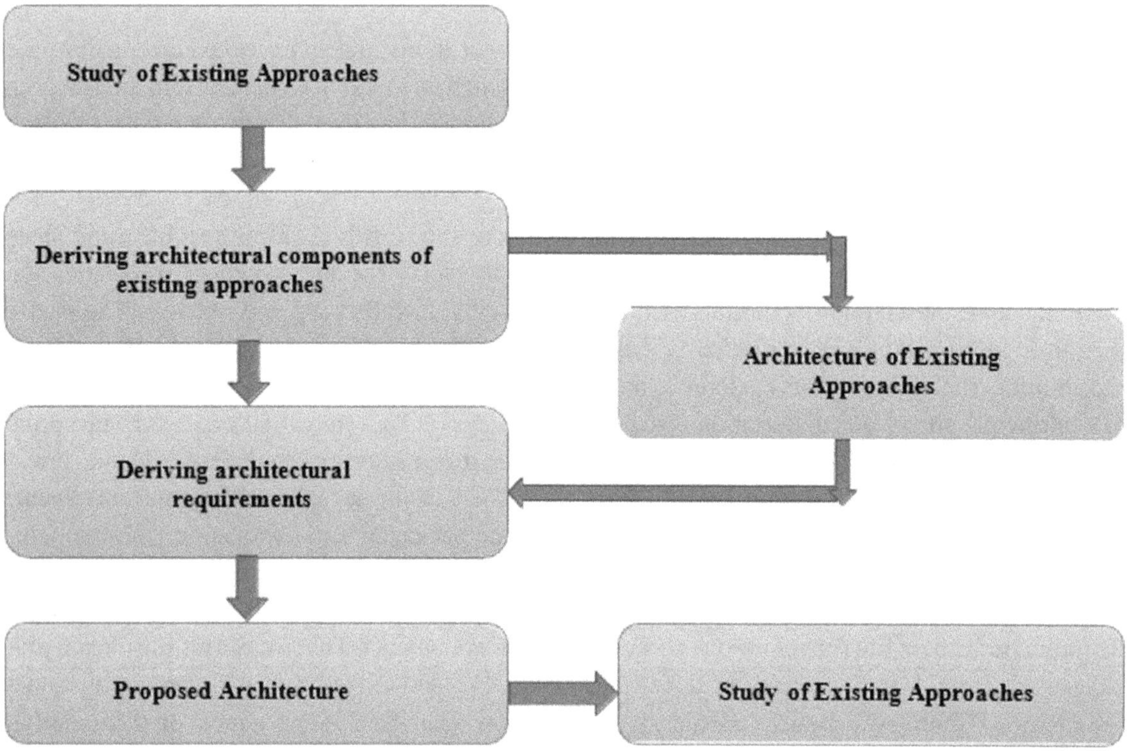

Since OAuth is an authentication protocol, there is plenty of research avenues and challenges that can be explored. These research areas are security, social-technical impacts, performance, etc. (Jason, B., Elie, B., Divij G. & John, C. M., 2010). However, this research will focus on the challenges facing security of user data stored in resource servers and transmit through the internet. In this paper we see the new OAuth security architecture for packaged web apps in comparison with other two approaches. If the flows of OAuth 2.0 are similar in user's visibility, so attackers might set response_type = token, and obtain the access token from the browser by a script except user's awareness. The receiver, excepting an attacker, of the code or access token might log in to RP, such as the victim, and obtain the victim's account. If the State parameter is spoiled, another countermeasure in opposition to CSRF attack (Yacin, N., Prateek, S. & Dawn, S., 2009) must be taken specified in the protocol. In practice, CSRF attack is mostly disregarded through developers. If obtaining Authorization Code and using it earlier on victim, the attacker might embody the victim and log in to RP, and that has a bigger risk if the code might be used for many RPs, e.g. the code sent to RP1 might be used to log in to RP2. Due to the DE cryptographic (William, K., Robertson, Giovanni V., 2009) design of OAuth 2.0, the access token is mostly performing as a bearer token (Jones & Hardt., 2012). That is to say, any bearer of the token, e.g. an attacker, might access or manage user's profile (Prateek, S., David M. & Livshits, B., 2011).

OVERVIEW OF OAuth 2.0 IN PACKAGED WEB APP

In the tradition client-server authentication design, the client requests an access-limited resource (protected source) on the server through authenticating with the server by using the resource owner credentials. So that, provide 3rd party applications access to limited resources, the resource owner shares its identity with the 3rd party (Mike, T. L. & Venkat, V., 2009). This creates different limitations and problems:

- 3rd party applications are necessary for storing the resource owner credentials for oncoming use, generally a password in clear-text.
- The servers are necessary to client password authentication, notwithstanding of security vulnerabilities inherent in passwords.
- 3rd party applications acquire too broad access to the resource owner preserved resources exploit resource owners except any ability to limit duration or access to a restricted subset of resources.
- The resource owners may not call access to an individual 3rd party except calling access to all 3rd parties and must do therefore by changing the 3rd party's password. Compromise of any 3rd party application outcomes in trade-off the end user's password & all of the data out of danger through that password.

Instead of utilizing the resource owner's credentials to access restricted resources, the client gains an access token a string denoting a lifetime, scope, and alternative access properties. Access tokens are issued to 3rd party users through the authorization server with the grant of the resource owner. The user uses the access token to get the restricted resources reserved within the resource server.

Roles

OAuth defines 4 prefaces:

Resource owner: An entity equipped for giving access to a reserved resource. once the resource owner may be a person, it's spoken as an end-user.
Resource server: The server facilitating the preserved resources, worthy of acceptive & reacting to preserved resource requests utilizing access tokens.
Client: The application creating preserved resource requests in favor of the resource owner and with its approval. The term "client" does not indicate any specific implementation properties.
Authorization server: The server giving access tokens to the user afterward effectively authenticating the resource owner & gaining authorization.

The interaction within the resource server & authorization server is on the far side of the Opportunity of this specification. The authorization server might be the very same server as much the resource server or a different entity (Adam, B., Juan, C. & Dawn S., 2009). An individual authorization server might issue access tokens adopted by many resource servers.

Protocol Flow

The summary of OAuth 2.0 flow (Hardt Ed., 2012) painted in Figure 2 recounts the interaction among the four roles and included bellow steps:

A. The user requests authorization from the resource owner. The authorization request maybe is also created on direct to the resource owner (as shown), or ideally indirectly by the authorization server as a mediate. The user requests authorization from the resource owner. The authorization request may be is created on direct to the resource owner (as shown), or ideally indirectly by the authorization server as a mediate.
B. The user gets an authorization to allow, that is a credential representing the resource owner's authorization, published utilizing one in every of four grant types outlined during this specification or utilizing an expansion allow type. The authorization allow type to rely on the tactic employed by the user to request authorization & also the types Through authorization server.
C. The user asks for an access token through validating with the authorization server & exhibiting the authorization grant.
D. The authorization server confirms the user and approves the authorization grant, and if legitimate, issues an access token.
E. The user asks for the preserved resource from the resource server and confirms through exhibiting the access token.
F. The resource server approves the access token, and if legitimate, serves the demand.

The favored strategy for the user to gain an authorization grant from the resource owner (portrayed in steps (A) and (B)) is to utilize the authorization server as a mediate.

Figure 2. Abstract Protocol Flow of OAuth 2.0

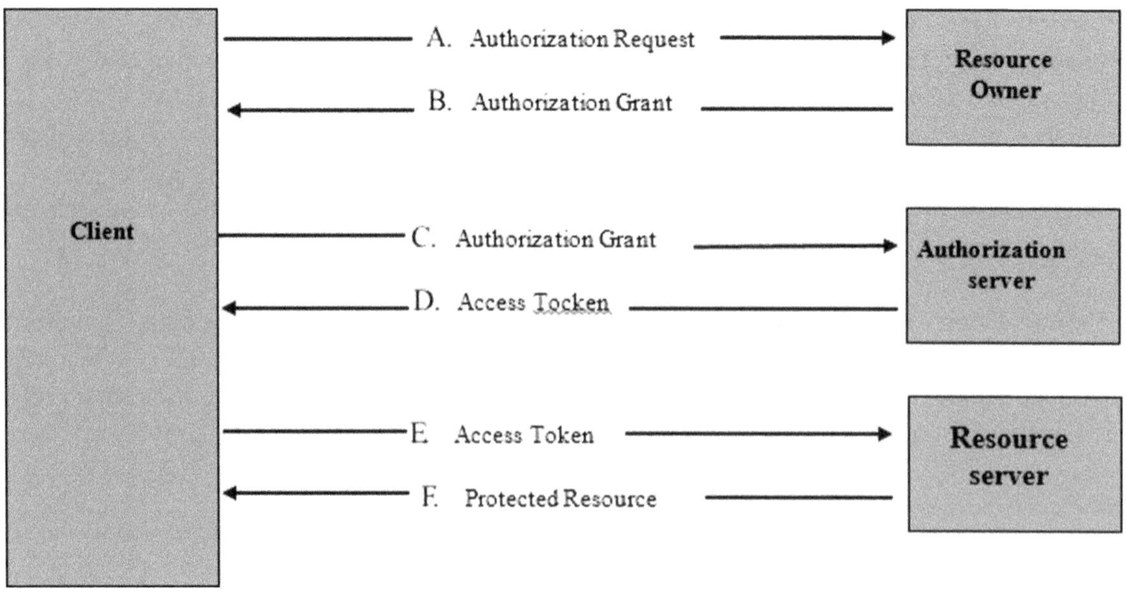

Authorization Grant

An authorization grant may be a credential representing the resource owner's approval (to get its preserved resources) utilized through the user to gain an access token (Richer, Mills & Tschofenig., 2012). This detail characterizes four grant varieties –

- Authorization code
- Implicit
- Resource owner password credentials and
- Client credentials.

Authorization Code

The authorization code is gained by utilizing an authorization server as a mediate within the resource owner & the user. Rather than asking for approval straightforwardly from the resource owner, the user guides to the resource owner to an authorization server that so coordinates the resource owner back to the user with the authorization code.

Implicit

The implicit grant is a rearranged authorization code stream upgraded for users implemented in a browser utilizing a scripting language for example JavaScript. In the implicit flow, rather than issuing the user authorization code, the user is issued an access token specifically (as the consequence of the resource owner authorization). The grant type is implicit, as no halfway credentials (for example an authorization code) are issued (and later used to gain an access token).

Resource Owner Password Credentials

The resource owner password credentials (i.e. password and username) may be utilized directly as an authorization grant to gain an access token. The credentials should just be utilized when there is a high level of trust within the user & the resource owner (e.g. the client is a piece of the device operating system or a profoundly advantaged application), and when several authorizations grant types are not accessible (like an authorization code).

Client Credentials

The user credentials (or other forms of user authentication) may be utilized as an authorization grant once the authorization scope is restricted to the preserver resources under the control of the user, or to preserved resources antecedently organized with the authorization server.

Access Token

Access tokens are credentials accustomed to accessing preserved resources. An access token could be a string speaking to an authorization issued to the user. The string is sometimes opaque to the user. Tokens represent specific Opportunities and span of access, granted by the resource owner, and executed by the resource server & authorization server. The access token gives an abstraction layer, substitution totally different authorization constructs (for example username and password) with one token understood through the resource server.

Refresh Token

Refresh tokens are credentials accustomed get access tokens. Refresh tokens (San-Tsai S., Kirstie, H. & Konstantin, B., 2012) are issued to the user through the authorization server and are accustomed gain a new access token whereas the current access token becomes void or expires, or to get extra access tokens with uniform or narrower Opportunity. Issuing a refresh token is volitional at the prudence of the authorization server. On the off chance that the authorization server issues a refresh token, it is comprised whereas issuing an access token.

The stream showed in Figure 3 incorporates the accompanying advances:

A. The client application asks for an access token through authenticating with the authorization server & introducing an authorization grant.
B. The authorization server verifies the client application & approves the authorization grant, and if legitimate, issues an access token & a refresh token.

Figure 3. Refreshing an expired access token

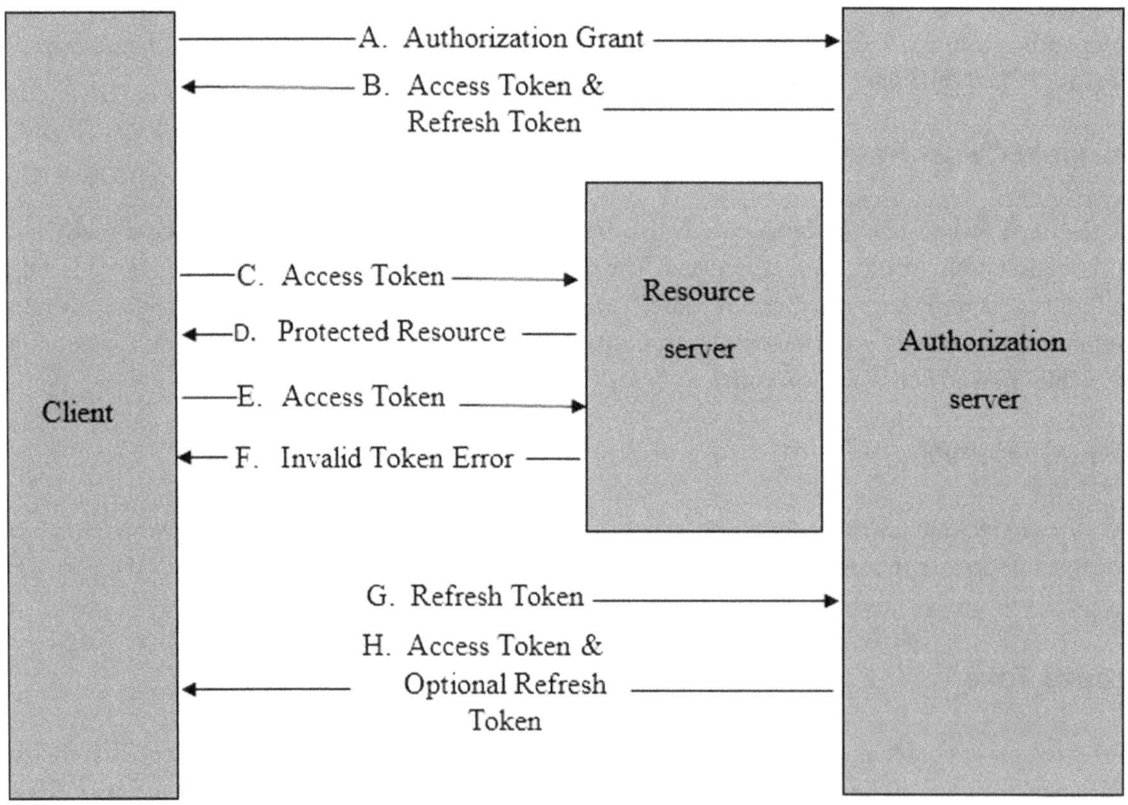

C. The client application makes a preserved resource demand to the resource server by showing the access token.
D. The resource server approves the access token, and if legitimate, serves the demand.
E. Steps (C) and (D) reiteration until the access token expired. On the off chance that the client application knows the access token lapsed, it skips to step (G); else, it makes another secured resource ask.
F. Since the access token is void, the resource server restores a void token error.
G. The client application asks for another access token through authenticating with the authorization server & introducing the refresh token. The client application authentication necessities depend on the user type & on the authorization server arrangements.
H. The authorization server authenticates the client application & verifies the refresh token, and if legitimate, issues another access token (and, alternatively, another refresh token).

Interoperability

OAuth 2.0 furnishes a rich authorization framework with very much characterized security properties. In any case, as a rich and profoundly extensible framework with numerous discretionary parts, all alone, this particular is probably going to deliver an extensive variety of non-interoperable executions.

OAuth Endpoints

OAuth 2.0 utilizes two endpoints, The Authorization endpoint, and the Token endpoint.

Authorization Endpoint: The Authorization endpoint is utilized to collaborate with the resource owner and obtain the authorization to access the preserved resource. To begin with, the service will divert you to Google in order to authenticate (in the event that you are not as of now signed in) and afterward you will get an assent screen, where you will be requested to approve the service to access some of your information, e.g., your email address and your list of contacts.

Token Endpoint: The Token endpoint is utilized through the application to gain an Access Token or a Refresh Token. It is utilized through all grant types, outside of Implicit grant (whereas an Access Token is issued straightforwardly). In the Authorization Code grant, the application alternates the authorization code it obtained from Authorization endpoint for an Access Token (Hardt Ed., 2012).

SECURITY ISSUE ON OAuth 2.0 FOR PACKAGED WEB APP

As a versatile and extensible framework, OAuth's 2.0 security concerns rely upon several factors. The subsequent sections give implementers with security rules concentrated on the three user profiles, native application, and user-agent-based apps. An extensive OAuth 2.0 security model and analysis, likewise as background for the protocol diagram, is given through the OAuth-Threat model (Nazmul, H., Md., A. H., Md., Z. H., Md., H. I. S. & Shawon, R., 2018).

Client Authentication

The authorization server builds up user credentials with web application user with the end goal of user authentication. The authorization server is Motivated to consider potential user authentication implies than a user password. Web application users must guarantee the privacy of user passwords and other client credentials. The authorization server should not issue user passwords or different user credentials to native applications (e.g. Facebook, Google, etc.) (Kitura., 2018) Or user-agent-based application users for the motive of user authentication. The authorization server can issue a user password or different credentials for the installation of a native application user on a particular device.

Client Impersonation

A malevolent user may impersonate another user and gain access to preserved resources if the impersonated user fails to, or is unable to, keep its user credentials confidential (Suresh, C., Charanjit, J. & Arnab R., 2011). The authorization server should authenticate the user whenever possible. In the event that the authorization server can't authenticate the user because of the user's nature, the authorization server should need the registration of any redirection URI utilized for receiving authorization responses and must use different intends to shield resource owners from like potentially malicious users.

Access Tokens

Access token credentials (and in addition any private access token properties) must be kept secret in transit & storage, and just apportioned among the resource servers, the authorization server, the access token is legitimate for, and the user the access token be issued to whom. Access token credentials should just transmit to utilize TLS (Wanpeng, L. & Chris J. M., 2014) with server authentication. When utilizing the Intrinsically grant type, the access token is sent in the URI section, which may open it to unapproved parties.

Refresh Tokens

Authorization servers can issue revitalize tokens to the web application users and local application users. Refresh tokens definitely kept privet in transit & storage and shared just only the authorization server and the user the refresh tokens were issued to whom. The authorization server must sustain the binding within a refresh token & the user to whom it was issued. Refresh tokens should just be transmitted utilizing TLS with server authentication.

Authorization Codes

The transmission of authorization codes ought to be made over a safe channel, and the user ought to require the utilization of TLS with its redirection URI if the URI recognizes a network resource. Since authorization codes are sent by means of user-agent redirections, they might possibly be revealed by user-agent history & HTTP referrer headers (Rui, W., Shuo, C. & Xiaofeng, W., 2012).

Authorization codes work as plaintext carrier credentials used to confirm that the resource owner who allowed authorization at the authorization server is a similar resource owner coming back to the user to complete the procedure.

Authorization Code Redirection URI Manipulation

When asking authorization utilizing the authorization code grant type, the user may specify a redirection URI by means of the "redirect_uri" parameter (Lodderstedt, McGloin & Hunt., 2013). In the event that an attacker may control the value of the redirection URI, it may because of the authorization server to divert the resource owner client operator to a URI under the control of the attacker with the authorization code. An attacker may make an account at a valid user and start the authorization flow. At the point when the attacker's user-agent is transmitted to the authorization server to allow access, the attacker possession the authorization URI provided through the valid user and replaces the user's diverted URI with a URI under the control of the attacker. The attacker at that point traps the victim into following the manipulated link to approve access to the valid user.

Resource Owner Password Credentials

The resource owner password credentials allow type is frequently utilized for legacy or migration reasons. It lessens the overall risk of storing passwords and usernames through the user but doesn't wipe out the need to uncover exceptionally favored credentials to the user. This grant type conveys higher jeopardy than different grant types as a result of it maintains the password anti-pattern this protocol seeks to get off. The user might misuse the password, or the password might surreptitiously be published to an attacker (for example, by means of log files or different records kept through the user).

Request Confidentiality

Refresh tokens, Access tokens, Resource owner passwords, and user credentials certainly not be sent in the clear. Authorization codes would not be sent in the clear. The "scope" and "state" parameters shouldn't include sensitive user or resource owner data in plain text, as they may be sent over shaky channels or stored shakily.

Ensuring Endpoint Authenticity

In order to obstruct man-in-the-middle attacks, the authorization server certainly needs the utilizes of TLS (Transport Layer Security) with server authentication for any asks transmitted to the authorization and token endpoints. The user certainly approves the authorization servers.

Credentials-Guessing Attacks

The authorization server certainly obstructs attackers from guessing refresh tokens, access tokens, resource owner passwords, authorization codes and client credentials (San-Tsai, S. & Konstantin, B., 2012). The possibility of an attacker guessing produced tokens (and different credentials not intentional for handling through end-users) certainly be less than or equal to 2-128 and could be less than or equal to 2-160. The authorization server certainly uses different means to defend credentials desired for end-user usage.

Phishing Attacks

Wide deployment of this and analogous protocols can cause end-users to become satisfied with the practice of being redirected to websites where they are inquiring to enter their passwords. If end-users aren't cautious to validate the authenticity of these websites earlier on entering their credentials, it will be feasible for attackers to utilize this practice to snitch resource owners' passwords. Service providers would try to educate end-users about the risks phishing attacks posing and would give mechanisms that create it simple for end-users to ensure the authenticity of their sites (Daniel, F., Ralf, K. & Guido, S., 2016).

Cross-Site Request Forgery

Cross-site request forgery (CSRF) (Caimei, W., Yan, X., Wenchao, H., Huihua X., Jianmeng, H. & Cheng S., 2017) is an adventure in which an attacker because the user- agent of a prey end-user to take after a malevolent URI (for example, gave to the user-agent as a deceptive link, redirection or image) to a confiding in server (generally established by means of the presence of a substantial session cookie. Once authorization has been gained from the end-user, the authorization server diverts the end- user's user-agent back to the user with the needed binding value included in the "state" parameter. The binding value enables the user to confirm the validity of the request by coordinating the binding value to the user agent's confirmed state. The binding value utilized for CSRF protection certainly holds a non-guessable value, and the user agent's confirmed state (for example, HTML5 local storage, session cookie) certainly be kept in a place accessible just once to the user and the user-agent (i.e., preserved through the same-origin policy). A CSRF attack in opposition to the authorization server's authorization endpoint may outcome in an attacker gaining end-user authorization for a bitchy user except alerting or involving the end-user. The authorization server certainly developed CSRF protection for its authorization endpoint and confirm that a malicious user cannot gain authorization except the awareness and evident consent of the resource owner.

Clickjacking

In a clickjacking attack, an attacker records a valid user and then built a malicious site in which it stacks the authorization server's authorization endpoint web page in a diaphanous iframe overlaid over a set of artificial buttons, that's are gingerly build to be placed specifically under significant buttons on the authorization page. When an end-user clicks a deceptive seen button, the end-user is really clicking an unseen button on the authorization page (e.g. an "Authorize" button). This permits an attacker to ruse a resource owner into allowing its user access except for the end-users knowledge.

Code Injection and Input Validation

A code injection attack happens when an input or in a different way the exotic variable is utilized through an application unsanitized and because of shuffling to the application logic. This can grant an attacker to obtain access to the application device or its information, because of denial of service, or introduce a spacious range of malicious side-effects. The authorization server and user certainly sanitize (and affirm when possible) any value accepted in individual, the value of the "redirect_uri" and "state" parameters.

Open Redirectors

The authorization server, client redirection endpoint, and authorization endpoint can be wrongfully configured and handle as free redirectors (OpenID Authentication 2.0 - Final., 2007). An open redirector is an endpoint conducting a parameter to automatically redirect a customer-viceregent to the position specified by the parameter value except any affirmative. Open redirectors can be applied in phishing invasions, or by a charger to catch end-users to visit malicious sites by applying the URI authority material of a homely and believable target (Victoria, B., 2016).

Misuse of Access Token to Impersonate Resource Owner in Implicit Flow

For public clients using implicit streams, this specification does not give anyway for the client to determine what client an access token was issued to. A resource proprietor may agreeably surrogate entree to a resource by offering an access token to an attacker's bitchy client. This may be payable to phishing or some other plea. An attacker may also snitch a token through several mechanisms. An aggressor may then try to incarnate the resource proprietor by giving the access token to a legal public client. Servers connecting with local applications that count on being adopted an access token in the backchannel to define the customer of the client may be likewise compromised by an attacker making a compromised application that can impel unrestricted robbed access tokens. (Francisco, C. & Karen, P. L., 2011)

EXISTING APPROACHES

OAuth is an uncovered security measure that enables customers to give appointed and time cramped rights to an application to entrance preserved customer's resources, gathered on several exotic resource servers, except needing them to divide their credentials, with this application. Handling OAuth, a client application has one ingress token for more use with an HTTP redirect reaction from the resource server sometimes the user authenticates the resource entry (lhshaoren., 2010). OAuth 2.0 stream is, particularly for user authorization. It is planned for applications that can hoard private news and sustain the realm. A rightly authorized web server application can entrance an API while the customers communicate with the application or afterward the user has left the application.

Approach A: OAuth 2.0 to Access Google APIs

The approach A explained in (Google Identity Platform, 2018) suggests specifying the redirect URL as a loopback IP address like http://127.0.0.1::port or http://::1::port and requires the client app to start listening to a randomly available HTTP port. To receive the access token using a loopback IP address, the app must be listening to the local webserver.

The Basic Pattern of OAuth 2.0 to Access Google APIs

To initiate, get OAuth 2.0 client certificates from the Google API Console. Then the client application entreats an ingress token from the Google Authorization Server, gist's a token from the reaction and transmits the token to the Google API that the user wants to access. Every application chases a fundamental pattern when engrossing a Google API conducting OAuth 2.0. At an exalted level, follows four steps.

- **Obtain OAuth 2.0 credentials from the Google API Console:** The Google API Console to make OAuth 2.0 certificates like a client ID and client covert that are acquainted with both Google and user applications. The group of values changes founded on what form of application is constructing. For example, a JavaScript application does not claim a hidden, however, a web server application does.
- **Obtain an access token from the Google Authorization Server:** Before an application can penetration personal data using a Google API, it must gain an entrance token that gives access to that API. An individual access token can grant asymmetrical degrees of ingress to multiple APIs. A changeable parameter named scope monitors the group of resources and actions that an access token permits. During the access-token request, an application gives one or more values in the scope parameter.
- **Send the access token to an API:** After an application gain an access token, it deputizes the token to a Google API in an HTTP authorization header. It is allowable to give tokens as URI query-string parameters because URI parameters can end up in log records that are not fully secure. Access tokens are valid only for the group of activities and resources narrated in the scope of the token request.
- **Refresh the access token, if necessary:** Access tokens have restricted lifetimes. If an application wants access to a Google API behind the lifetime of an individual access token, it can favor a refresh token.

Scenarios of OAuth 2.0 to Access Google APIs

Generally, OAuth provides clients (e.g. Mobile app) a 'secure confer penetration' to server funds on the part of a resource owner. The Web-server applications scenario is applied to authenticate a web application with a third-party service. There are many scenarios of OAuth 2.0 to access Google APIs.

Web Server Applications

The Google OAuth 2.0 endpoints backings web server applications that usage languages and molds such as PHP, Java, Python, Ruby, and ASP.NET (Android WebView., 2018). The authorization series starts when an application redirects a browser to a Google URL; the URL involves query parameters that imply the type of ingress being petition. Google manages user authentication, session selection, and user consent. The outcome is an authorization code, which the application can interchange for an access token and a refresh token.

Figure 4. Abstract Protocol Flow of Web server applications OAuth 2.0

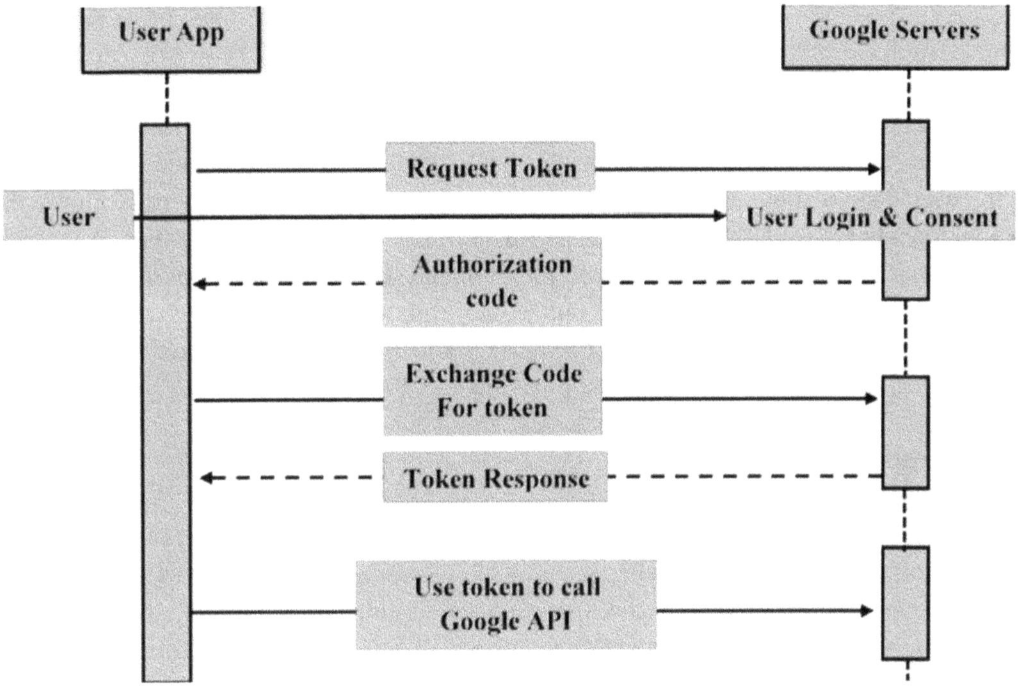

Figure 5. Abstract Protocol Flow of Installed applications OAuth 2.0

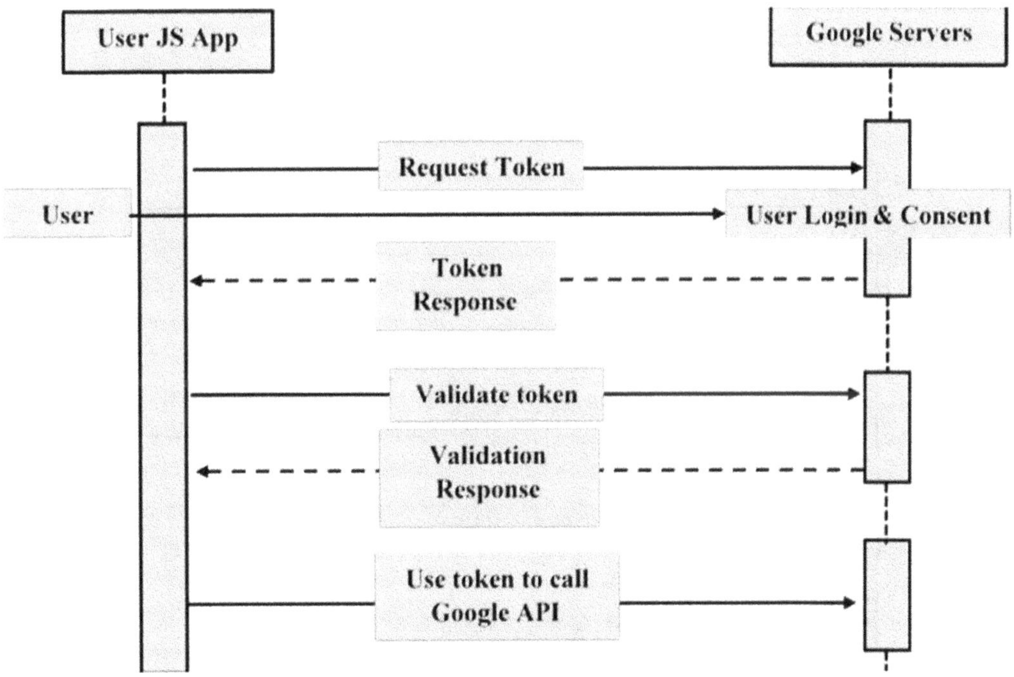

Installed Applications

The Google OAuth 2.0 endpoints clinch applications that are installed on devices like computers, mobile devices, and tablets. When user creates a client ID from the Google API Console, specify that this is an installed application, then choose Chrome, iOS, Android, or "Other" as the application type.

The procedure output in a client ID and, in some law, a client covert, which user embed in the source code of the user's application. The authorization sequence begins when user's application redirects a browser to a Google URL; the URL added query parameters that introduce the type of ingress being asked. Google maintains the session selection, user authentication, and user consent. The output is an authorization code, which the application can interchange for a refresh token and an access token. The application should hoard the refresh token for next usage and usage the access token to ingress a Google API. At one time the access token died, the application uses the refresh token to gain a new one.

Client-Side Applications

The Google OAuth 2.0 endpoint supports client-side applications that run in a browser.

The authorization sequence begins when the user's application redirects a browser to a Google URL; the URL adds query parameters that select the sampling of entree being asked. Google manages the session selection, user consent, and user authentication. The output is an access token, which the client should verify before adding it in a Google API request. When the token departs, the application repetitions the procedure.

Figure 6. Abstract Protocol Flow of Client-side applications OAuth 2.0

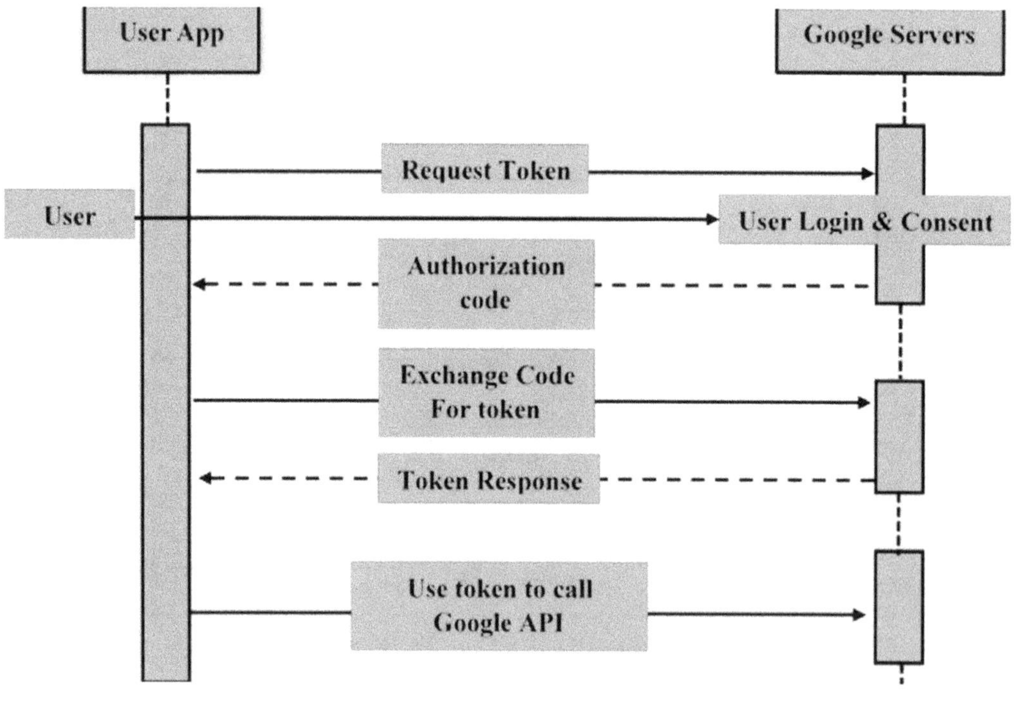

Applications on Limited-Input Devices

The Google OAuth 2.0 endpoint helps applications that move on limited-input instruments such as video cameras, game consoles, and printers. The authorization order starts with the application creating a web service that asks a Google URL for an authorization code. The reaction contains different parameters, adding a URL and a code that the application displays to the user.

The user gains the URL and code from the apparatus, then switches to individual instruments or computers with larger input capabilities. The user opening a browser navigates to the specified URL, logs in, and enters the code. Hitherwards, the application peaks a Google URL at a specified interim. After the customers approve ingress, the answer from the Google server holds an access token and a refresh token.

The application should gather the refresh token for the next usage and usage of the access token to ingress a Google API. At one time the access token dies, the application conducts the refresh token to gain a new one

Figure 7. Abstract Protocol Flow of limited-input devices applications OAuth 2.0

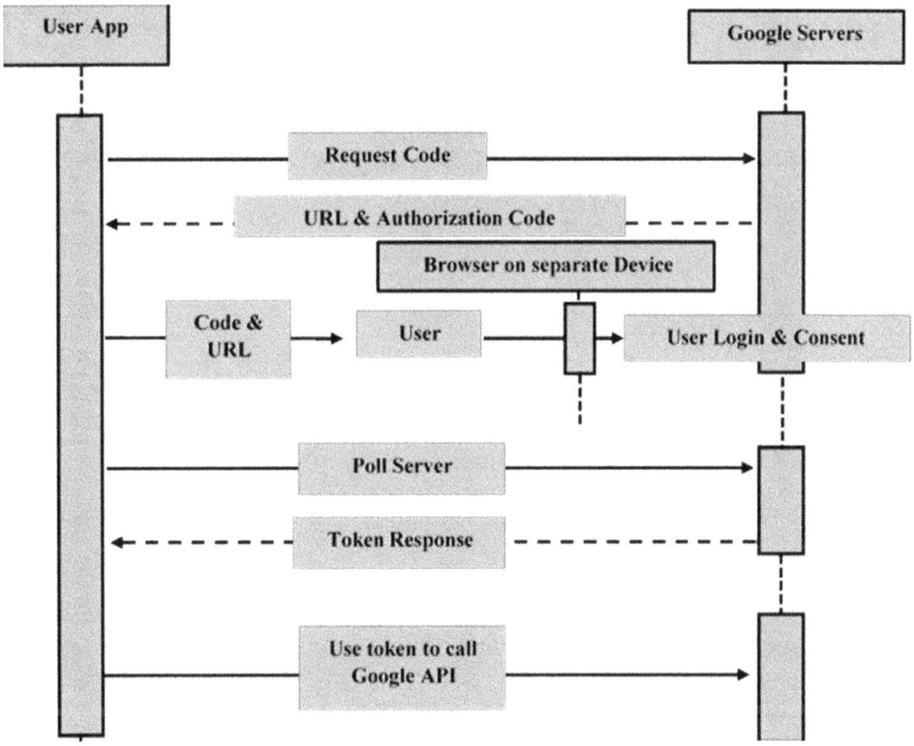

Service Accounts

A service account's credentials, which user obtains from the Google API Console, involve a propagated email address that is identical, a client ID, and at least one public/private key couple. A user uses the client ID and a private key to make a signed JWT and build access-token solicitation in the opposite

Figure 8. Abstract Protocol Flow of Service account applications OAuth 2

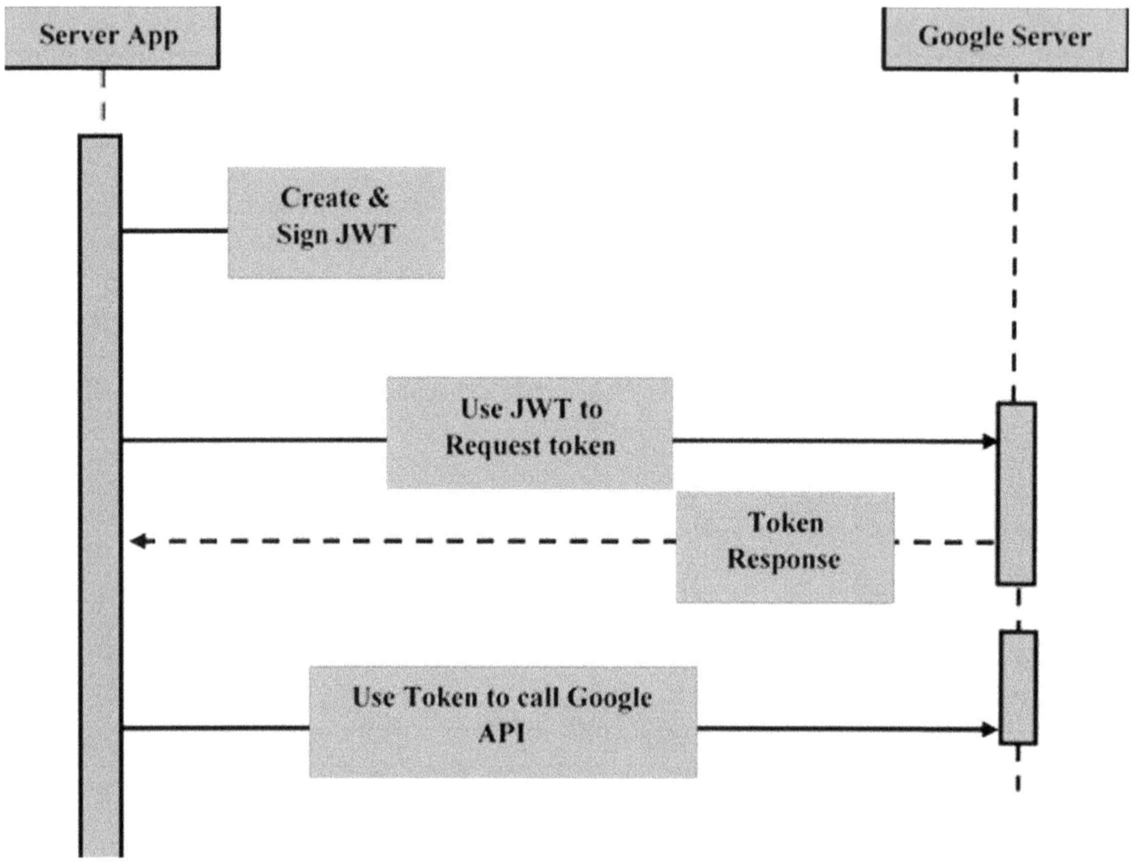

ordination. The application then gives the token indictment to the Google OAuth 2.0 Authorization Server, which retires an access token. The application demeanor the token to entrance a Google API. When the token kicks off, the application repetitions the methods.

Token Expiration

Users must write their code to relate the probability that an approved refresh token might no longer active. A refresh token might pause working for one of these reasons:

- The user has rejected the app's ingress.
- The refresh token has not been demeanor for six months.
- The user altered passwords and the refresh token take on the Gmail tract.
- The user account has surpassed the greatest number of inflicted (live) refresh tokens.

There is now a border of 50 refresh tokens per customer account per client. If the boundary is touched, making a new refresh token automatically defeats the ancient refresh token without inflaming.

Limitations of OAuth 2.0 to Access Google AP

In this approach it receives the access token using a loopback IP address, the app must be listening to the local webserver. This approach recommends that the client app, upon receiving the token response, should show an HTML page instructing the user to close the browser tab and return to the client app. This approach also allows using localhost in place of loopback IP, if that does not lead to any firewall related issue in the underlying platform. In this approach, it is possible to successfully get the access token, but OAuth flow initiated from a packaged web app is exposed to security vulnerability due to week redirection endpoint definition and handling.

Approach B: Facebook App in Tizen

This approach (Facebook App in Tizen, 2018) suggested for performing OAuth with Facebook on the Tizen platform, is to provide a legitimate HTTP URL as redirect endpoint and continuously poll for redirection to the given redirect URL. To complete the Facebook Authentication method user has to follow three steps.

Initiate a Redirect to an Endpoint

The app must begin a redirect to an endpoint having the according to parameters:

- client_id: The ID of the app, established in the app's fascia.
- redirect_uri: The Site URL found in the "Website with Facebook Login" chapter of the app's fascia.
- scope: A comma differentiated list of Permissions to ask from the customer of the app.

The endpoint will show the Facebook login window. After admitting the credentials, a dialogue will look inquiring user to grant entry to his friend list, public profile, and any extra Permissions asked by the app. If the user selects "OK" on the dialogue, the app will achieve entree to specified permissions.

Child Window

The "Child Window" idea is applied to open the Facebook login window (Erika, C., & David W., 2014). At one time the user finished the login procedure, the authentication URL will give the access code to redirect_uri. Instructor the authentication URL to recover the 'access code' and regain to the application by sending the child window. It starts a timer that will monitor inside timeout callback the change in the URL for the child window due to redirection to retrieve the 'access code'.

Collecting the Access Code

After collecting the access code use Facebook Graph API to generate the access token. It retrieves 'access code' and initiate an AJAX request and in the success callback of AJAX request parse the access token and close the child window.

Limitations of Facebook App in Tizen

This approach provides a legitimate HTTP URL as redirect endpoint and continuously poll for redirection to the given redirect URL. It is possible to successfully get the access token, but OAuth flow initiated from a packaged web app is exposed to security vulnerability due to week redirection endpoint definition and handling.

NEW OAuth SECURITY ARCHITECTURE FOR PACKAGED WEB APPS

The Internet today provides us a collaborated environment of various types of software platforms and social networking systems (Mohamed, S. & Fadi, M., 2014) (Marian, H., Markus, H., Susanne, W. & Matthew, S., 2014) Application (henceforth referred to as "app") developers, especially client web app developers, want to utilize this medium to access various user resources, such as photos or videos maintained in some external resource server, in order to provide a richer and connected experience through their apps (Hao, H., Vicky, S. & Wenliang, D., 2013). Nowadays OAuth 2.0 is very widely used on most websites, and there is a correspondingly rich infrastructure of identity providers (IdPs) providing identity services using OAuth 2.0. Use of OAuth 2.0 (Fadi, M. & Mohamed, S., 2016) by Facebook, Google and Microsoft have previously been studied, and issues have been identified. However, security efficient OAuth implementations are not available for locally installed packaged client web apps (Tara, W. & Kori I. Q., 2005). In this section, we propose a novel OAuth implementation for these kinds of apps. We developed a packaged web app with Web Runtime environment.

After analyzing the existing approaches of OAuth 2.0 for packaged web apps we have found some similarities and dissimilarities. To develop an architecture capable of addressing the existing architectural problems, several architectural requirements have been derived and discussed.

Packaged Web App Development

To implement our own new OAuth authentication protocol, we need to develop a Packaged Web App. A Packaged Web application (Dongwan, S. & Rodrigo, L., 2011) includes all the backup files that are wanted by the Web application. Hence, a Web application may run without network connectivity or any other external resources after finishing the installation. The packaged web application model supports a big amount of standard W3C/HTML5 characteristic, which involves many JavaScript APIs besides extra HTML markups and CSS features. Those features, besides the Device APIs and UI framework support, may be used to make a rich Web application in a diversity of categories, such as device information access, multimedia, graphics, contacts, messaging, and games. For packaged web apps the installation descriptor is a very important component of the package. It declares various properties of the app that should be manifested in the system during its lifecycle. It includes the icon and name to be shown when the app is installed.

Web Runtime Environment

The Web Runtime performs a body of the core behavior of any programming language and allows it to be changed via an API or inlaid domain-specific language. The web runtime (Web Runtime – Tizen developer., 2018) is like it uses web-based programming languages like Java-script which utilizes the core behavior a computer language. Another example of a Web Runtime is JsLibs which is a stand able JavaScript development runtime environment for using JavaScript as a common all-round scripting language. JavaScript is used to make responsive interfaces that enhance the user experience and give dynamic functionality except having to wait for the server to respond and direct to another page. It supports:

- Management of packages (installation, update, etc.)
- Lifecycle and Execution (launching, pause, resume, etc.)
- Runtime security (API/network access, sandboxing, etc.)
- Platform and device integration (access local device and platform resources)

OAuth 2.0 Authentication Flows

The OAuth-SSO systems (Fadi, M. & Mohamed, S., 2016) are based on browser redirection that an RP redirects the Clint browser to an IP which collaborates with the Clint before redirecting the Clint back to the RP website. The IP authenticates the Clint, recognizes the RP to the Clint, and asking for consent to offer the RP access to the services and resources on the side of Clint. Once the implored consents are granted, the Clint is redirected back to the RP with an access token which illustrates the granted permissions.

With the approved access token, the RP then invokes web APIs revealed by the IP to access the client's profile properties. The OAuth 2.0 specification which defines two flows for RPs to gain access tokens: server-flow (called as the Authorization Code Grant), designed for web applications which gain the access tokens from their server-side program logic; and client-flow (called as the \Implicit Grant") for JS applications which runs in the web browser.

Authentication & Resource Server

An authentication server offers a network service that applications use to certify the credentials, typically account names and passwords, of their users. Major authentication algorithms embrace passwords, Kerberos, and public-key cryptography.

The Resource Server could be a huge collection of libraries and the web application engaged towards a lot of economical inclusion of static resources like CSS and JavaScript in Java web applications. The tip goal of those utilities is to scale back the number of distinct resource URLs the browser should recover the content from for any given page and to require advantage of browser facet caching to scale back the number of requests the browser should build.

Access Tokens

Access tokens area unit credentials used to access safe resources. associate access token could be a string describing an authorization issued to the client. The string is sometimes obscure to the client. Tokens represent special scopes and durations of access, approved by the resource owner, and insisted by the resource server and authorization server. Then the token could denote an identifier that will not retrieve the authorization info or could self-contain the authorization info in an exceedingly verifiable manner (i.e., a token string consisting of some information and a signature). further authentication credentials, that area unit on the far side the scope of this specification, is also needed for the client to use a token.

Fulfilling Architectural Requirements

The proposed framework fulfills the requirements derives in the previous section as follows:

- The architecture ensures that access token if delivered is delivered to the right, intended, a legitimate target. The feature Web Runtime ensures that.
- The architecture makes our approach handle OAuth response faster than another two existing approaches.
- The proposed architecture performs better than the other two approaches.

PROPOSED ARCHITECTURE

Description of Proposed Architecture

Securing the OAuth Flow for Packaged Web Application

In our proposed system, web runtime (WRT), W (Web View), is the key component that ensures a secure environment for OAuth flow execution from a packaged web app. In our system, we have considered packaged web apps because those are more aligned to W3C standards. With packaged web apps, as mentioned already, each app mentions its lifecycle properties inside installation descriptor or configu-

Figure 9. Architecture of new OAuth flow with WRT

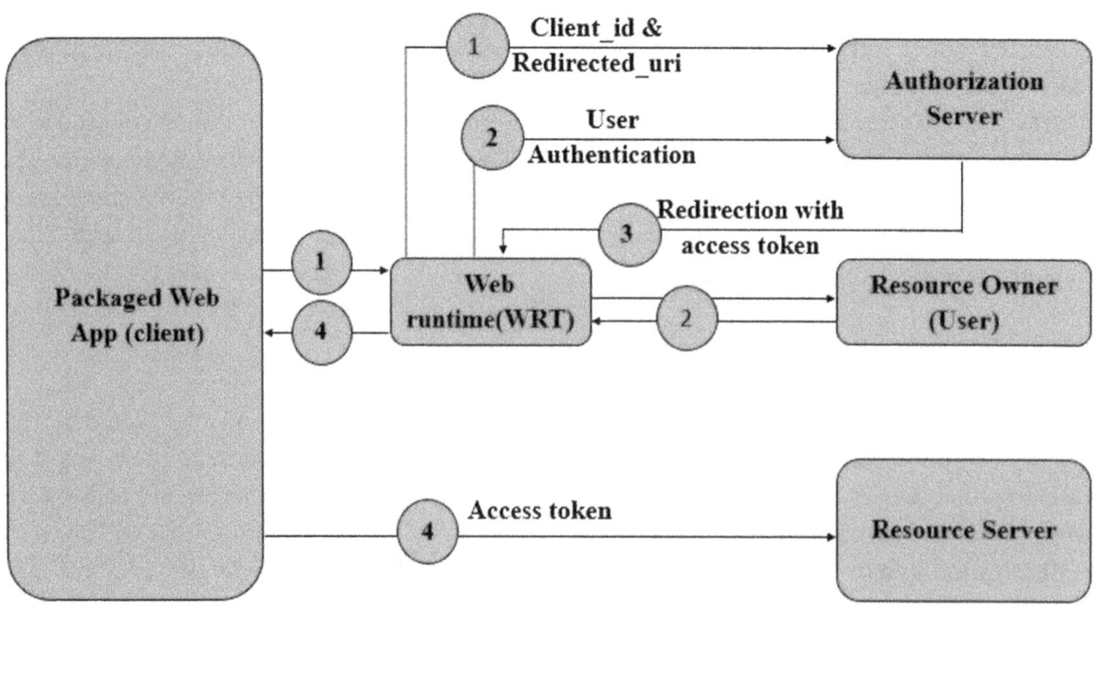

ration document, config.xml, by using different XML elements. In some of these XML elements are taken directly from W3C packaging specification (Dongwan, S., Huiping, Y. & Une, R., 2013) with its custom XML elements. Each app that wants to execute OAuth flow should specify special OAuth related information in its config.xml file (Kaushik, D., Prabhavathi, P. & Joy, B., 2017).

Custom XML Element OAuth

For the previously defined purpose, we have introduced a special custom XML element <OAuth> for the config.xml file (Custom Elements- W3C Working Group Note., 2018). Listing 1 depicts the part of the XSD showing this custom element. This custom element, if present, must include its mandatory redir-url attribute. The client apps need to specify in config.xml the set of redirection URLs that are already registered with the authorization server using one or more <OAuth> elements by specifying each redirected URL as the value for redir-url attribute. If there be multiple redirection URLs used by the client app, equal number of <OAuth> elements should be added to the config.xml file. The order in which the elements are specified in the config.xml file is not important. Each value of the redirection URL specified as the value for redir-url attribute must be of the form http://localhost/[optional_path_components]/<app_id>, where <app_id> is the unique id that is used by W to identify an installed packaged app. This requires the path component of the redirection URL to end with <app_id>.

Listing 1: XML Element OAuth (Custom)

```
<xs:element name="OAuth">
<xs:complexType>
<xs:attribute name="redir-url"
type="xs:anyURI"
use="required"/>
</xs:complexType>
</xs:element>
```

The Packaging on the Web

A packaged web application includes all the support files required by the web application. Hence, a web application will run with none further resources or network property when installation. connected data the application model supports an upscale set of standard W3C/HTML5 options, that embrace varied JavaScript App is similarly as Understanding further HTML mark-ups and CSS options (McGloin & Hunt., 2011). These options, besides the Device App, is and UI framework support, can Programming be wont to produce made web applications during a type of classes, like contacts, messaging, device info access, multimedia, graphics, and games. Applications within the same package follow an equivalent installation life-cycle, handled by the application package manager.

Application Package Manager: The application package manager is one in all the core modules of our application framework, and is liable for installing, uninstalling, and change packages, and storing their data. using the package manager, we will able to additionally retrieve data associated with the packages that can be installed on the device. the application package manager module supports numerous kinds of applications, and selected installation modules may be added to that.

Web Package Configuration: Every package of a Web application consists of a configuration file, config.xml, which follows a packaging format and corresponding data for the application. The Web application needs to follow the characteristics of W3C Widget Packaging format and XML Configuration and additional configuration (in mobile and wearable applications)

Figure 10. Application package manager

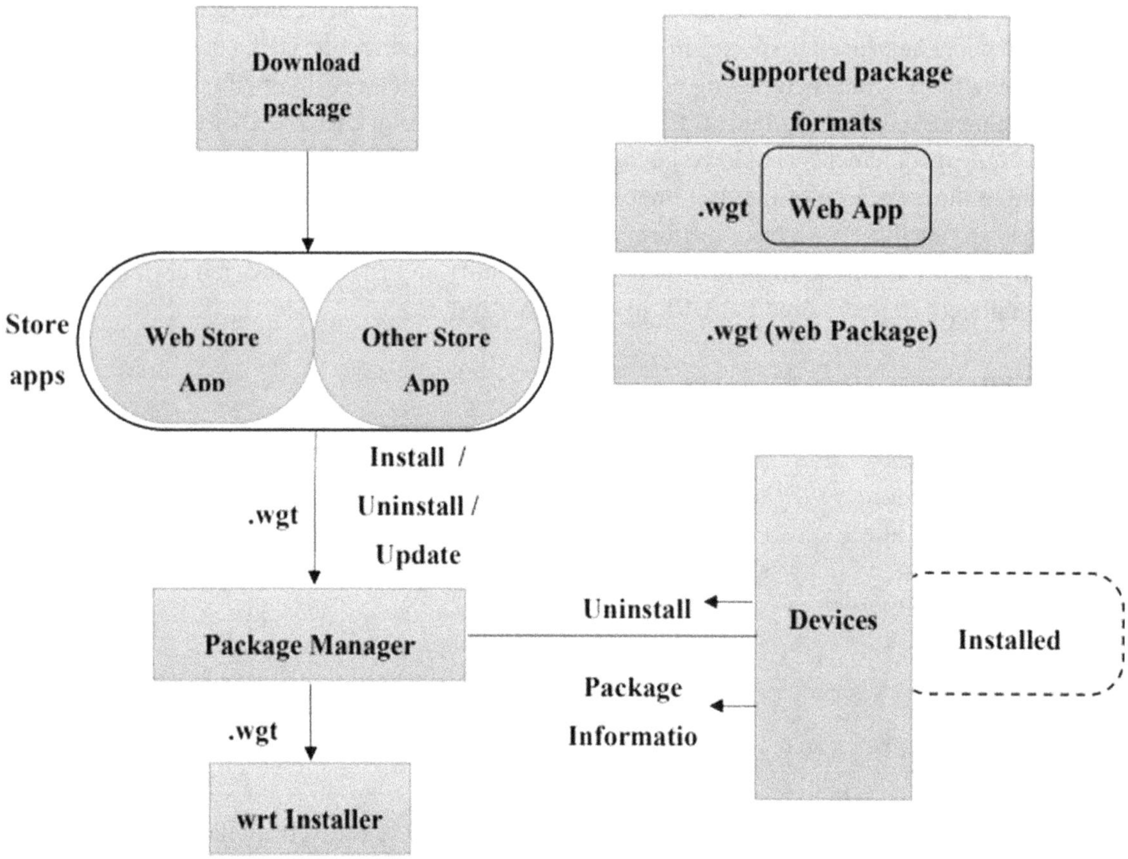

Web Application Package: The software development kit packages the whole web application as a zipped file (McGloin & Hunt., 2011). Therefore, while packaged web application installation W 1st extracts the web app package to any location Lf in the file system. After that it parses the config.xml file from the desired location Lf. The web platform supports HTML, JavaScript, and CSS based Web applications and packaged following the W3C specification. The platform additionally gives device APIs to gain platform capabilities, which enables a rich Web apps development environment. An application package of Web (Hardt., 2011) must keep to the following conventions:

- The file extension and Package format

File extension: *.wgt* (for ex. *abcde.wgt*)
File format: ZIP
MIME type: *abcde/widget*

- Application ID
  After publishing the application, the App ID can't be changed.
- Contents of the Package
- File and foldcrs: The root of the Web package the ZIP file and it contains numerous files and folders, some of these are reserved. The contents of a package show in table 1.
- Directory hierarchy (after installation on device)
- The following table 2 illustrates the Web application package directory structure.

Table 1. Package content

Name	Type	Description
config.xml	File	Application configuration document
icon.gif	File	Application default icon
icon.ico	File	Application default icon
icon.jpg	File	Application default icon
icon.png	File	Application default icon
icon.svg	File	Application default icon
index.html	File	Application default start file
index.htm	File	Application default start file
index.svg	File	Application default start file
index.xhtml	File	Application default start file
index.xht	File	Application default start file
locals	Folder	Container for localized content

WRT Operations

We already know from the previous sections that; Web Runtime environment performs part of the core behavior of any programming language. Web runtime on the OS follows:

- Managing package (install, update, etc.)
- Lifecycle and Execution (launch, pause, resume, etc.)
- Security in Runtime (API/network access, etc.)
- Platform and device integration (access platform resources and local devices)

Web Runtime is the component that allows web applications to run outside the browser. The users able to install the

web apps and use web apps as a standalone application. It manages the installation, update, life cycle, system calls, API/network access, resources, platform integration, and access control of the desired web

Table 2. Web application package structure

Package	Root Directory	Application ID	Core XML file
app	*home/owner/apps_rw/<Package ID>* For example: *home/owner/apps_rw/qik37po9ck*	*<Package ID>.<Name>* For example: *qik37po9ck.sample*	*opt/share/packages/<Package ID>.xml* For example: *opt/share/packages/qik37po9ck.xml*

Figure 11. Web runtime internal block

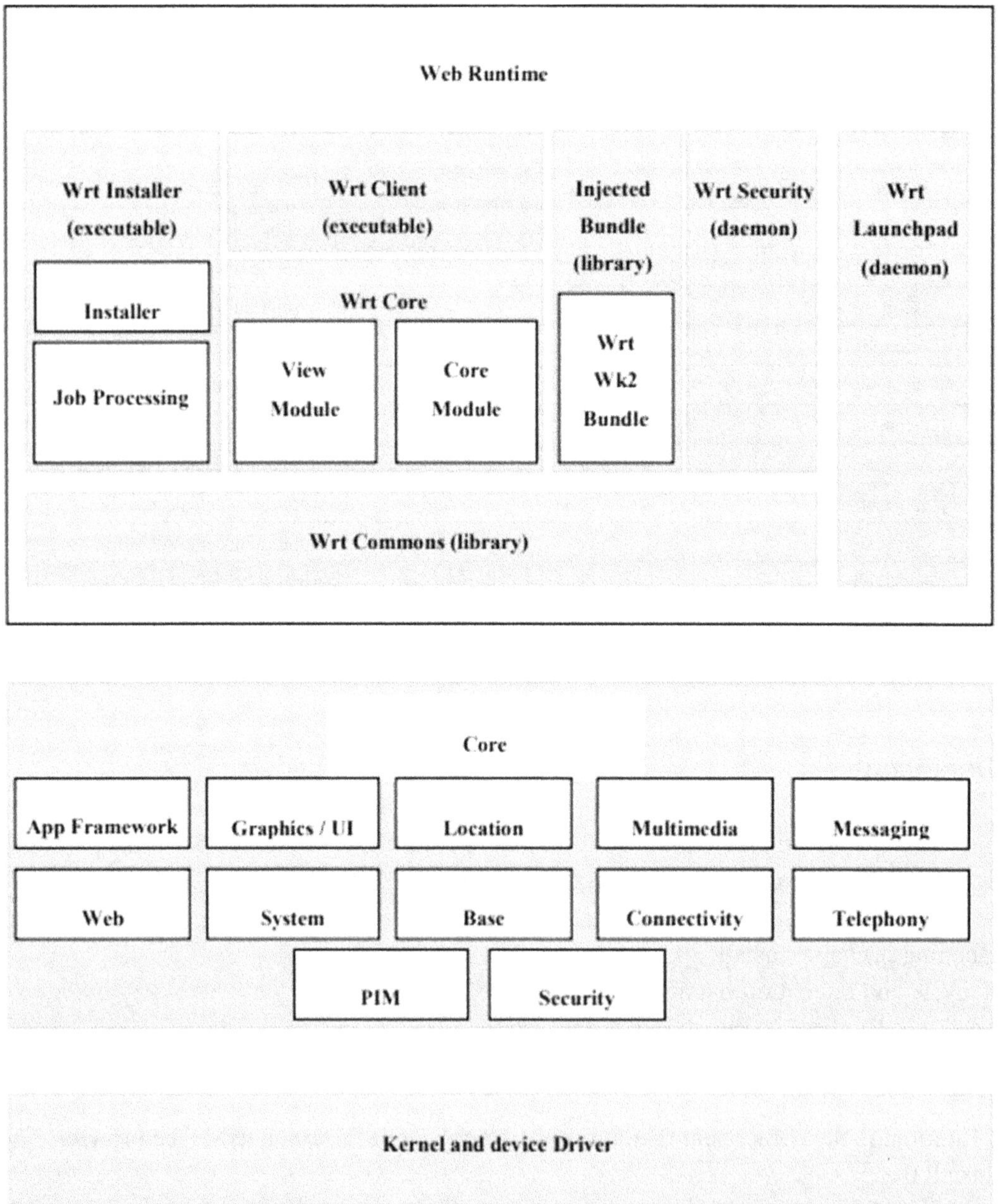

application. Figure 11 illustrates the Internal Blocks of Web Runtime. The Web structure furnishes the best Web involvement with Browser and packaged Web Applications by means of Web Runtime. The Web Runtime is the Execution condition for packaged Web Application. It has the accompanying highlights:

- Focusing on functionality, performance, Standard Compliance(W3C)
- More gadget include availability through Tizen Device API
- jQuery Mobile based Tizen Web UI FW enables easy Web Application development

Parsing OAuth Element

While parsing the components present in the config.xml record, the parser segment of W approves every component against the specified components. The parser extracts relevant information from the element only if the element is valid and it packs with appropriate values related in-memory data structures.

Some of these data structures are also serialized securely to a database in the local file system. When a packaged web app is undergoing any lifecycle operation at some later point in time, refer to the values stored in the database for appropriate actions. Figure 12 explains these operations pictorially.

If the type of an element is *ELEMENT-OAuth* (denoting an *<OAuth>* element) the function checks if it has the mandatory *redir-url* attribute. It also checks if the value of the *redir-url* attribute follows the expected format. If yes, this function populates redirect URL member of *OAuth* element object with the value of *redir-url* attribute for further processing.

The function *PARSE-URI-FOR-APP-ID()* parses the underlying redirect URL to retrieve the app id from it. If the retrieved app id successfully matches with the app id of the client app, the redirect URL is appended to a list of redirect URLs S maintained for this app. This list S is stored permanently in a database table for this client app and is used for OAuth flow control when validating an initiated OAuth flow, as explained in the next section. This database is maintained securely so that it cannot be tampered by any means.

Each time the parser component of W encounters *an <OAuth>* element it executes the algorithm shown in Listing 2. The *ELMNT-VALID ()* function checks the validity of a *config.xml* element against its type.

Listing 2: Algorithm implemented by W to parse an <OAuth> element encountered in config.xml

```
PARSEOAuthELMNT() // Parse OAuth element
w = a packaged web app; // Input
ε = Current OAuth element in w.configXml; // Input
S = Set of redirection URLs retrieved already; // Input
o = empty OAuth element; // local object
if (true == ELMNT-VALID(ε, ELMNT-OAuth, o))
// get app id;
D = PARSE-URI-FOR-APP-ID(o.redirUrl);
if (D == w.app_id) {
S = S  U { o.redirUrl };
return TRUE;
else
return FALSE;
```

Figure 12. WRT operations during installation of packaged web application. Dotted lines indicate data flow and control flow. Solid lines indicate data flow.

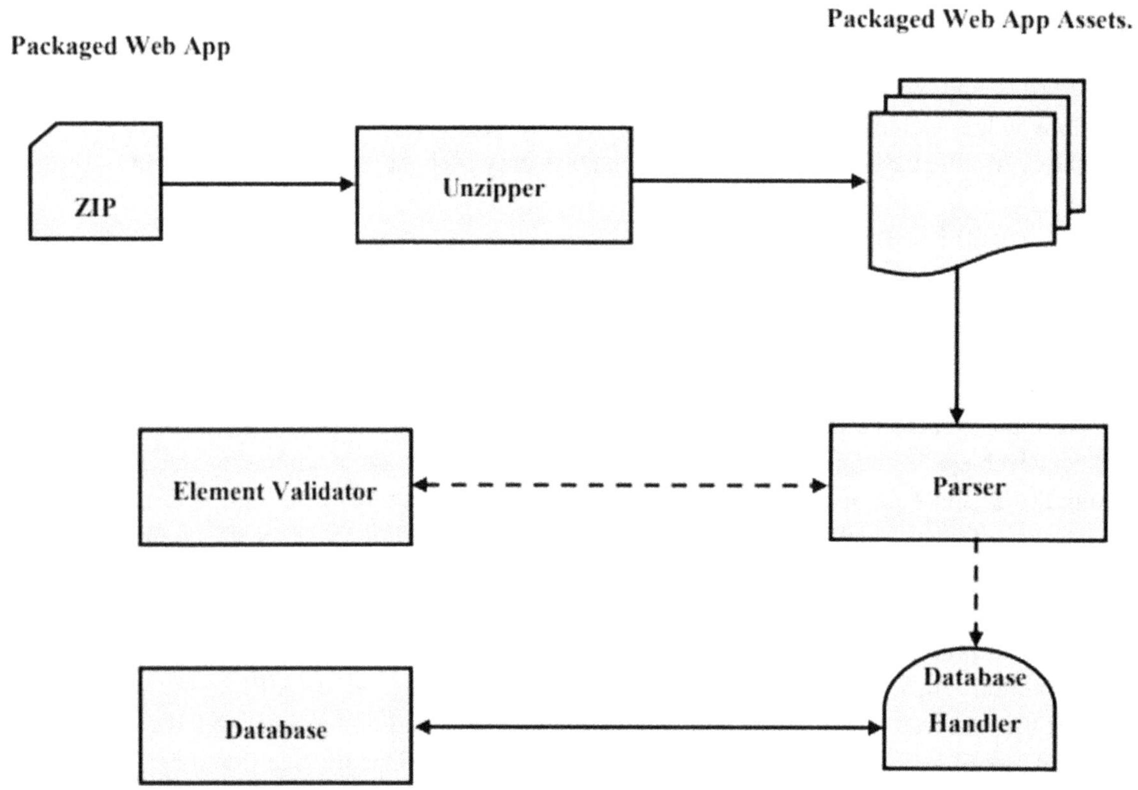

Algorithm for OAuth Flow Control

In our architecture, W provides a JavaScript API to enable a packaged web app create a web view which is a mini window within the current view window and start a browsing context when it is executing. A packaged web app can create such a view only if it declares a custom privilege inside its configuration document. If an app tries to create a web view, W checks if the app has got necessary privileges as per the security model and its configuration document declaration. If allowed to create the web view, the client app loads authorization server URL in the web viewers requesting the access token. With this step, the login page of the authorization server gets loaded within the web view. If user grants the access the authorization server redirects the web view to already registered redirect. *W* intercepts the redirection from the authorization server. *W* fetches the *URL U* authorization server has redirected to and compares the same against the list of the redirection URLs declared inside the configuration document of the underlying packaged web app. *W* retrieves from database the list *S* of redirection URLs that it created during the installation of the app. Next it compares *U* against each string entry *u* in *S*.

Given three finite-length strings a, b and c created using characters from an alphabet Σ of finite number of characters, i.e., a, b, c, Ω, Σ*. We denote equality of two such strings x and y as x = y. If all the entries in S fail to satisfy W aborts the OAuth execution process. If for any entry u in S, W finds u = U satisfying, W stops checking remaining entries in S, if any, and extracts the last part β of the path

Figure 13. Installation/Update flow for Parsing OAuth Element

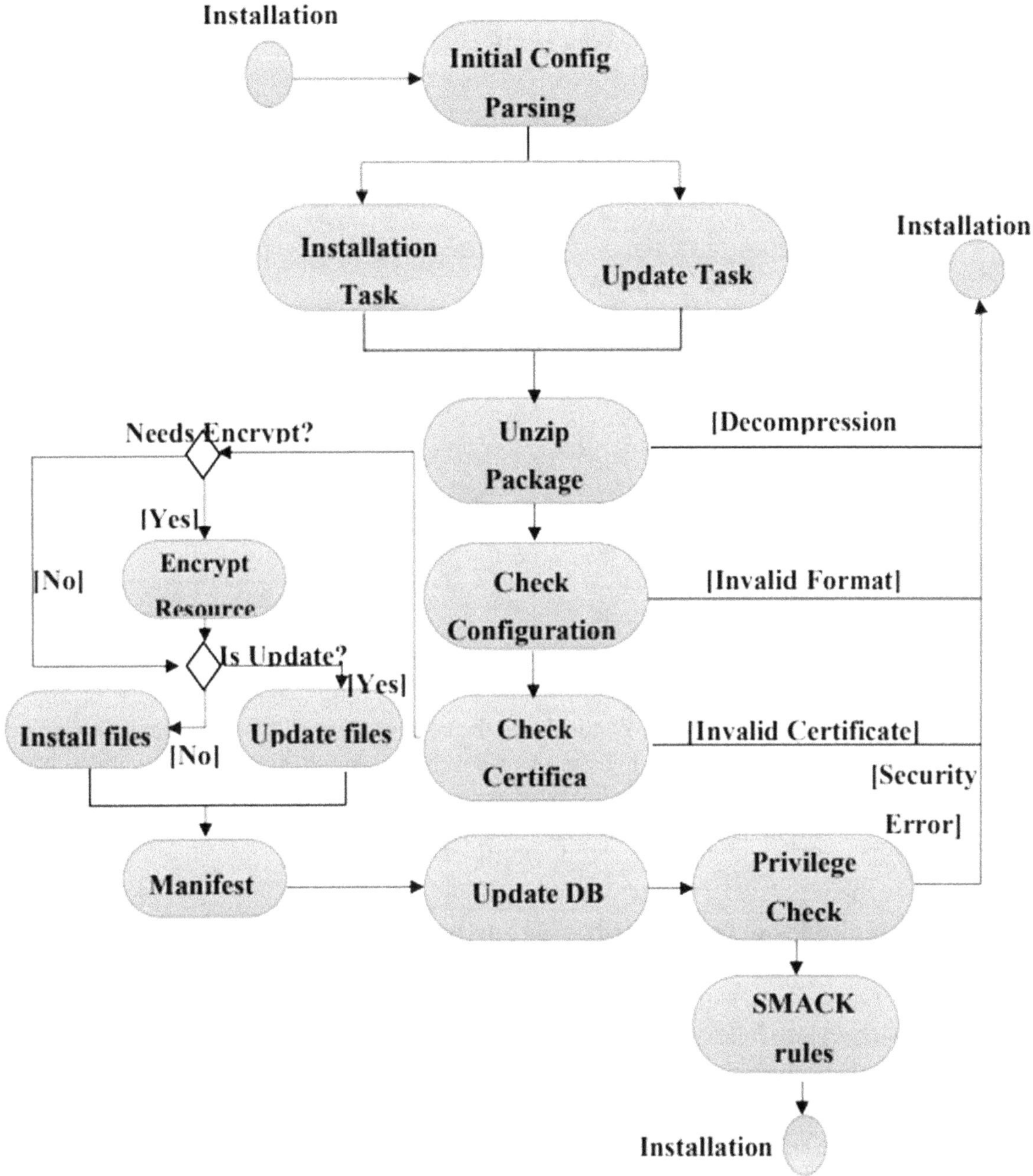

component of the redirection URL U. W matches β (last part of re) with app_id of the destination app using (6) to ensure if the OAuth is getting executed to the right, legitimate app. Upon successful verification of β = app_id, W extracts the access token ξ from U by fetching the value of the query parameter access_token in U. W passes the access token ξ to packaged app for further use.

Listing 3: OAuth flow control algorithm implemented by W.

```
OAuthFLOWCONTROL() // OAuth Flow Controlled by WRT
w = a packaged web app; // Input
ξ = OAuth access token to be used by w; // Output
S = Ø; // set of redirection URIs for each redirection URL u in W.configXml
S = S U {u};
v = CREATE-OAuth-WEBVIEW(auth_server_url);
U = INTERCEPT-REDIRECT-RESP(); // redirect URL
if (true == MATCH-REDIRECT-URI(U, S))
D = PARSE-URI-FOR-APP-ID(U); // get app id
if (D == w.app_id)
ξ = PARSE-URI-FOR-ACCESS-CODE(U);
CLOSE-OAuth-WEBVIEW(v);
return ξ;
else
// OAuth flow terminated by WRT
OAuth-EXCEPTION(Target_Mismatch);
else
// OAuth flow terminated by WRT
OAuth-EXCEPTION(Redirection_Mismatch);
```

It does not satisfy for β with app_id, W concludes the current OAuth flow as illegal and aborts the OAuth execution flow raising exception to the packaged web app. In either case W closes the web view created for OAuth flow execution. Listing 3 shows the pseudocode for the algorithm followed by W to ensure secure OAuth flow initiated by a packaged web app. Figure 14 illustrates the flow chart of OAuth flow control using *W*.

Proposed Attack Model to Detect the Security of OAuth Flow

It occupied that 1) customer's browser and computer are impervious 2) both RP (Relying Parties) and IdP (Identity Provider) are not harmful, and 3) the forthright contacts between RP and IdP are assured. However, owing to the universality of web vulnerabilities (Hammer., 2010), IdP or RP may have several vulnerabilities, such as Cross-Site Scripting (XSS), Cross-Site Request Forgery (CSRF). The objective of an invader is to reach an unauthorized access customer's funds hosted on IdP or RP. In concise, the power of the invader is restricted as follows:

- The charger is a web user who could consecrate a website, issue some links by the blog, microblog, and email, and start an attack on an attackable website.
- The invader is even a smaller who could smell the unencrypted traffic on the Internet, and bribe simply with the traffic.

Our analytical framework is to give a normal approach to find the security of OAuth service. It is a two-stage procedure adding protocol analysis and empirical analysis, as manifested in Figure 15. In this

Figure 14. Flow chart of OAuth flow control using W

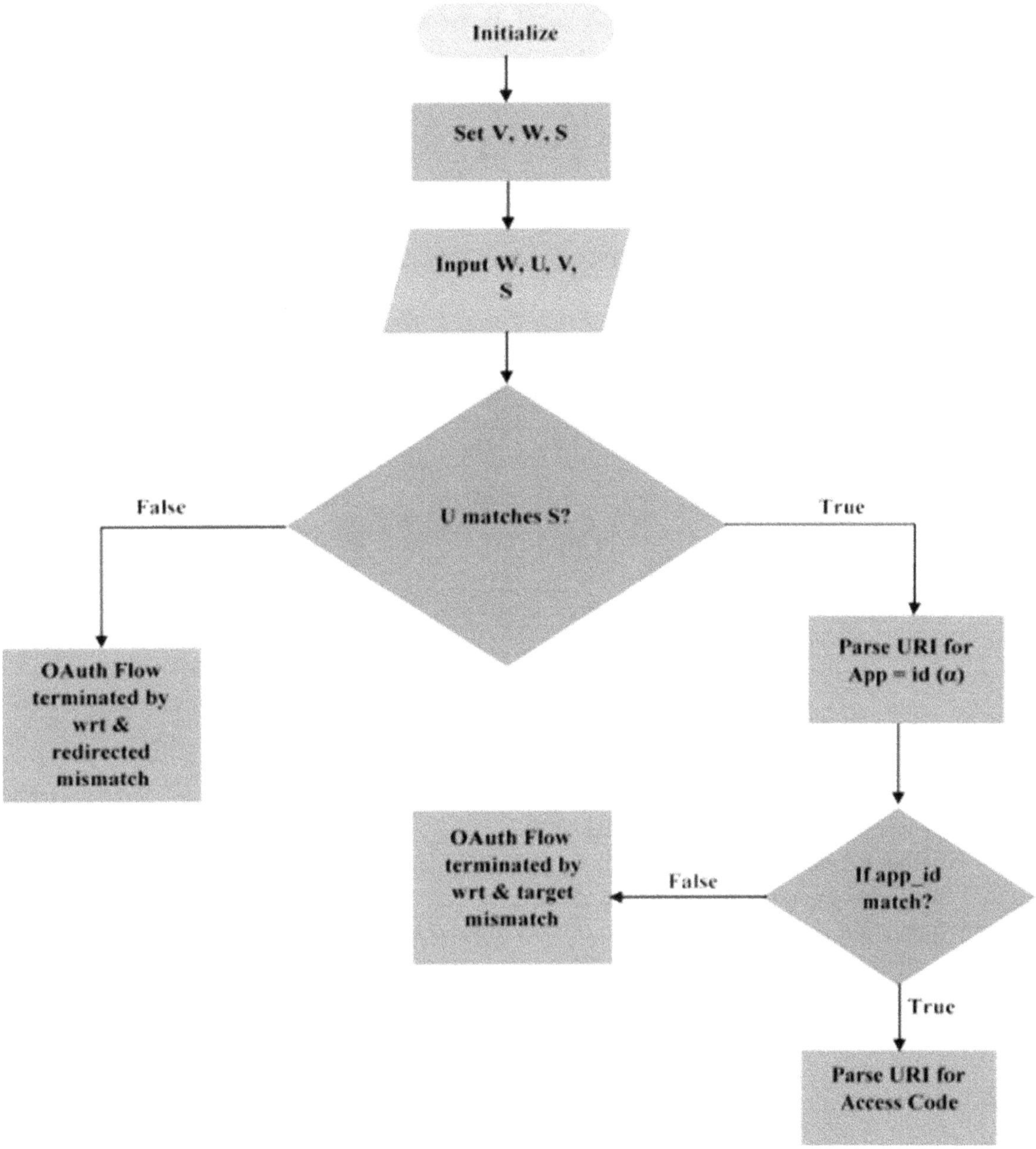

period of protocol analysis, we investigate the protocol gingerly to publish the usages, requirements and strong jeopardies of the five alterable parameters and session S.

Five attacks (recognized by A1 to A5) (lhshaoren., 2010) in view of the previous are given in the last stage, all of which have been demonstrated accessible in the accompanying test. What's more, the investigation on the presuppositions of the five assaults uncovers that we require just to execute a discovery on thing I1 to I9 as takes after to assess the security of OAuth services:

Figure 15. Analytic framework of detecting the security on parameters and session S

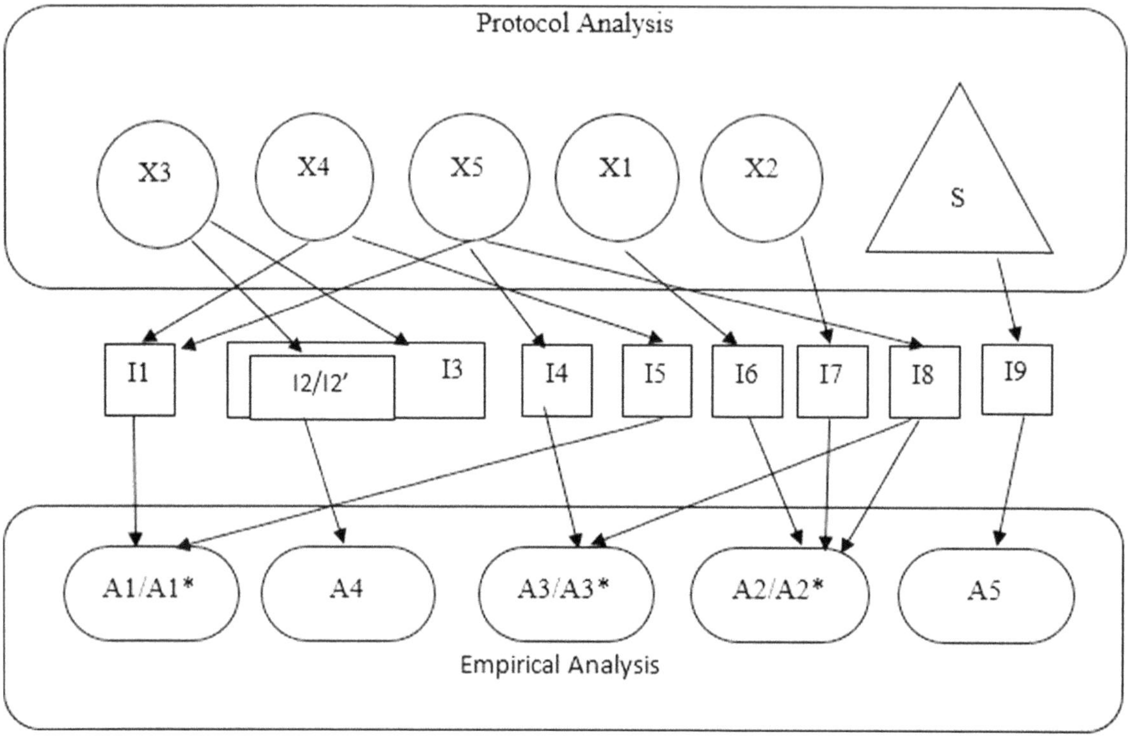

- I1: Whether HTTPS protection is conveyed by RP.
- I2: Whether an unusual state parameter is utilized by RP.
- I3: Whether RP is resisted against CSRF attack.
- I4: Whether RP stores the access token in the URI or cookie.
- I5: Whether the code used in the cross.
- I6: Whether IdP underpins the two reactions composes all the while and unpredictably.
- I7: Whether any redirection URI in the domain of RP could pass IdP's checking.
- I8: Whether the token is a bearer token.
- I9: Whether a strategy to end the session S is provided.

Based on the analytic framework, the proposed approach can be described with the pseudocode as follows:

Listing 4: Pseudo-code for Attack List Detection on Parameters and Session S

```
Attack List Detection on Parameters and Session S
AttackList Detection() {
// Parameters X[1-5] and S are defined in Section 3.3;
// Attacks A[1-5] are defined in Section 3.4;
// The symbol * means the attack has greater damage. AttackList attacks = NULL;
// all the potential attacks. if(CHECK(X4.I1: HTTPS is employed by RP) = TRUE)
```

```
{
if(CHECK(X4.I5: The code is cross in use) = TRUE) attacks.Add(A1*);
else
attacks.Add(A1);
}
if(CHECK(X1.I6: Two response types are supported by IP) = TRUE AND CHECK(X5.
I8: The token is a bearer token) = TRUE)
{
if(CHECK(X2.I7: The realm URIs are checked out by IP)= TRUE) attacks.Add(A2*);
else
attacks.Add(A2);
}
if(CHECK(X5.I4: The access token is stored incorrectly by RP) = TRUE
ANDCHECK(X5.I8: The token is a bearer token) = TRUE)
{
if (CHECK(X5.I1: HTTPS is employed by RP) = FALSE) attacks.Add(A3*);
else
attacks.Add(A3);
}
if(CHECK(X3.I3:Protection against CSRF is employed by RP) = FALSE)
{
attacks.Add(A4); // check Item X3.I2': Whether state is invalid. if(CHECK(X3.
I2: The state parameter is used by RP)= TRUE)
Warning(The state parameter(X3) is INVALID);
}
if(CHECK(S.I9:Ending session S is provided by RP)= FALSE) attacks.Add(A5);
return attacks;
}
```

PERFORMANCE ANALYSIS OF NEW OAuth SECURITY ARCHITECTURE FOR PACKAGED WEB APPS

When we make our Packaged Web application, we marked the following options, which all used the Web-native languages like HTML5, CSS, and JavaScript. We follow the Web application life-cycle from commodity concept via upliftment and redemption to end-of-life application retirement.

Planning and Designing the Application

The first step in creating a Web application is scheming and designing it using the design instruments of our option. Sometimes we have closed the application scheme and design, we are prepared to make the application project.

Creating the Application Project

We create our application using Visual Studio 2012. The Visual Studio gives several project templates that create it easier for us to initiate coding our application. When we make a new project, we can choose an earmarked template or specimen. Founded on the choice, the Visual Studio Web Project Wizard automatically builds basic functionalities that the application must apply to be capable to run. The default project files and folders are also made.

Figure 16. Packaged Web application development process

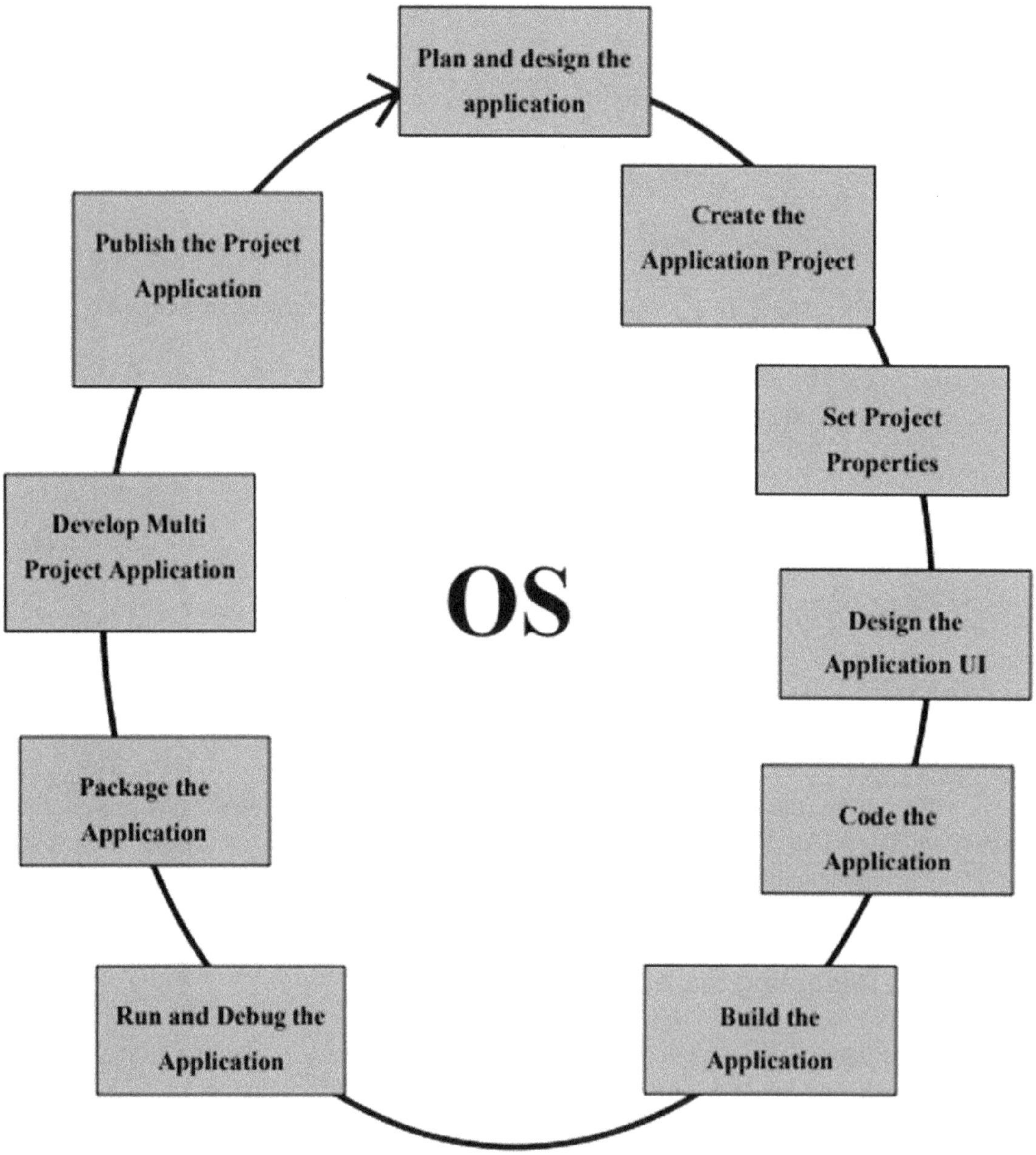

Setting Project Properties

After making the application project, we may configure the characteristics of the project and the Web application to acquire the need functionality and properties for our application.

Designing the Application UI

We designed the application UI utilizing the UI materials defined in the Visual Studio Advanced UI Framework. It gives tools, like effects, UI components, events, and animations, for Web application development. We leverage these tools by just selecting the required screen elements and creating the applications.

Coding the Application

The Visual Studio Web Device API based on JavaScript gives advanced ingress to the device's platform capabilities. We can develop rich Web applications using the Visual Studio Web Device APIs. We can, for example, control the application life-cycle, manage our schedules, exchange data, or make payments using NFC. Sometimes we have ended coding our application, we are ready to construct our application.
Building the Application

When the Visual Studio creates an application then the following procedure is executed:

- Validation check for:
    - JavaScript, CSS
    - Privilege
- Compile for:
    - Coffee script
    - Less

If the project has some errors, they are displayed in the Problems and Project Explorer scenes after the construct. Anyone can create a Web application automatically or manually. We build our apps manually. In the Visual Studio menu, select Project > Build Project. We may construct our project at our benefit. If we need to usage the manual set up and confirm that the Project > Build Automatically choice is not picked.

Running and Debugging the Application

We use Google Chrome to run our application. The running of our application illustrated pictorially in below:

When the Visual Studio runs or debugs the application, the following procedure is executed:

- Construct automatically if no construct has been made yet.
- Package: The optimization procedure is only executed when we execute the packaging procedure.
- Execute the application to the imitator or destination device.

Comparison Table of Existing Approaches With Proposed Approach

The comparison of existing approaches with a proposed approach is given below in table 3:

Packaging the Application

Our Web application packaging procedure is founded on the W3C packaging and configuration.

When the Visual Studio packages the application, the following procedure is executed:

- Construct automatically if no construct has been made yet
- Optimize resources:
- Obfuscation (for JavaScript)
- Minification (for HTML, CSS, JavaScript, and PNG)
- Make the frame formation (for hybrid core applications).
- Create resources (for font, hybrid core, and UI framework applications).
- Maintain signing

We may bunch a Web application behaving the web-packaging mandate in the Command Line Interface (CLI) and where it is a working instrument in the Visual Studio: web-packaging project.wgt project/. By default, the Web application bundle is made at one time. We may display the pack volume at any drop of the application development procedure by double-clicking the project.wgt sfile in the Project Explorer display. All the files existent in the application project are viewed in a chart.*Certifying and Publishing the Application*

After we have packaged our application, we are ready to certify and publish our application. To certify and publish our application:

- Upload our Web application to the localhost.
- After the application is uploaded then the application is signed as an attested application installer bundle and the <Application_name>.wgt archive ordering, which holds the distributor signature and it is attached by the applicable hoard.

If we need to pick off your application from parceling and action, you want to request for application retirement from the hoard.

Comparison Between Existing Approaches With Proposed Approach

In our experimental setup, we compared our approach with two other approaches for OAuth flow execution – Approach A: the client app runs a web server, and Approach B: the client app creates a child window and initiates OAuth flow from the child window. In Approach A the OAuth access token is received by the web server and in Approach B the client keeps polling the HTTP redirect by running a timer with 100 ms timeout. 100 ms timeout has been considered after several trials optimizing the performance for OAuth response handling using Approach

B. In timer callback the client app checks if it has received a valid HTTP redirect with access code. If not, it again starts the timer with 100 ms timeout and continues waiting till the next timeout.

This process is repeated until the client app receives a valid HTTP redirect with access code in it or the number of attempts exceeds a pre-determined number. If the client app receives a valid HTTP redirect with the access code in it, it closes the child window and makes an AJAX request to the authorization server to get the access token. The client app uses the access code received in the response to its earlier AJAX request. We used packaged web app for accessing resources from Facebook. We also created malicious versions of these apps impersonating the actual apps. We compared the system behavior while executing OAuth flow from legitimate and malicious apps. We also measured the time to handle OAuth redirection response and average power consumed during OAuth flow execution.

Characteristics

We have some common characteristics among the approaches. They also have some different characteristics. Now we mention the characteristics of the approaches:

- Supported Platform
- CSRF Attack
- XSS Attack
- Illegal OAuth Flow
- Response Handling Delay
- Power Consumption

Table 3. Comparison table of existing approaches with proposed approach

Parameter	Approach A	Approach B	Proposed Approach
Supported Platform	Windows 7,8,10, MAC OS, iSO 6.0 and later and Android 2.0 and later	Windows 7,8,10, MAC OS, iSO 6.0 and later and Android 2.0 and later	Windows 7,8,10, MAC OS, iSO 6.0 and later and Android 4.0 and later
CSRF Attack	Suffer Slightly	Suffer Mostly	Don't Suffer
XSS Attack	Suffer Slightly	Suffer Mostly	Don't Suffer
Illegal OAuth Flow	Some of these are blocked	Some of these are blocked	Aborts All
Response Handling Delay	Medium	High	Low
Power Consumption	Medium	Worst	Low

Performance Analysis and Results

In this section, we compared our approach with two other approaches for OAuth flow execution –Approach A: the client app runs a web server, and Approach B: the client app creates a child window and initiates OAuth flow from the child window. We compared the system behavior while executing OAuth flow from legitimate and malicious apps. We also measured the time to handle OAuth redirection response and average power consumed during OAuth flow execution. The OAuth flow is executed from our apps in 5 iterations.

System Setup

In this section, the application will be examined. Further, the testing environment is described.

- Application Tools: Microsoft Visual Studio 2012, Microsoft SQL Server and Sublime Text is used to create the application.
- Testing Environment: The specifications of the computer used are:
- Operating System – Windows 7.
    - Processor – Intel(R) Core (TM) i7-6700 CPU @ 3.40GHz.
    - RAM – 16 GB (DDR-3).
    - Clock – Core speed 3401 MHz, 4 Core(s)
    - Bandwidth – 100 mbps

Analysis

In this section, the performance of new OAuth security architecture for Packaged Web apps is described with the corresponding figure, table and performance graph. At the initial stage, we analyze the security of our new OAuth security architecture for Packaged Web apps with the proposed attack model to detect the security of OAuth flow described in section 5.6. Five types of attacks (identified by A1 to A5) founded on the previous are given in the latter period, all of which have been verified available in the following research. And the resolution on the prolepsis of the five attacks publishes that we want only to execute a discovery on item I1 to I9.

Here we used DVWA (Damn Vulnerable Web App) web app tools to perform the previously defined attacks. Some of the attacks are listed below:

- Access token eavesdropping (A1)
- Access token theft via XSS (A2)
- Impersonation (A3)
- Session swapping (A4)
- Force-login CSRF (A5)

We have used the Website plus Tool for measuring the response handling delay (ms) for OAuth flow. We perform 5 iterations with our new developed Packaged Web app. Figure 18 to Figure 22 pictorially shows the iteration results for measuring the response handling delay.

We can calculate the Response Handling Delay as:

$$T_{RHD} = T_{CBF} - T_{ARL}$$

Here we denote TRHD as a time for response handling delay, TCBF as time for call back first and TARL as time for OAuth request last. By using the above equation, we got 53.3 ms as response handling delay from the first iteration. Then we got 49.7 ms as response handling delay from the second iteration.

We got 50.8 ms as a response handling delay from the third iteration.

We got 53.9 ms as a response handling delay from the fourth iteration.

Figure 17. Attacks tool - Damn Vulnerable Web App

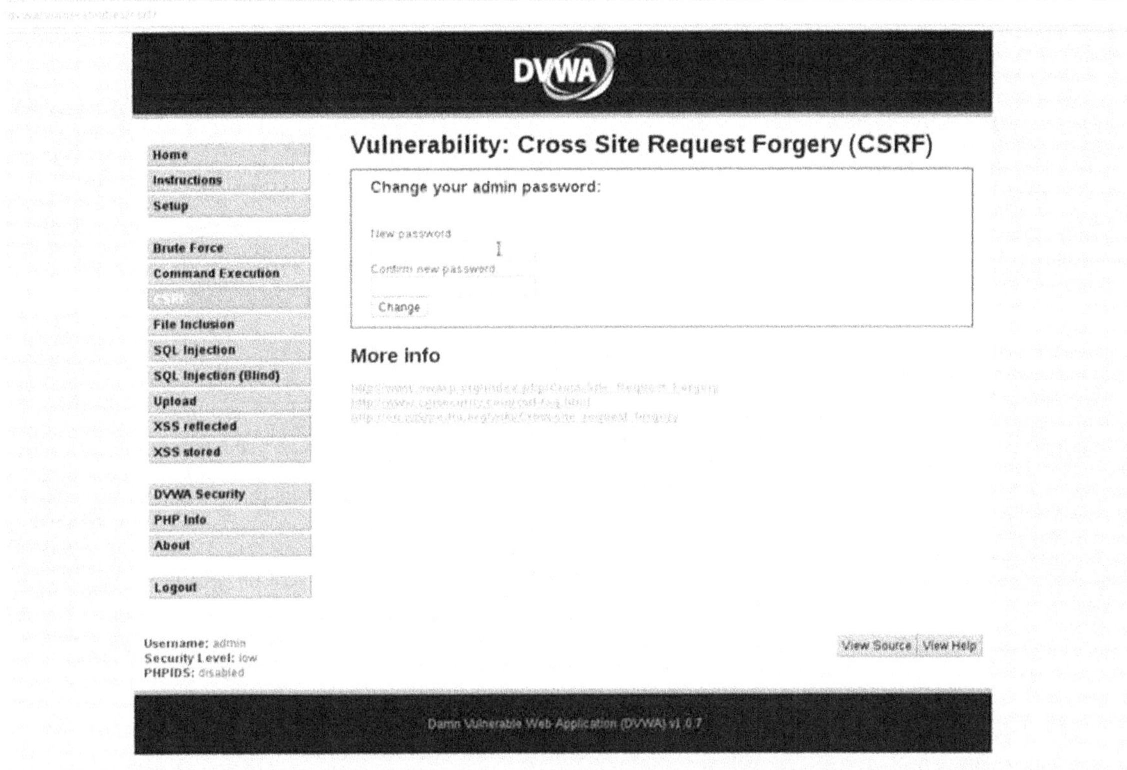

Figure 18. Capturing response handling delay (Iteration 1)

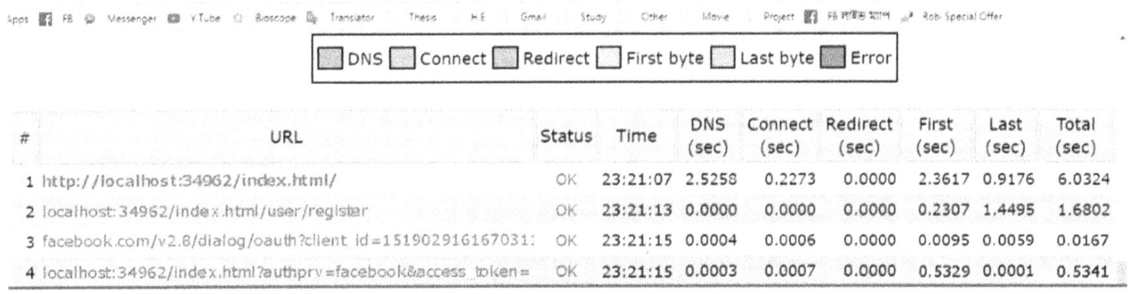

DNS Connect Redirect First byte Last byte Error

#	URL	Status	Time	DNS (sec)	Connect (sec)	Redirect (sec)	First (sec)	Last (sec)	Total (sec)
1	http://localhost:34962/index.html/	OK	23:21:07	2.5258	0.2273	0.0000	2.3617	0.9176	6.0324
2	localhost:34962/index.html/user/register	OK	23:21:13	0.0000	0.0000	0.0000	0.2307	1.4495	1.6802
3	facebook.com/v2.8/dialog/oauth?client_id=151902916167031	OK	23:21:15	0.0004	0.0006	0.0000	0.0095	0.0059	0.0167
4	localhost:34962/index.html?authprv=facebook&access_token=	OK	23:21:15	0.0003	0.0007	0.0000	0.5329	0.0001	0.5341

Figure 19. Capturing response handling delay (Iteration 2)

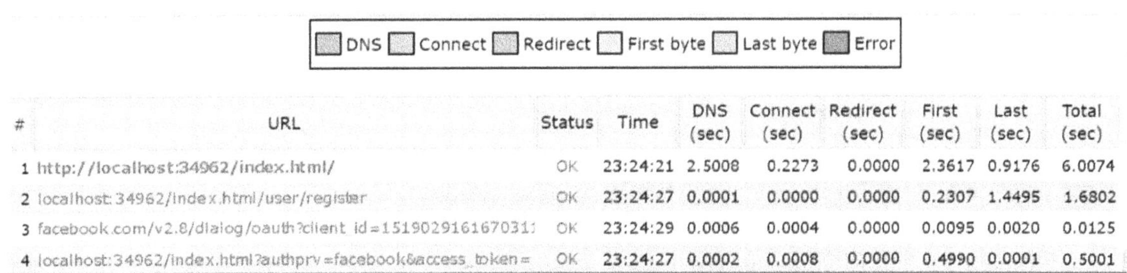

DNS Connect Redirect First byte Last byte Error

#	URL	Status	Time	DNS (sec)	Connect (sec)	Redirect (sec)	First (sec)	Last (sec)	Total (sec)
1	http://localhost:34962/index.html/	OK	23:24:21	2.5008	0.2273	0.0000	2.3617	0.9176	6.0074
2	localhost:34962/index.html/user/register	OK	23:24:27	0.0001	0.0000	0.0000	0.2307	1.4495	1.6802
3	facebook.com/v2.8/dialog/oauth?client_id=151902916167031	OK	23:24:29	0.0006	0.0004	0.0000	0.0095	0.0020	0.0125
4	localhost:34962/index.html?authprv=facebook&access_token=	OK	23:24:27	0.0002	0.0008	0.0000	0.4990	0.0001	0.5001

Figure 20. Capturing response handling delay (Iteration 3)

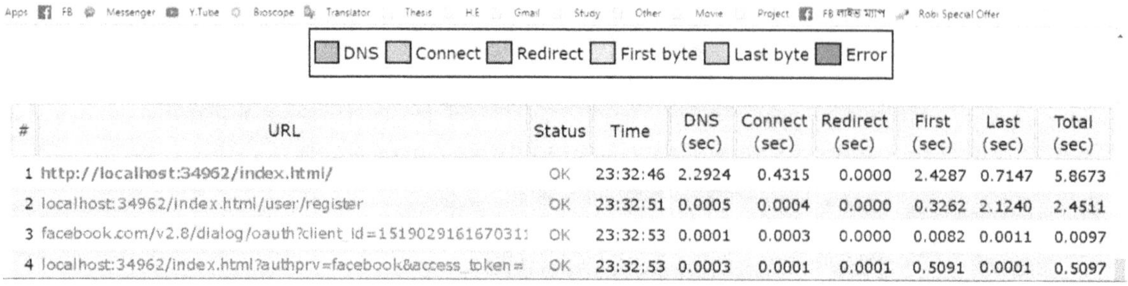

#	URL	Status	Time	DNS (sec)	Connect (sec)	Redirect (sec)	First (sec)	Last (sec)	Total (sec)
1	http://localhost:34962/index.html/	OK	23:32:46	2.2924	0.4315	0.0000	2.4287	0.7147	5.8673
2	localhost:34962/index.html/user/register	OK	23:32:51	0.0005	0.0004	0.0000	0.3262	2.1240	2.4511
3	facebook.com/v2.8/dialog/oauth?client_id=151902916167031	OK	23:32:53	0.0001	0.0003	0.0000	0.0082	0.0011	0.0097
4	localhost:34962/index.html?authprv=facebook&access_token=	OK	23:32:53	0.0003	0.0001	0.0001	0.5091	0.0001	0.5097

Figure 21. Capturing response handling delay (Iteration 4)

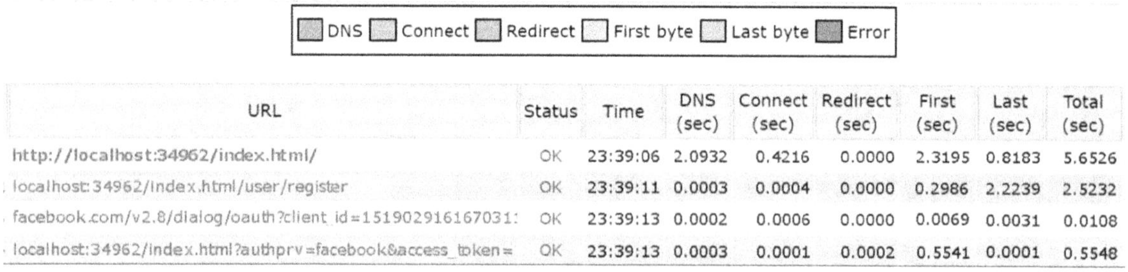

URL	Status	Time	DNS (sec)	Connect (sec)	Redirect (sec)	First (sec)	Last (sec)	Total (sec)
http://localhost:34962/index.html/	OK	23:39:06	2.0932	0.4216	0.0000	2.3195	0.8183	5.6526
localhost:34962/index.html/user/register	OK	23:39:11	0.0003	0.0004	0.0000	0.2986	2.2239	2.5232
facebook.com/v2.8/dialog/oauth?client_id=151902916167031	OK	23:39:13	0.0002	0.0006	0.0000	0.0069	0.0031	0.0108
localhost:34962/index.html?authprv=facebook&access_token=	OK	23:39:13	0.0003	0.0001	0.0002	0.5541	0.0001	0.5548

Figure 22. Capturing response handling delay (Iteration 5)

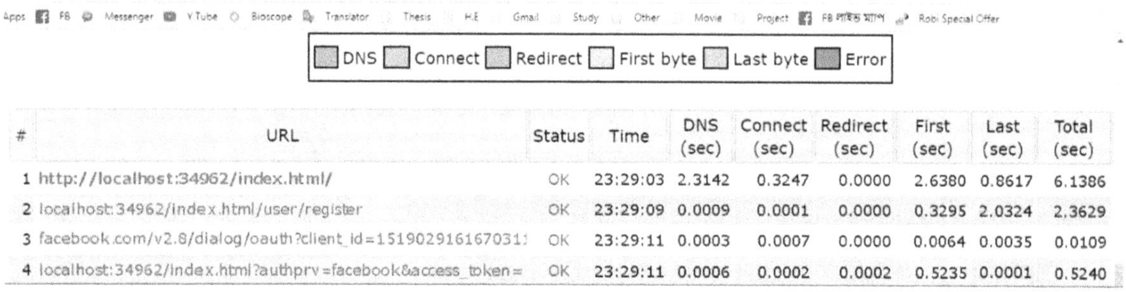

#	URL	Status	Time	DNS (sec)	Connect (sec)	Redirect (sec)	First (sec)	Last (sec)	Total (sec)
1	http://localhost:34962/index.html/	OK	23:29:03	2.3142	0.3247	0.0000	2.6380	0.8617	6.1386
2	localhost:34962/index.html/user/register	OK	23:29:09	0.0009	0.0001	0.0000	0.3295	2.0324	2.3629
3	facebook.com/v2.8/dialog/oauth?client_id=151902916167031	OK	23:29:11	0.0003	0.0007	0.0000	0.0064	0.0035	0.0109
4	localhost:34962/index.html?authprv=facebook&access_token=	OK	23:29:11	0.0006	0.0002	0.0002	0.5235	0.0001	0.5240

We got 52.0 ms as a response handling delay from the fifth iteration. The average response handling delay of our new developed Packaged Web app is 51.94 ms.

Then we used the Joulemeter Tool for measuring the power consumption (mW) for OAuth flow. We perform 5 iterations with our new developed Packaged Web app. Figure 23 pictorially shows the iteration results for measuring power consumption.

We can calculate the Response Handling Delay as:

$$E_{(mW/time)} = P_{(nW)} * t_{(time)} / 1000_{(W/mW)}$$

Here we denote E mW/time as energy, P mW as power and t as time.

We got 173 mW as power consumption from the second iteration. Then we got 175 mW as power consumption from the third iteration. We got 170 mW as power consumption from the fourth iteration.

Figure 23. Capturing power consumption (Iteration 1 to 5)

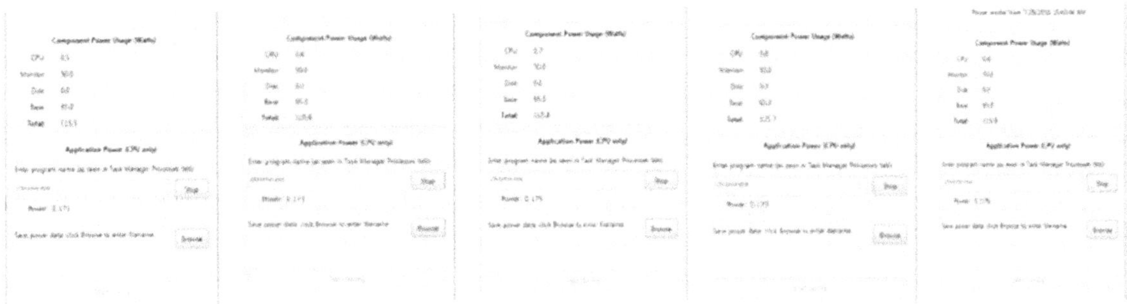

And then we got 175 mW as power consumption from the fifth and final iteration. The average power consumption of our new developed Packaged Web app is 173 mW.

Handling Illegal OAuth Flow

In this section we discuss handling capacity of our new OAuth based Packaged Web Application. Since our prime goal was to check if our approach can block malicious OAuth, we first checked the behavior with malicious OAuth flow initiated by impersonating client web apps. All these malicious web apps create OAuth web view and attempt OAuth flow using app id of some other legitimate packaged web app. Results reveal that our approach blocks all malicious OAuth flow executions which are summarized in table 4. Since in our approach W checks the app id present in the redirection URL against the app id the app executing the OAuth flow W aborts OAuth flows from malicious impersonating apps considering those flows as illegal flows. Since an app cannot tamper its app id and W also not allow it to use app id of an already installed legitimate app, W always successfully aborts all illegal OAuth flows.z

Table 4. OAuth Flow execution results from our apps

Attack No.	Attack Type	Resource Server	Attacks (Attempted/ Blocked)
A1	Access token eavesdropping		1/1
A2	Access token theft via XSS		3/3
A3	Impersonation	Facebook	1/1
A4	Session swapping		2/2
A5	Force-login CSRF		5/5

The algorithm OAuthFLOWCONTROL() executed by W is the key component of our approach. In the other two approaches, there is no means to check if the access token is getting delivered to right, intended, a legitimate target. But unlike, those two approaches by means of this algorithm our method ensures that access token if delivered, is delivered to the right, intended legitimate target. W ensures that by executing this algorithm OAuthFLOWCONTROL().

Evaluation Table of OAuth Response Handling Delay of Proposed Approach

Our approach gives better OAuth response handling delay compared to the other two approaches across all five iterations. In the case of Approach A, first the authorization code is received and the using the same access token is received next. Hence, getting access token requires a greater number of message exchanges with the authorization server and hence its OAuth response handling delay is more compared to our approach. The Evaluation Table of OAuth Response Handling Delay of Proposed Approach is given in Table 5:

Table 5. Evaluation table of OAuth response handling delay of proposed approach

Iteration No.	Resource Server	Response Handling Delay (ms)	Average Response Handling Delay (ms)
1	Facebook	53.9	52.94
2		53.3	
3		49.7	
4		52.0	
5		50.8	

Graphical Representation of Response Handling Delay of Proposed Approach

The graphical representation of Response Handling Delay of the proposed approach is given below

Graphical Representation of Response Handling Delay with Existing Two Approaches

The graphical representation of response handling delays with existing two approaches is given below:

Evaluation Table of Power Consumption of Proposed Approach

Since three approaches handle OAuth flow involving different software and hardware components, we thought of capturing the average power consumption during OAuth flow execution with each of these approaches. These approaches possess the access token by executing different techniques and require additional processing to gain the possession of the access token, employing OAuth during execution web app will cause additional CPU overhead. Therefore, power consumption by web apps due to OAuth execution is a key performance indicator to compare OAuth handling techniques. The Evaluation Table of OAuth Power Consumption of Proposed Approach is given in table 6:

Graphical Representation of Power Consumption of Proposed Approach

The graphical representation of power consumption of the proposed approach is given below for all five iterations:

Figure 24. Graphical representation of response handling delay of proposed approach

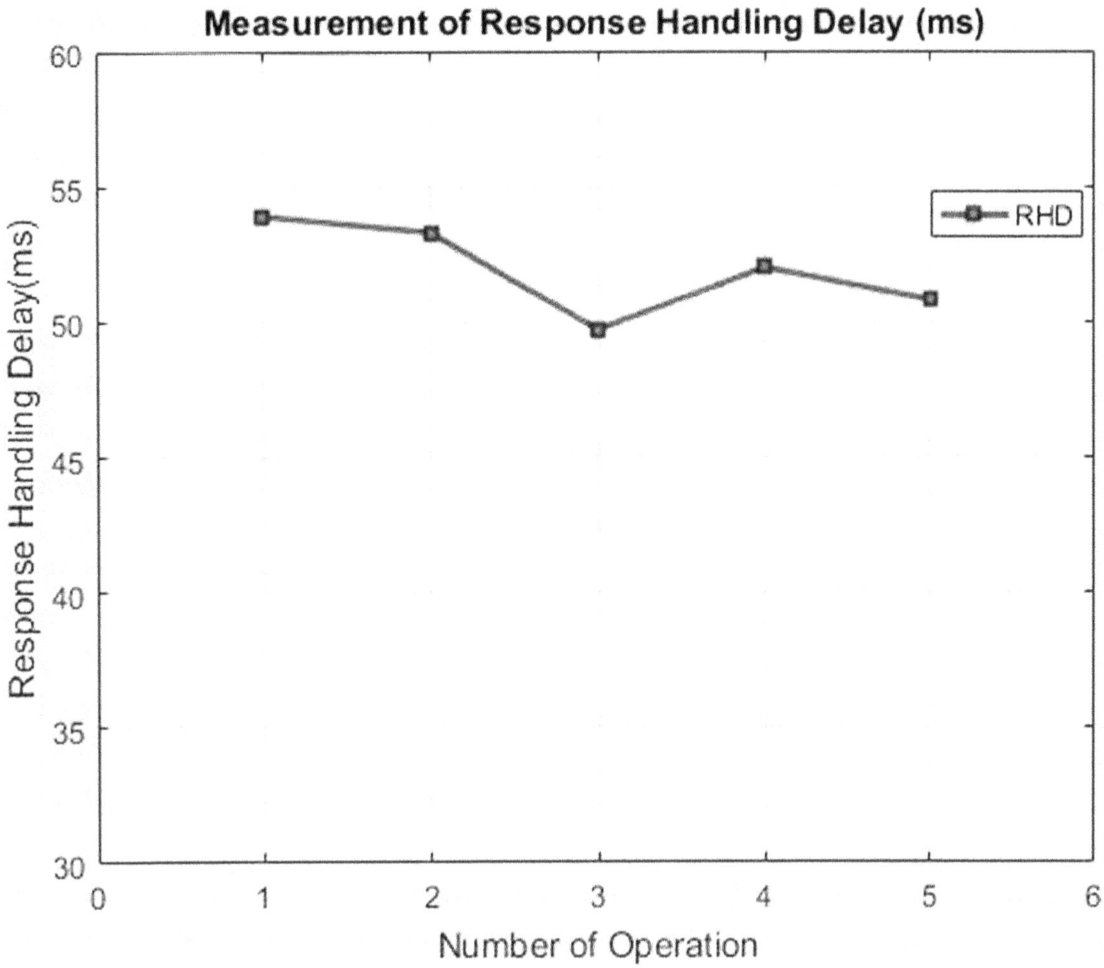

Graphical Representation of Power Consumption with Existing Two Approaches

The graphical representation of power consumption with existing two approaches is given below:

CONCLUSION AND FUTURE WORK

Unlike operable security protocols, OAuth 2.0 is designed except sound cryptographic protection like a digital signature, a random nonce, and encryption. The deficiency of encryption in the protocol needs RPs to employ SSL, but many appreciated websites do not follow this exercise. Compared to server stream (Adrienne, P. F., Elizabeth, H., Serge. E., Ariel, H., Erika, C. & David, W., 2012) client-flow is inherently insecure for OAuth. Based on these acumens, we trust that OAuth 2.0 at the hand of maximum developers without a profound understanding of web security is perhaps to produce precarious implementations.

Figure 25. Graphical representation of response handling delay with existing two approaches

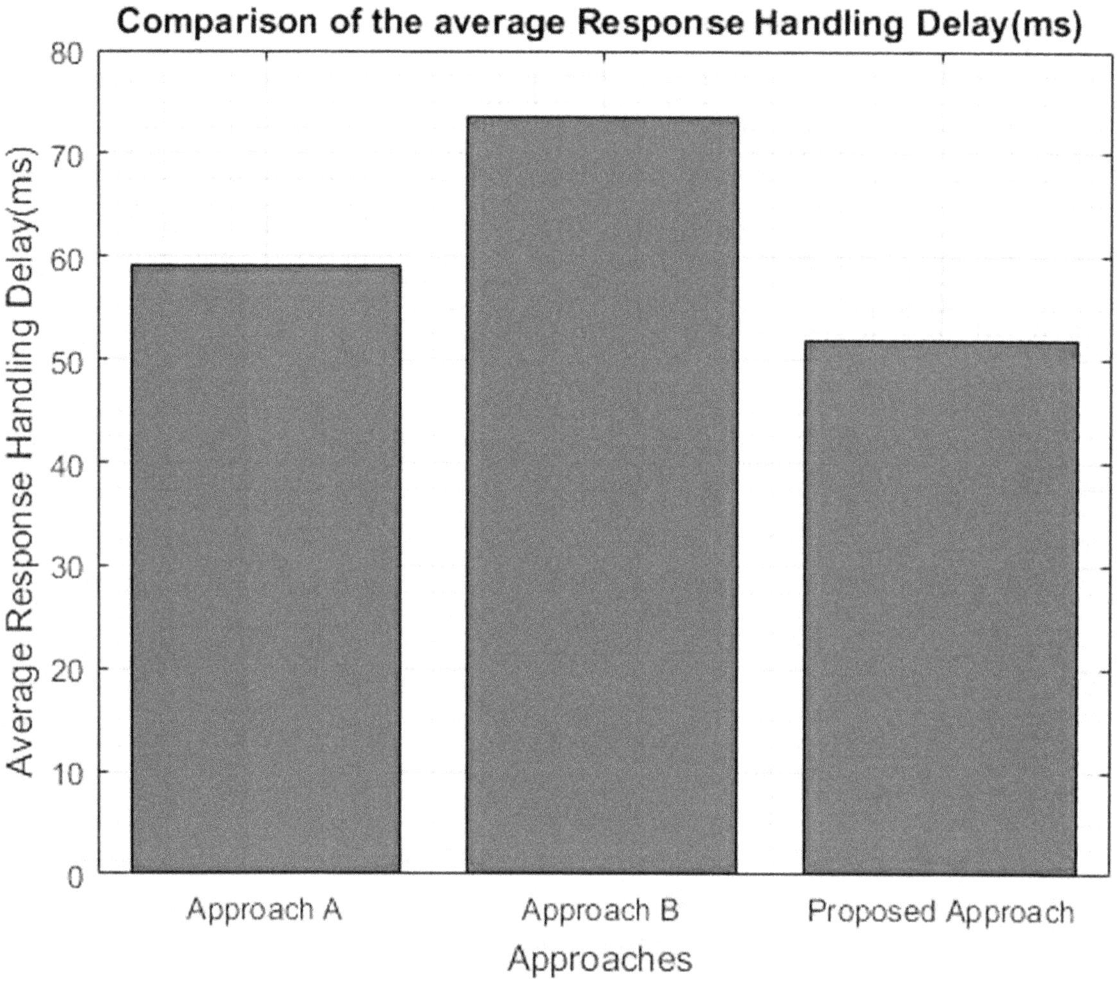

Table 6. Evaluation table of OAuth power consumption of proposed approach

Iteration No.	Resource Server	Power Consumption (mW)	Average Power Consumption (mW)
1	Facebook	171	173
2		173	
3		175	
4		170	
5		173	

Figure 26. Graphical representation of power consumption of proposed approach

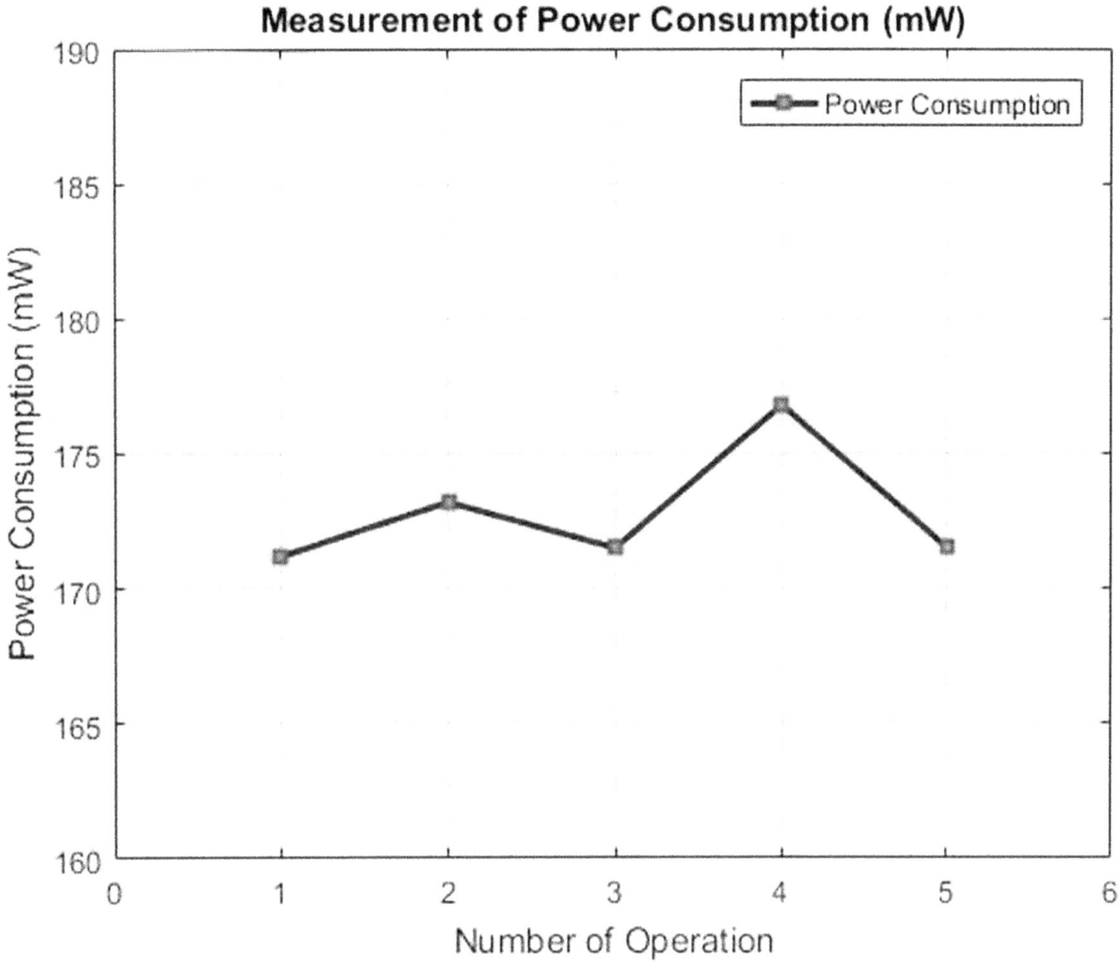

In our system any malicious app cannot execute OAuth flow impersonating as a legitimate app even after gaining the access to proper client_id, redirect_uri and other secretes this is ensured because the OAuth verification logic is executed by W that looks for the presence of the app id in the redirect URL. App id of a packaged app is unique across all apps and any app cannot tamper that. Hence, the strategy of embedding the app id in the redirect URL ensures that the redirect response will reach only to the legitimate app even though the app cannot define the redirect endpoint using a valid HTTP URL like a website.

Our approach still requires the developers to generate the redirection URLs to be registered and declared in the configuration document manually. But that opens the scope of manual errors, which if present, will stop a legitimate valid app from getting OAuth access token. In the future, we intend to wire logic of providing client app credentials through app stores reducing the possibility of developer errors. We also intended to use the Encryption method in future.

Figure 27. Graphical representation of power consumption with existing two approaches

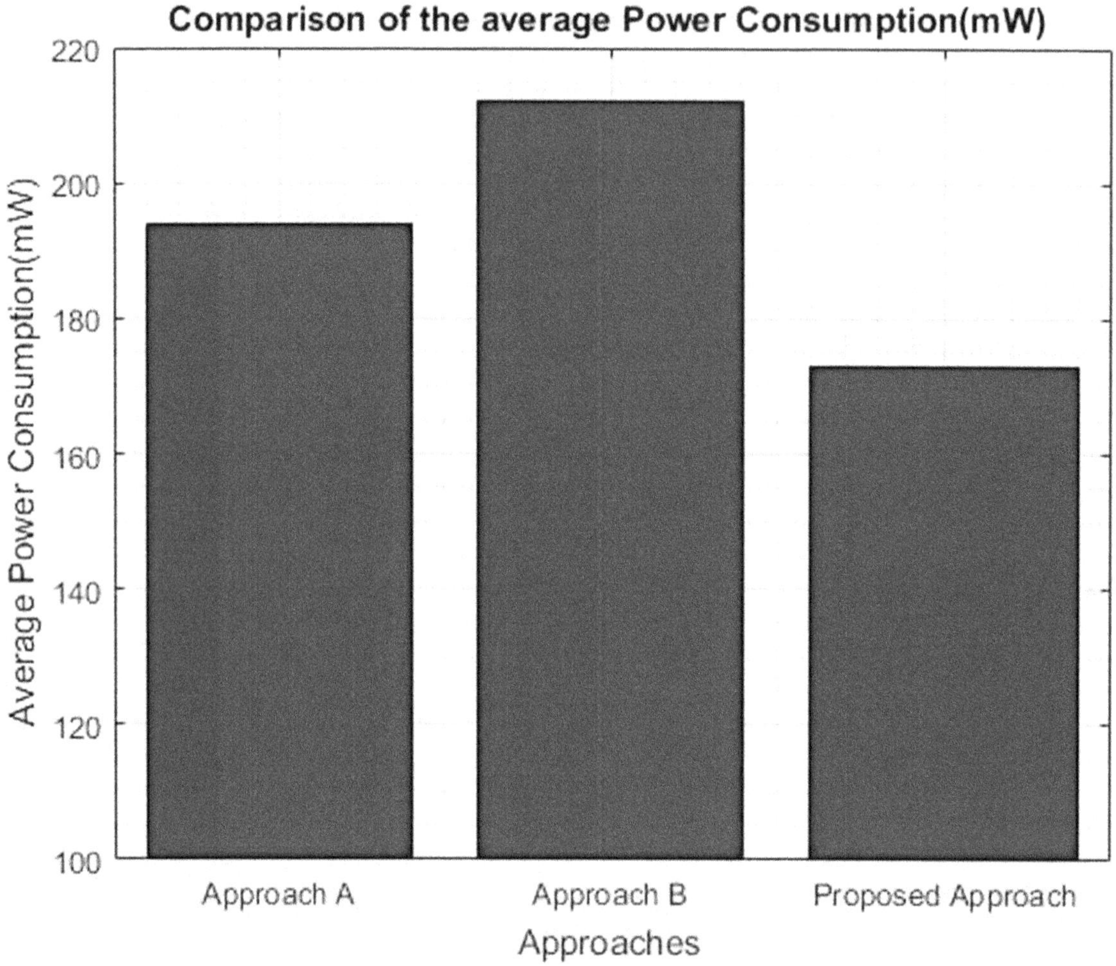

REFERENCES

Adam, B., Juan, C., & Dawn, S. (2009). Secure Content Sniffing for Web Browsers, or How to Stop Papers from Reviewing Themselves. *IEEE Symposium on Security and Privacy, 30*(8), 360-371. doi: 10.1109/SP.2009.3

Adrienne, P. F., Elizabeth, H., Serge, E., Ariel, H., Erika, C., & David, W. (2012). Android permissions: user attention, comprehension, and behavior. *Eighth Symposium on Usable Privacy and Security*, (3). doi: 10.1145/2335356.2335360

Android WebView. (2018). Retrieved from https://developer.android.com/reference/ android/webkit/ WebView.html

Caimei, W., Yan, X., Wenchao, H., Huihua, X., Jianmeng, H., & Cheng, S. (2017). A Verified Secure Protocol Model of OAuth Dynamic Client Registration. *3rd International Conference on Big Data Computing and Communications (BIGCOM),* 106 – 110. doi: 10.1109/BIGCOM.2017.50

Charlie, C., Ben, L., Benjamin, G. Z., & Christian, S. (2011). ZOZZLE: Fast and precise in- browser JavaScript malware detection. *Proceedings of the 20th USENIX Conference on Security*. Retrieved from https://usenix.org/conference/usenix-security-11/zozzle-fast-and-precise-browser-javascript-malware-detection

Daniel, F., Ralf, K., & Guido, S. (2016). A Comprehensive Formal Security Analysis of OAuth 2.0. *CCS '16 Proceedings of the 2016 ACM SIGSAC Conference on Computer and Communications Security,* 1204-1215, doi: 10.1145/2976749.2978385

Dongwan, S., Huiping, Y., & Une, R. (2013). Supporting visual security cues for WebView-based Android apps. *28th Annual ACM Symposium on Applied Computing, ser. SAC '13. New York,* 1867–1876. doi: 10.1145/2480362.2480709

Dongwan, S., & Rodrigo, L. (2011). An empirical study of visual security cues to prevent the ssl stripping attack. *27th Annual Computer Security Applications Conference, ser. ACSAC '11. New York,* 287–296. doi: 10.1145/2076732.2076773

Elements, C. - W3C Working Group Note. (2018). Retrieved from https://www.w3.org/TR/custom-elements

Erika, C., & David, W. (2014). Bifocals: Analyzing webview vulnerabilities in android applications. *Information Security Applications - 14th International Workshop, WISA*, 8267. doi: 10.1007/978-3-319-05149-9_9

Facebook App in Tizen. (2018). Retrieved from https://developer.tizen.org/ko/development/articles/ facebook-apptizen

Fadi, M., & Mohamed, S. (2016). Hardening the OAuth-WebView Implementations in Android Applications by Re-Factoring the Chromium Library. *2nd IEEE International Conference on Collaboration and Internet Computing*, 196–205. doi: 10.1109/CIC.2016.036

Francisco, C., & Karen, P. L. (2011). *Security Analysis of Double Redirection Protocols*. Retrieved from https://bit.ly/2o5eggP

Google Identity Platform. (2018). *Using OAuth 2.0 to Access Google APIs*. Retrieved from https://developers.google.com/identity/protocols/OAuth2

Hammer. (2010). The OAuth 1.0 protocol. *The Internet Eng. Task Force RFC 5849*. Retrieved from https://tools.ietf.org/html/rfc5849

Hao, H., Vicky, S., & Wenliang, D. (2013). On the effectiveness of API-level access control using bytecode rewriting in Android. *8th ACM SIGSAC Symposium on Information, Computer and Communications Security*, 25–36. 10.1145/2484313.2484317

Hardt. (2011). The OAuth 2.0 authorization framework. *The Internet Eng. Task Force RFC 6749*. Retrieved from https://tools.ietf.org/html/rfc6749

Hardt. (Ed.) (2012). *The OAuth 2.0 authorization framework*. IETF Std., RFC6749. doi:10.17487/RFC6749

Jason, B., Elie, B., Divij, G., & John, C. M. (2010). State of the Art: Automated Black-Box Web Application Vulnerability Testing. *IEEE Symposium on Security and Privacy, 31*(1), 332 - 345. doi:10.1109/SP.2010.27

Jones & Hardt. (2012). *OAuth 2.0: Bearer token usage*. Available: https://tools.ietf.org/html/draft-ietf-OAuth-v2-bearer-23

Kaushik, D., Prabhavathi, P., & Joy, B. (2017). Security Mechanism for Packaged Web Applications. *2017 IEEE International Conference on Web Services (ICWS)*. doi: 10.1109/ICWS.2017.72

Kitura. (2018). *OAuth 2.0 authentication with Facebook/Google*. Retrieved from https://kitura.io/docs/authentication/fb-google-OAuth2.html

lhshaoren. (2010). *A vulnerability on weibo.com (sina.com)*. Retrieved from http://www.wooyun.org/bugs/wooyun-2010-039727

Lodderstedt, McGloin, & Hunt. (2013). *OAuth 2.0 Threat Model and Security Considerations*. IETF Std., RFC6819. doi:10.17487/RFC6819

Marian, H., Markus, H., Susanne, W., & Matthew, S. (2014). Using personal examples to improve risk communication for security & privacy decisions. *32nd Annual ACM Conference on Human Factors in Computing Systems,* 2647–2656. doi: 10.1145/2556288.2556978

McGloin & Hunt. (2011). *OAuth 2.0 Threat Model and Security Considerations*. Retrieved from http://tools.ietf.org/id/draft-ietf-OAuth-v2-threatmodel-00. txt/

Mike, T. L., & Venkat, V. (2009). Blueprint: Robust Prevention of Cross-site Scripting Attacks for Existing Browsers. *IEEE Symposium on Security and Privacy, 30*(1), 331-346. doi: 10.1109/SP.2009.33

Mohamed, S., & Fadi, M. (2014). Towards Enhancing the Security of OAuth Implementations in Smart Phones. *IEEE Third International Conference on Mobile Services*, 9–46. doi: 10.1109/MobServ.2014.15

Nazmul, H., Md, A. H., Md, Z. H., Md, H. I. S., & Shawon, R. (2018). OAuth-SSO: A Framework to Secure the OAuth-Based SSO Service for Packaged Web Applications. *IEEE International Conference on Big Data Science and Engineering (TrustCom/BigDataSE), 12*(1), 1575-1578. doi:10.1109/TrustCom/BigDataSE.2018.00227

OpenID Authentication 2.0 - Final. (2007). Retrieved from https://openid.net/specs/openid-authentication-2_0.html

Prateek, S., David, M., & Livshits, B. (2011). ScriptGard: Automatic Context-Sensitive Sanitization for Large-Scale Legacy Web Applications. *ACM Conference on Computer and Communications Security, 18*(2), 601-614. doi: 10.1145/2046707.2046776

Richer, Mills, & Tschofenig. (2012). *OAuth 2.0 message authentication code (MAC) tokens*. Retrieved from https://tools.ietf.org/html/draft-ietf-OAuth-v2-http-mac-00

Rui, W., Shuo, C., & Xiaofeng, W. (2012). Signing Me onto Your Accounts through Facebook and Google: A Traffic-Guided Security Study of Commercially Deployed Single-Sign-On Web Services. *2012 IEEE Symposium on Security and Privacy*, 365-379. doi: 10.1109/SP.2012.30

San-Tsai, S., Kirstie, H., & Konstantin, B. (2012). Systematically breaking and fixing OpenID security: Formal analysis, semi-automated empirical evaluation, and practical countermeasures. *Computers & Security, 31*(4), 465–483. doi:10.1016/j.cose.2012.02.005

San-Tsai, S., & Konstantin, B. (2012). The Devil is in the (Implementation) Details: An Empirical Analysis of OAuth SSO Systems. *CCS '12 Proceedings of the 2012 ACM conference on Computer and communications security,* 378-390. doi: 10.1145/2382196.2382238

Suresh, C., Charanjit, J., & Arnab, R. (2011). Universally Composable Security Analysis of OAuth v2.0. *IACR Cryptology ePrint Archive*. Retrieved from https://eprint.iacr.org/2011/526

Tara, W., & Kori, I. Q. (2005). Gathering evidence: use of visual security cues in web browsers. *Proceedings of the Graphics Interface 2005 Conference,* 137–144. doi: 10.1145/1089508.1089532

Victoria, B. (2016). Characterization of web single sign-on protocols. *IEEE Communications Magazine, 54*(7). doi:10.1109/MCOM.2016.7514160

Wanpeng, L., & Chris, J. M. (2014). Security Issues in OAuth 2.0 SSO Implementations. *International Conference on Information Security, 8783*(1), 529-541. doi: 10.1007/978-3-319-13257-0_34

Web Runtime – Tizen developer. (2018). Retrieved from https://developer.tizen.org/development/training/web-application/understanding-tizen- programming/web-runtime

William, K., & Robertson, G. V. (2009). Static Enforcement of Web Application Integrity Through Strong Typing. *USENIX Security Symposium, 18*(6), 283-298.

Yacin, N., Prateek, S., & Dawn, S. (2009). Document structure integrity: A robust basis for cross-site scripting defense. *Proceedings of the Network and Distributed System Security Symposium (NDSS), 16*(1), 1-20.

Chapter 6
A Technological Solution for Communication With Deaf Students in the Institutions of Kuwait

Wadhah A. AlKhaldi
Kuwait University, Kuwait

Muhammad Sarfraz
https://orcid.org/0000-0003-3196-9132
Kuwait University, Kuwait

ABSTRACT

The deaf category represents a significant class at the global level, and it is therefore necessary to pay attention to the provision of prominent educational programs and services to help them achieve self-realization and full integration into society. This study seeks to propose a new translation system for deaf students in faculties, universities, and higher institutes in Kuwait. It can help deaf students to obtain better educational services and improve their communications with their hearing students and teachers in the State of Kuwait. Let us call such a system a video relay service (VRS); it can also be adopted in institutions in other parts of the world. The overall research goal is to propose a new translation system that can help students obtain better educational services and improve their communication with their hearing peers and teachers in the State of Kuwait. This chapter proposes a new translation system for deaf students;, a detailed study and methodology of the targeted research has been explained with all its essential elements.

INTRODUCTION

No one can deny the critical importance of communication in the life of individuals. It is the main method through which we can convey messages and understand each other. Without communication, there is always a space for misunderstanding and misconception among individuals. Sometimes, people may

DOI: 10.4018/978-1-7998-3355-0.ch006

have certain problems that prevent them from effectively communicating and engaging in the society. Hearing impairments are among these problems. In order to overcome the problem of communication among individuals with hearing impairments, sign languages have been introduced. They can be regarded as important means of communication with people who have hearing impairments. Moreover, they can help in improving the quality of the educational services being provided to deaf students in particular all over the world.

Marschark et al. (2005) explain that over the past several decades, there has been a modification in the face of deaf education. Legislation such as National Law 118/71 in Italy and Ley Orgánica 10/2002 de Calidad de la Educación in Spain has sought to progress equity and access for people with disabilities. In the United States, Section 504 of the Rehabilitation Act of 1973 prohibits discrimination against individuals with disabilities, who are qualified, in any program receiving federal funding, as well as academic institutions. The Education for All Handicapped Children Act (PL 94–142), passed in 1975, guaranteed free and suitable public education for children with disabilities. Largely as a result of such legislation, the amount of deaf students in integrated or "mainstream" classrooms has improved dramatically.

Johar et al. (2014) believe that sign language is a language which uses manual communication and body movements to convey meaning instead of using sound patterns by combining hand movement, body reactions and facial expressions to convey meaning.

Antonakos et al. (2015) assure that around 70 million deaf people worldwide use Sign Languages (SLs) as their native languages to communicate and interact with the surrounding environments. Sign languages have appeared as a strategic solution through which a person can convey messages by the help of the body's movements and facial expressions. The recent years have witnessed a great focus on how to use computers and electronic tools to develop sign languages to communicate with deaf students in different communities (Parton, 2005).

Microsoft has provided many products and services to ease the process of deaf communication such as TTY/TDD service (Microsoft Corporation, 2013). In addition, Samsung has designed a video call center for deaf individuals that can be accessed through an application installed on mobile phones Technology Usage for Deaf Community. There are also many iCloud services available for deaf people to improve the process of communication (Australian Web Industry Association, 2014). Kaufmann (Kaufmann, 2011) assures that virtualization and virtual environments can play critical roles in improving the process of teaching and learning for deaf students at different educational levels.

A sign language can be processed for translation using cameras or flex sensors (Arif et al., 2016). Sign language translation is considered as essential issue. It has the ability to bridge the communication gap between deaf people and their hearing counterparts as it enables communication without any writing or typing challenges. These translation tools in fact have the ability to improve the quality and speed of the educational process (Baldassarri et al., 2009). The process of sign language translations involves three steps (Martins et al., 2015) represented in recognition as follows: (1) interpretation of all gestures, movements and facial expressions; (2) representation (reflected by avatars or animations); (3) translation (from text/speech to sign language and vice versa).

At the Arab level, we can notice that Arabic Sign Language (ArSL) is limited to two basic classes; nouns/adjectives and verbs. Although there are a lot of attempts that have been made to reach a unified sign language, many of these attempts have not been successful and do not achieve the required level of success (Abdel-Fattah, 2005). This means that the signs used in one language will differ from the other one (Abdo et al., 2015). This in fact has made many researchers believe that the Arabic Sign Language is still under development (Assaleh et al., 2010).

In addition to the challenges mentioned above, we must admit that although sign languages have witnessed increased attention and undergone development in recent years. Arabic Sign Languages are still in their developmental stages. Abdel-Fattah (2005) states that Arabic sign languages will not enable students to reach the level of excellence in materials being taught. Also, there are certain limitations related to traditional videos especially those concerned with bandwidth constraints (Boulares and Jemni, 2012).

Halawani (2008) proposes an Arabic Sign Language Translation System (ArSL-TS) for translating text into sign language animations for deaf users with mobiles based on a standard Arabic Sign Language. The system is provided with instant feedback about the meaning of the Arabic text to help deaf people understand the massage more clearly.

A close look at the educational process in Kuwait shows that there are many problems associated with teaching students with disabilities represented in inappropriate curriculum that focuses on quantity alone and discourages hands-on activities that are considered as necessity to fulfill students' needs. In addition, Kuwaiti teachers have negative attitudes towards inclusion of students with different impairments into regular classrooms (Aldaihani, 2010).

The researchers have some personal observations which are being expressed here. Kuwait University used to admit students with special needs and major disabilities like blindness. Recently, the university started accepting deaf students too. The university, as yet, is not fully aware of the deaf students' needs. The university does not have any fulltime translators. Students are allowed to bring translators from the Ministry of Education to fulfill the special needs of the students. These part time translators are usually hired by various schools and colleges. They are full time translators in different places. The translators are very selective and very efficient in understanding the needs of the deaf students. There are no specific organizations to organize the specializations in the translators' groups. Sometimes, the translators lack good abilities in sign language that allow him/her to contact deaf students by writing messages through WhatsApp, etc. Deaf students are allowed to visit the teacher's office where they can contact other translators when needed by FaceTime call through the student's phone. In Kuwait University, every deaf student has a translator. Translators are allowed to write for deaf students at lectures and during examination.

The Kuwaiti society realizes that deafness is a phenomenon that cannot be ignored. Joshua Project (2019) reports that the number of deaf people in the State of Kuwait is not exactly known. However, it roughly estimates that there are around eight thousand deaf people in the State of Kuwait. Al-Meqbel and Al-Baghli (2015) sought to identify the prevalence of hearing impairments among high-risk infants within the Kuwaiti State. The results showed that the prevalence of hearing impairment was 11.5%. About 43.48% had severe prematurity that represented a main risk factor of hearing loss. Al-Sebeih et al. (2014) revealed that the high percentage of marriages between relatives among Kuwaitis was a critical factor in the development of hereditary hearing impairment.

According to the Administration of Schools of Special Education in Kuwait, the first school for deaf males was established in 1959–60, followed by another school for deaf females in 1960–61 (Al-Hilawani, 2009). The president of the Women's Institute for Development and Peace in Kuwait called for the provision of efficient translators who can effectively use sign language for deaf community in the field of education for its prominent significance in improving communication and social inclusion of those people in the community.

Literature Review

This section handles two major parts; the first part tackles the following theoretical related issues: an overview of the deaf category, sign language, Video Relay Service (VRS), and video call centers. The second part presents a review of the empirical studies on the incorporation of the video relay service into the education process of the deaf category.

Theoretical Literature

Chen (2004) and Vervloed et al. (2006) agree that the deaf category represents a significant class at the global level. They are always seen as exceptional individuals with different challenges to be faced in academics, society and also in emotional balance. They seek for opportunities to better utilize the gap that co-exists alongside their fellow students to be seen as a socially, emotionally, academically, and physically recognized student. It is therefore necessary to pay attention to the provision of prominent educational programs and services to help them achieve self-realization and full integration into society. This category needs deaf education. That is, the teaching of students with any level of hearing loss or deafness which addresses their variances and individual needs. This procedure involves individually-planned, adaptive materials, systematically-monitored teaching methods, accessible settings and other interventions developed to help students reach an advanced level of self-sufficiency and success in the school and community better than any typical classroom education. Some countries focus on developing and training teachers to educate deaf students with a variety of methods and create organizations to aid deaf students. Among those countries are France, Great Britain, The United States, Canada, Nepal, New Zealand, and The Netherlands. We must admit that signed languages have witnessed increased attention and undergone major developments in recent years.

Sign language based communication is a vital type of correspondence in the hard of hearing group. It has been a basic part of correspondence all through mankind's history. Since the start of human correspondence, communication via gestures has changed and developed into the framework that individuals see today. Earlier humans used basic translation mechanism. Ahead of schedule in mankind's history, people utilized basic sign based communication to express fundamental thoughts.

("The history of sign language," 2013) reports that sign language is an essential form of communication in the deaf community. Without sign language, deaf people would have difficulty speaking and learning languages but with sign language they are able to communicate as seamlessly and efficiently. On the other hand sign language has been a necessary part of communication throughout human history. From the start of human communication, sign language has improved and evolved into the system that people can see today. Humans in early history used simple sign language to express basic ideas. Even when vocal communication became the mainstream form of interaction, hand and facial gestures were used to enhance ideas in communication.

Moreover, Reagan (2019) adds that the history of sign language has an remarkable past, it was the first form of communication in early man. Sign language then continue to help finish the discrimination of deaf people, and assisted the deaf to become educated similar to their hearing peers. This phase started in France and then moved to the United States. Now worldwide, lots of sign language schools and different sign languages exist.

American Sign Language (ASL) on the other hand was not the only sign language established. All over the world, many sign languages developed, including in England (BSL) and Australia (Auslan).

Although speakers of English language can understand American, British, and Australian people equally. However, with some informal differences, signers in America, England, and Australia would be incapable to understand each other because the signs are not similar. The differences in these signs are based on nuances within the deaf communities of that area, which causes an interesting evolution of sign language worldwide. It can be supposed that there are many sign languages in the world as much as there are spoken languages.

Abdel-Fattah (2005) reports that Arabic sign languages (ARSLs) are still in their developmental stages and as a result they will not enable students to reach the level of excellence in materials being taught.

Arab Sign-language family (2013) clarifies that the Arab Sign Language family is a family of sign languages that spreads across the Arab Mideast. Its scope is not yet identified, because only few of the sign languages in the region have been compared. A project for a single Arabic Sign Language is in progress, with much of the vocabulary voted on by regional Deaf associations. Yet, so far only a dictionary has been compiled; grammar has not been addressed, so the outcome cannot be considered a language.

Several studies have been conducted in order to provide assistive technologies that can help in enhancing the process of learning and integrating deaf people into the community (Martins et al., 2015). These technological tools in fact have the ability to improve the quality of life of the deaf community (Halawani and Zaitun, 2012).

Halawani (2008) comments that Sign Language Translation Systems that translate text into sign language animations have the ability to improve deaf people lives especially in aspects related to communication and accessing information in educational institutions.

A Video Relay Service (VRS), also known as a Video Interpreting Service (VIS), is a video telecommunication service that allows deaf, hard-of-hearing and speech-impaired (D-HOH-SI) individuals to communicate over video telephones and similar technologies with hearing people in real-time, via a sign language interpreter. VRS is a newer system of telecommunication service to the D-HOH-SI community, which had, in the United States, started earlier in 1974 using a simpler non-video technology called telecommunications relay service, also known as "TRS", or simply as "relay service". VRS is a form of Telecommunications Relay Service (TRS) that allows individuals with hearing disabilities who use ASL to communicate with voice telephone users via video equipment, rather than through typed text. Video equipment links the VRS user with a TRS operator – called a Communications Assistant (CA) so that the VRS user and the CA can see and communicate with each other in signed conversation. Because the conversation between the VRS user and the CA flows much faster than with a text-based TRS call, VRS has become very popular form of TRS. ("Video relay service," n.d.).

VRS, same as other forms of TRS, allows individuals who are deaf or hard-of-hearing to communicate through the telephone system with hearing people. The VRS caller, using a television or a computer with a video camera device and a broadband high speed Internet connection, contacts a VRS CA, who is a qualified sign language interpreter. They communicate with each other in sign language via a video link. The VRS CA then places a telephone call to the party the VRS user needs to call. The VRS CA relays the conversation back and forth between the parties - in sign language with the VRS user, and by voice with the called party. No typing or text is involved. A voice telephone user can also start a VRS call by calling a VRS center, usually through a toll-free number.

The VRS CA can be reached through the VRS provider's Internet site, or through video equipment attached to a television. At this time, around ten providers offer VRS. Like all TRS calls, VRS is free to the caller. VRS providers are compensated for their costs from the Interstate TRS Fund, which the FCC oversees.

Cromartie et al. (2012) report that Video Call centers have succeeded in attracting the attention of many researchers in the field of improving communication for the deaf community. For example, in Australia, there is a National Relay Service (NRS) that depends on a call center which provides the relay service component for deaf community in the country. Moreover, Colorado State has developed a relay call for deaf or hard of hearing people who can write a text using a computer or a wireless devise and the relay operator will read it aloud so that a hearing person can understand, the same operator transforms words into text so that a deaf person can read them.

VRS services have become well developed nationally in Sweden since 1997. Except Sweden, For only a few years VRS has been provided in Europe since the mid-2000s, and as of 2010 has not been established in many European Union countries with many European countries still requiring the legislation or the financing for large-scale VRS services, and to offer the necessary telecommunication equipment to deaf users. Germany and the Nordic countries are among the other leaders in Europe, while the United States is another world leader in the provisioning of VRS services.

Halawani (2008) believes that there is lack of researches that are dedicated to investigate the possibility of developing video call centers in Arab countries in general and in the state of Kuwait in particular.

López-Ludeña et al. (2012) describe a methodology for developing a speech into sign language translation system using a user-centered strategy. The methodology consisted of four basic steps: analysis of technical and user requirements, data collection, technology adaptation and evaluation. The results showed that many aspects can be updated automatically from the generated parallel corpus.

Cromartie et al. (2012) point out that video call centers have succeeded in attracting the attention of many researchers in the field of improving communication for the deaf community. In Australia, there is a National Relay Service (NRS) that depends on a call center which provides the relay service component for deaf community in the country. Colorado State has developed a relay call for deaf or hard of hearing people who can write a text using a computer or a wireless devise and the relay operator will read it aloud so that a hearing person can understand, the same operator transforms words into text so that a deaf person can read (Herr, 2004).

Taking into account that Sign Language Translation Systems that translate text into sign language animations have the ability to improve deaf people lives especially in aspects related to communication and accessing information in educational institutions. The statement of the problem revolves around introducing an advanced translating system to help in improving the quality of the educational services being provided to deaf students in Kuwait by keeping in mind the necessity of overcoming video limitations and enhancing the capacity of the proposed video call center.

Empirical Literature

Deaf students' academic challenges are associated to differences in language comprehension, cognition, and learning strategies, not only language modality. Deaf and hearing children do not have the same knowledge, backgrounds, experiences, and learning strategies. Cognitive differences can be strengths, weaknesses, or just differences, but all add to diversity in the classroom. Students and teachers must identify those differences. Deaf students can learn as much as hearing peers when taught by skilled teachers of the deaf. Deaf children are not hearing children who cannot hear.

Stein et al. (2007) developed a new approach to recognize sentences of continuous sign language speaker independently. The robust tracking algorithm based on dynamic programming was required.

The used tracking system achieved very good results as the incorporation of the tracking data for deixis words helped to properly interpret the meaning of the deictic gestures.

Introduction to VRS Interpreting Curriculum (2017) exposes interpreting students to the environment of the VRS and Video Remote Interpreting (VRI) settings, including the inherent challenges and interpreting skills needed for deaf students in this type of work. The aim of this work is to provide students with sufficient information and practical experience to be capable of making an informed decision about future work in these settings. It is not designed as a training course for working in VRS or VRI settings, nor is it intended to imply to students that they will be prepared to work in these settings upon completion of the course. The curriculum was designed with flexibility in mind. It can be used in 1, 2, or 3 credit hour courses that meet for a variety of time blocks, and a variety of number of days per week.

Lessons may also be used independently of the entire curriculum, although certain scaffolding components will be lost if used in this fashion. There is enough material in each lesson to satisfy 3 credit hour courses. Instructors teaching 1 or 2 credit hour courses will have to make decisions about which activities to eliminate based on their time constraints.

The instructor of such course needs to be an experienced video relay interpreter and a trained, experienced instructor. In some cases, where the instructor is not an experienced VRS interpreter, it is suggested that at the very minimum, he needs to obtain permission from a local VRS call center manager or director to spend two to three hours observing VRS calls. In cases where instructors lack experience in specific lesson topics, it is recommended that they invite guest speakers with expertise in that area to supplement or provide the lesson.

Marschark et al. (2005) explain that four experiments investigated classroom learning by deaf college students receiving lectures from instructors signing for themselves or using interpreters. Deaf students' prior content knowledge, scores on post lecture assessments of content learning, and gain scores were compared to those of hearing classmates. Consistent with prior research, deaf students, on average, came into and left the classroom with less content knowledge than hearing peers, and use of simultaneous communication (sign and speech together) and American Sign Language (ASL) apparently were equally effective for deaf students' learning of the material. Students' self-rated sign language skills were not significantly related to performance. Two new findings were of particular importance. First, direct and mediated instruction (via interpreting) was equally effective for deaf college students under the several conditions employed here. Second, despite coming into the classroom with the disadvantage of having less content knowledge, deaf students' gain scores generally did not differ from those of their hearing peers.

Proposed Research

Here is the contribution towards the approach and methodology of the proposed system. First of all, a statement of the problem is presented. It is followed by purpose of the study and then the significance of this study. The section closes by the major components of research methodology.

Statement of the Problem

As a result of the recent enrollment of deaf students in universities and colleges of Kuwait, a new sign language translating system should be introduced to ease the process of education for these students. To the researcher's best knowledge, an online sign language translating system for deaf students to support faculty members, universities and higher institutes in Kuwait has never been introduced and studied.

In response to this problem, this study aims to investigate deaf students' needs in terms of improving the translation system through introducing new technology of Video Relay Service. It is the technology that will be utilized and examined to measure the level of effectiveness, it can have, on the learning process of deaf students.

Purpose of the Study

This study attempts to answer few Research Questions (RQ): The main research question of the current study can be formulated as "To what extent can a new sign language translating system help in improving the educational system for deaf students in Kuwait?" Under this main research question, there are certain other questions that can be reviewed as follows:

RQ1. What is the expected service standard of the deaf students from the video relay system in Kuwait?
RQ2. What is the expected service standard of the teachers from the video relay system in Kuwait?
RQ3. Does the expectation match the deliverables by the Video Relay Service (VRS) provider?
RQ4. What is the real time application environment of the VRS?

Significance of the Study

The work seeks to study the need of a new translation system that can help students obtain better educational services and improve their communication with their hearing peers and teachers in the State of Kuwait. This study sets the stage for further research on the importance of the use of translator sign language for university students. It is potentially significant for the following parties:

- The Kuwaiti Ministry of Education: since it may design training sessions to introduce the translator sign language principles and instructional practices for educators of different specializations.
- Curriculum designers: since they may attempt to incorporate the translator sign language in the different subjects being taught in Kuwait.
- Researchers: since they may derive insights for different research studies stemming from translator sign language.

Research Methodology

For this study, a qualitative research approach is proposed. The data collection process will consist of several essential methods. The first data collection method will be the in-class observations of the interactions between the deaf students and their teachers, peers, and translators. The second method will be case studies of deaf students, teachers, and translators. The third method is going to be interviews done with deaf students, teachers, and translators that will depend on three main questions RQ1, RQ2 and RQ3. The fourth method that will be used for acquiring data is the Video Relay System experiment. This part will rely on the fourth research question RQ4 in its administration.

The researchers will adopt the definitions of the following terms in the study:

1. Deaf Students: Deaf students can be defined as those students who use sign language as a main method of communication (ACCESS-Disability Support Unit, 2016).

2. Sign Languages: Stein et al. (2007) defined sign languages (SLs) as "natural, indigenous and independent means of communication for deaf and hard-of-hearing communities worldwide".
3. Sign Language Translation: Atwood et al. (2012) defined sign language translation as an important application for "vision-based gesture recognition methods, in which highly structured combinations of static and dynamic gestures correlate to a given lexicon" .
4. Video Relay Services (VRS): VRS is a video telecommunication service that helps deaf, hard-of-hearing and speech-impaired (D-HOH-SI) individuals to communicate via video telephones and similar technologies with hearing people in real-time, through a sign language interpreter.

This study is going to be limited to deaf students studying in institutions, universities and higher institutes in Kuwait. Figure 1 represents the diagram of this study on effectiveness of VRS implementation for deaf students in Kuwait.

Proposed System

This section describes the practical implementation of video relay service facility and its scope for the deaf students. It also discusses the methodology involved pertaining to Research Questions 1, 2, 3 and 4 described in Sections 1.3 and 1.5. It discusses the proposed model, objectives, hypothesis, and approach for answering the research questions. Moreover, it also discusses the issue of validating the study samples for further analysis. The sampling technique and sample size justifications are provided together with methodology administered for all the research questions.

Practical System

We have proposed a practical implementation of video relay service facility and its scope for the deaf students. This will give a practical insight on how this service can be extended in education platforms in Kuwait. Besides that, we have discussed the expressions of the technology involved in the system to analyze the degree of usefulness for the people, in-depth review of the business model, and prototype of this model.

Figure 2 explains the process of Traditional Telecommunication Relay System in Figure 2(a) and the Contemporary Video Relay System in Figure 2(b). The technology of Video Relay Service (VRS) is built upon the technology of Telecommunication Relay Service (TRS). Hoisting from such communication medium, the technology has grown benefiting many areas and sectors worldwide. Usually, TRS users relay back or forth between the text telephones and communication assistants. The major input is carried inside the technology of VRS service. In VRS call, communication assistants are regarded as Video Interpreter (VI) who has a provincial knowledge on the sign languages. The VI builds greater importance to the communication medium and builds a wide scenario of interaction between the parties.

The benefit of VRS is more spontaneous and carries forwards in par with TRS. The sign languages are easily transmitted and carried in real time without any difficulties. Though there is backlog that may rise because of the technology infrastructure or lack of proper network connection, the carry forward of the VRS is so effective and it extends to various fields especially for deaf students. Beside this, the Internet Protocol Relay (IPR) is another technology which is similar to traditional methods. But, the main difference between VRS and IPR is the spontaneity that VRS acclaims are higher than that of IPR. This anticipates to yield great results overall for the VRS system.

Figure 1. Effectiveness of VRS implementation diagram

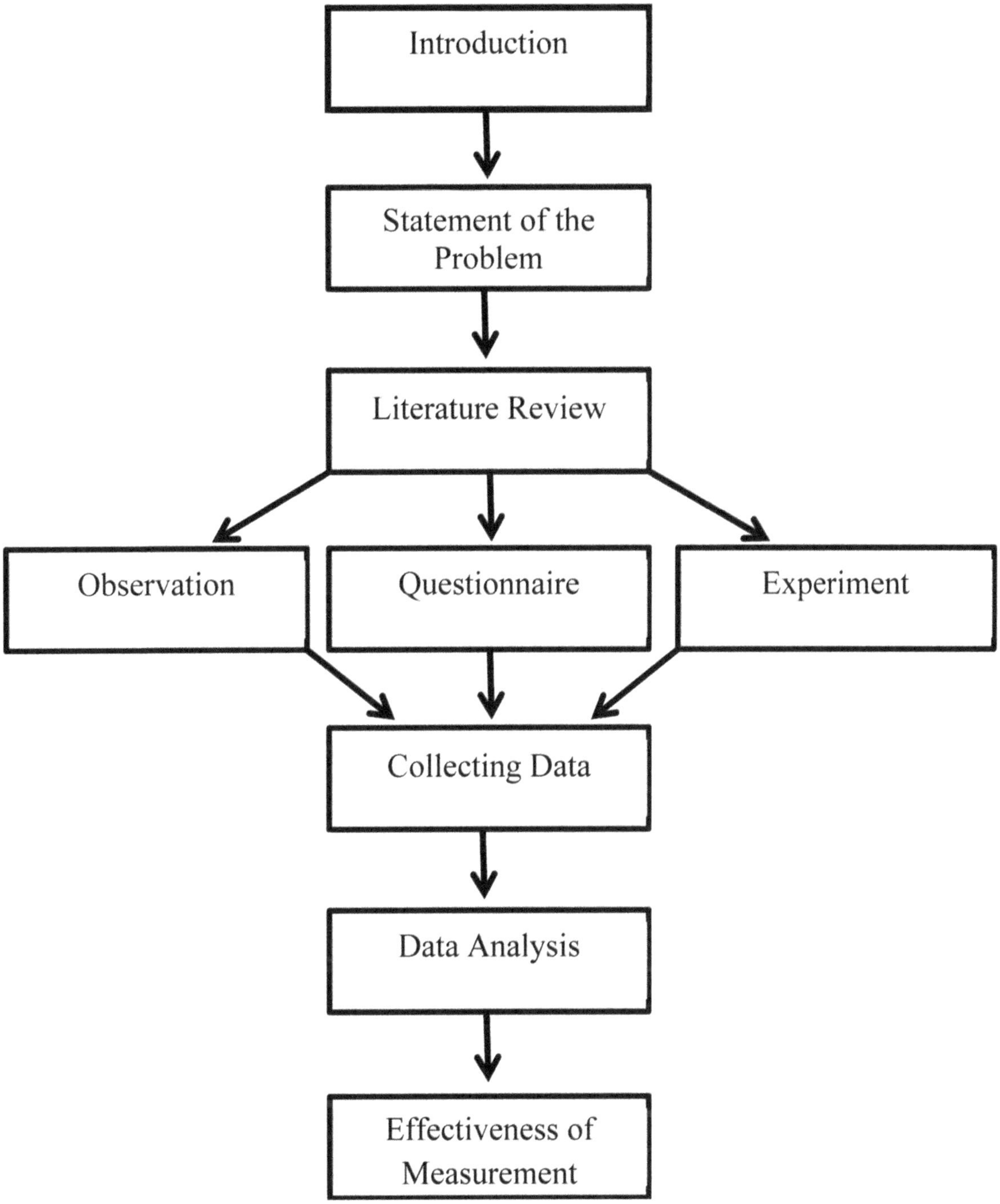

Figure 2. The process of Video Relay System

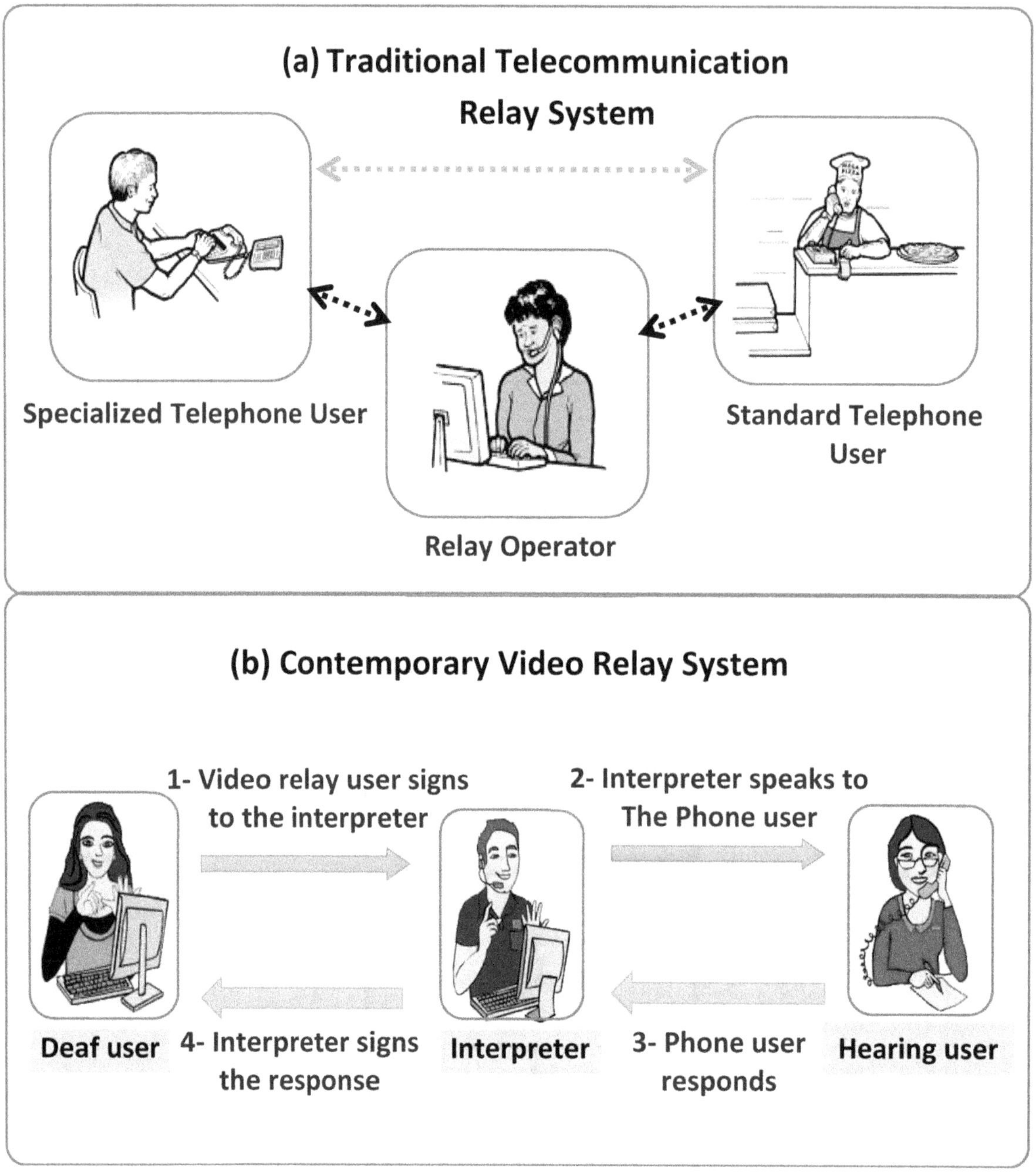

Outline of Research

The research questions RQ1, RQ2 and RQ3 will be answered using an exploratory methodology with an open ended questionnaire circulated among the deaf students. The questionnaire will be prepared for student, Teachers and Translators. The questionnaire also contains questions pertaining to investigating on RQ4. This study administers a recursive depth interview design to identify the variables that influence students to choose video relay service (RQ1), the variables that portray the expectations of the teacher (RQ2), and the variables that portray the expectations of the video relay translators (RQ3). The variables are then analyzed with content analysis technique to understand the underlying factors in the study pertaining to all research questions which simultaneously finds a valuable insight for the research question. Since the research design is exploratory by nature, this gives an insight and the enlargement to understand the conclusion is still void. Exploratory researches give insight and orientation towards future research in a bigger scale. Likewise, this study tries to dissect the importance and content pattern pertaining to the three research questions.

A non-probabilistic sampling method will be administered using a convenient or snowball effect. This sampling method is mostly administered when the sampling frame of the population is unknown. Most of the previous research has administered this research design with greater importance. Convenient sampling method conveys a greater meaning for the samples as long it does not deviate the representation of the population.

A reasonable amount of samples are planned to be collected of deaf students, teachers, and translators. The sample is convenient and representative of the investigating population. The sample interviews would require reasonable time to arrive at a meaningful understanding. The samples would be identified from Kuwait University, Kuwait International Law School, and the College of basic education.

An exploratory analytical approach would be chosen to analyze the variables respective to the study and the research design. More precisely this design facilitates a recursive model to ladder the questions accordingly and to build the study appropriately without deviating the study aspects. We would use tables and figures for each question of the questionnaire in order to illustrate the frequencies and the percentages of each teachers', translators' and students' responses. In this research, the research instruments to be adopted would be as follows:

- Direct observation of the samples in their real classes.
- A Questionnaire that would be answered by the respondents in personal interviews conducted.
- Experiments.

The questionnaire would answer all research questions. It would contain both parametric and non-parametric sections of questions, composing of nominal, ordinal, interval and ratio scale questions. The questionnaire would allow the use of multiple analyses throughout the study. The questionnaire would have three segments for RQ1, RQ2, and RQ3 questions to students, teachers, and translators. The questionnaire would be validated by experts to confirm the questions diversification and framing construct validity. All segments in the questionnaire would be mainly open ended segments except for couple of them. Of the two segments, one would measure the impact of service quality and the other would measure the rank order scaling of the items pertaining to RQ1 and RQ2 respectively.

Though it is a traditional mechanism to use frequency testing, it is always good to report the descriptive or inferential statistics using this mechanism. It has received high recognition among the past researchers and the induction of conducting it is understandable by all readers regardless of their occupation. This is the most comprehensive analysis for many industries and researchers to post their statistical views before performing any computational analysis. This analysis would be mainly chosen because of the exploratory nature of the study. The analysis would be performed on the five items measuring the service quality of VRS as perceived by the students. The mean values of the questions and the inter-correlation

Figure 3. Possible prototype for VRS

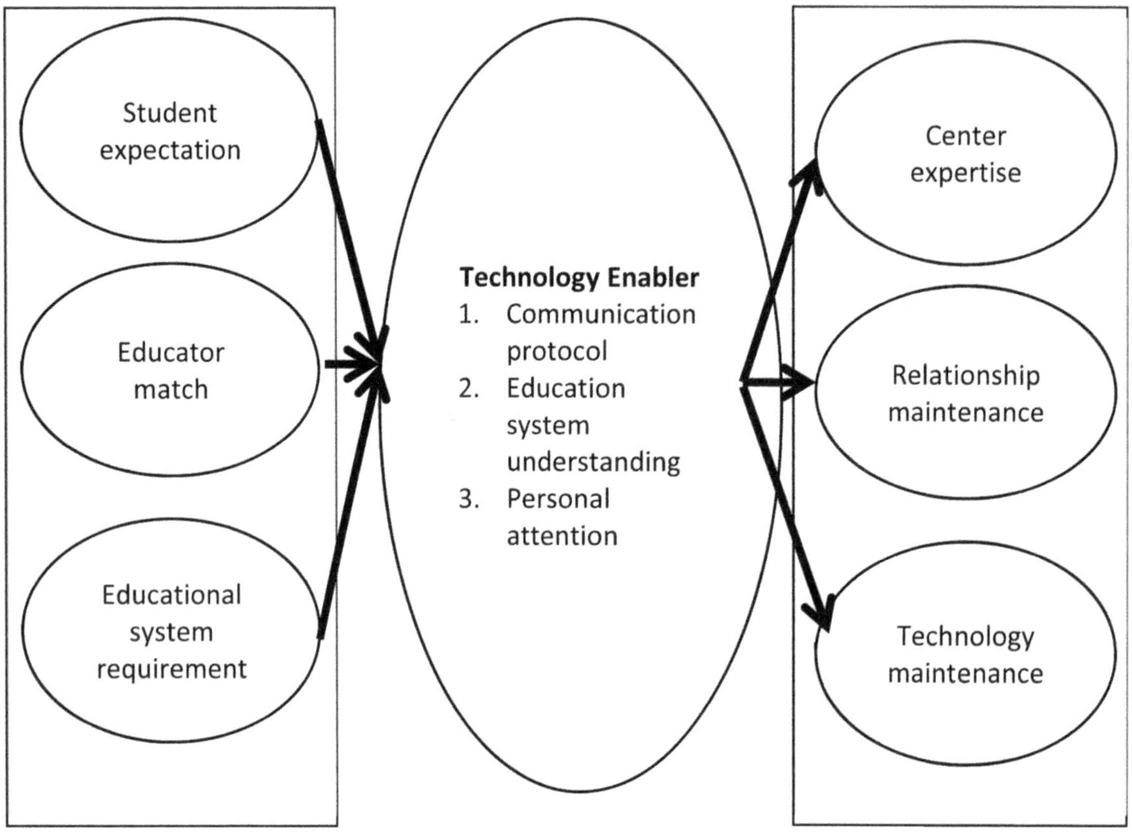

among the service quality items will be calculated. This will highlight the importance of the service quality requirements and expectations.

The methodology pertaining to RQ4 would be different from the other three research questions in a conditional format. It would be used as an experimental research design with the same sample in three levels namely, students, teachers and the translators to investigate the effectiveness of the VRS. Figure 3 refines the framework that would be investigated among the deaf students. This research question is expected to adopt a case methodology to complete the research question RQ4.

CONCLUSION

The objectives of the study are to focus the new virtual relay system with real time hearing technologies through sign language interpreter. This chapter has provided a concrete proposal to work on. It has introduced the structure of the study and explains the key problems that the deaf students have faced in the universities and other educational institutions. This chapter has discussed key areas of the study like the challenges of deaf students, their problems, why the study is needful for the society, the objectives of the study and the research used for the study.

Several studies have been conducted in order to provide assistive technologies that can help in enhancing the process of learning and integrating deaf people into the community. These technological tools in fact have the ability to improve the quality of life of the deaf community. It is worthy here to mention that teaching and learning are products of collective efforts. Teachers and hearing individuals can make important transformational changes in the process of teaching for deaf learners through providing support and generating new interactional patterns for easing the interaction process.

This chapter has discussed the methods and tools used in collecting the data. The research intends to use a qualitative research method and self-administrated questionnaire to collect the data. To estimate the performance of Video Relay Service to be realized by the users and providers, the data would be collected from the deaf students, teachers, and translators through deep interview questionnaire and the results and findings would be provided in a later publication by the authors.

REFERENCES

Abdel-Fattah, M. (2005). Arabic Sign Language: A Perspective. *Journal of Deaf Studies and Deaf Education, 10*(2), 212–221. doi:10.1093/deafed/eni007 PMID:15778217

Abdo, M., Hamdy, A., Salem, S., & Saad, E. M. (2015). Arabic alphabet and numbers sign language recognition. *International Journal of Advanced Computer Science and Applications, 6*(11), 209–214.

Access-Disability Support Unit. (2016). *Glossary of Conditions/ Impairments/ Disability*. Malta: University of Malta.

Al-Hilawani, Y. (2009). Perspectives and reviews on deafness in the State of Kuwait and the United Arab Emirates. *Deaf people around the world: Educational and social perspectives*, 119-132.

Al-Meqbel, A., & Al-Baghli, H. (2015). The prevalence of hearing impairment in high-risk infants in Kuwait. *Auditory and Vestibular Research, 24*(1), 11-16.

Al-Sebeih, K., Al-Kandari, M., Al-Awadi, S. A., Hegazy, F. F., Al-Khamees, G. A., Naguib, K. K., & Al-Dabbous, R. M. (2014). Connexin 26 gene mutations in non-syndromic hearing loss among Kuwaiti patients. *Medical Principles and Practice, 23*(1), 74–79. doi:10.1159/000348304 PMID:24080506

Aldaihani, M. (2011). *A comparative study of inclusive education in Kuwait and England* (Doctoral dissertation). University of Birmingham.

Amin, O., Said, H., Samy, A., & Mohammed, H. K. (2015, December). HMM based automatic Arabic sign language translator using Kinect. *Tenth International Conference on Computer Engineering & Systems*, 389-392. 10.1109/ICCES.2015.7393081

Antonakos, E., Roussos, A., & Zafeiriou, S. (2015, May). A survey on mouth modeling and analysis for sign language recognition. *11th IEEE International Conference and Workshops on Automatic Face and Gesture Recognition,* 1, 1-7. 10.1109/FG.2015.7163162

Aouiti, N., & Jemni, M. (2016, July). Classifiers in Arab Gloss Annotation System for Arabic Sign Language. *International Conference on Computers Helping People with Special Needs*, 400-406. 10.1007/978-3-319-41267-2_56

Arab sign-language family. (2013). Retrieved in 17 July 2019, from https://en.wikipedia.org/wiki/Arab_sign-language_family

Arif, A., Rizvi, S. T. H., Jawaid, I., Waleed, M. A., & Shakeel, M. R. (2016). Techno-Talk: An American Sign Language (ASL) Translator. *International Conference on Control, Decision and Information Technologies*, 665-670. 10.1109/CoDIT.2016.7593642

Assaleh, K., Shanableh, T., Fanaswala, M., Amin, F., & Bajaj, H. (2010). Continuous Arabic Sign Language Recognition in User Dependent Mode. *Journal of Intelligent Learning Systems and Applications*, *2*(01), 19–27. doi:10.4236/jilsa.2010.21003

Atwood, J., Eicholtz, M., & Farrell, J. (2012). *American Sign Language Recognition System. In Artificial Intelligence and Machine Learning for Engineering Design*. Dept. of Mechanical Engineering, Carnegie Mellon University.

Australian Web Industry Association. (2014). The accessibility of Cloud computing – Current And Future Trends. Author.

Baldassarri, S., Cerezo, E., & Royo-Santas, F. (2009, August). Automatic translation system to Spanish Sign Language with a virtual interpreter. In *IFIP Conference on Human-Computer Interaction*, 196-199. Springer. 10.1007/978-3-642-03655-2_23

Boulares, M., & Jemni, M. (2012, April). Mobile sign language translation system for deaf community. *Proceedings of the international cross-disciplinary conference on web accessibility*, 37. 10.1145/2207016.2207049

Bowe, F. G. (2002). Deaf and Hard of Hearing Americans' Instant Messaging and E-Mail Use: A National Survey. *American Annals of the Deaf*, *147*(4), 6–10. doi:10.1353/aad.2012.0251 PMID:12592800

Chen, D. (2004). Young children who are deaf-blind: Implications for professionals in deaf and hard of hearing services. *The Volta Review*, *104*(4), 273–284.

Deaf Australia Inc. (2010). *Next generation National Relay Service Campaign. How does VRS work?* Retrieved in May 2018 https://sites.google.com/site/nrscampaign/resources/how-does-it-work

Deaf in Kuwait. (2019). Retrieved in 17 July 2019, from https://joshuaproject.net/people_groups/19007/KU

Deafwebsites.com. (2013 Sep 26). *The History of Sign Language*. Retrieved in May 2018. http://www.deafwebsites.com/sign-language/history-sign-language.html

Florida Telecommunications Relay, Inc. (1990). *Federal Legislation Americans with Disabilities Act ADA 1990 Title 4 (Telecommunications) Deaf, Hard of Hearing, Deaf/Blind*. Speech Disabled Relay Service.

Gaffey, B. E., Cromartie, J. B., & Seaboldt, M. R. (2012). *Evaluating communication technologies for the deaf and hard of hearing*.

Halawani, S. M. (2008). Arabic sign language translation system on mobile devices. *IJCSNS International Journal of Computer Science and Network Security*, *8*(1), 251–256.

Halawani, S. M., & Zaitun, A. B. (2012). An avatar based translation system from Arabic speech to Arabic sign language for deaf people. *International Journal of Information Science and Education*, *2*(1), 13–20.

Herr, R. (2004). Telecommunications Relay Services: An Introduction to Communication Options. *Colorado Commission for the Deaf and Hard of Hearing E-Journal,* 1-6.

Hussana, J. R., Priyanka, A., Revathi, A. M., Suchitha, K., & Sumana, K. J. (2014). Multiple sign language translation into voice. *International Journal of Engineering and Innovative Technology*, *3*(10).

Introduction to video relay services. (2005). Sorenson's Communications.

Introduction to VRS Interpreting. A Curriculum Guide. (2017). Retrieved in 17 July 2019, from http://www.vrsii.com/assets/pdf/curriculum/VRS_Curriculum.pdf

Kaufmann, H. (2011). Virtual environments for mathematics and geometry education. *Themes in Science and Technology Education, 2*(1-2), 131-152.

Landed, J. (n.d.). *Language in Kuwait.* Retrieved in 17 July 2019, from https://www.justlanded.com/english/Kuwait/Kuwait-Guide/Language/Language-in-Kuwait

Lopez Ludeña, V., San Segundo Hernández, R., González Morcillo, C., López López, J. C., & Pardo Muñoz, J. M. (2012). *Methodology for developing a Speech into Sign Language Translation System in a New Semantic Domain.* Academic Press.

Marschark, M., Pelz, J. B., Convertino, C., Sapere, P., Arndt, M. E., & Seewagen, R. (2005). Classroom interpreting and visual information processing in mainstream education for deaf students: Live or Memorex®? *American Educational Research Journal, 42*(4), 727–761. doi:10.3102/00028312042004727 PMID:16628250

Martins, P., Rodrigues, H., Rocha, T., Francisco, M., & Morgado, L. (2015). Accessible options for deaf people in e-learning platforms: Technology solutions for sign language translation. *Procedia Computer Science, 67*, 263–272. doi:10.1016/j.procs.2015.09.270

Microsoft Corporation. (2013). *Administrator's Guide for Microsoft Application Virtualization (App-V) 5.0 MDOP Information Experience Team.* TechNet Library.

Modi, K., & More, A. (2013). Translation of Sign Language Finger-Spelling to Text using Image Processing. *International Journal of Computers and Applications, 77*(11).

Nayef, N. (2017). *7 thousands deaf in Kuwait and requests for generalization of sign language.* Localities.

Parton, B. S. (2005). Sign language recognition and translation: A multidisciplined approach from the field of artificial intelligence. *Journal of Deaf Studies and Deaf Education, 11*(1), 94–101. doi:10.1093/deafed/enj003 PMID:16192405

Reagan, T. (2019). Sign Language and the Deaf-World: 'Listening without hearing'. In Linguistic Legitimacy and Social Justice, (pp. 135-174). Palgrave Macmillan.

Shawnee. (2018). *Sprint National Relay Services. TTY/ASCII/Voice/Spanish/Speech-to-Speech/IP Relay/Federal Relay.* Author.

Stein, D., Dreuw, P., Ney, H., Morrissey, S., & Way, A. (2007). *Hand in hand: automatic sign language to English translation.* Academic Press.

Technology Usage for Deaf Community. (2017). *Communication*. Retrieved in October 25, 2017 at https://elbdie.nfshost.com/mod/page/view.php?id=145&lang=tr

The U.S. Census Bureau. (2010). *Americans with disabilities: 2010*. Retrieved in May 1, 2010, from https://www.census.gov/prod/2012pubs/p70-131.pdf

Trine, E. (2013). *ةمجرتم: a case study of an Arabic/Jordanian sign language (liu) interpreter in Jordan*. Academic Press.

Tsuladze, M. (2015). *Teaching and Learning of Deaf Students in Ordinary Vocational Education Setting-The Georgian Case* (Master's thesis).

Vervloed, M. P., Van Dijk, R. J., Knoors, H., & Van Dijk, J. P. (2006). Interaction between the teacher and the congenitally deafblind child. *American Annals of the Deaf*, *151*(3), 336–344. doi:10.1353/aad.2006.0040 PMID:17087444

What is video relay service. (2009). *Convo and Convo Relay are registered trademarks*. The Federal Communication Commission.

Wikipedia Contributors. (2018, July 12). Video relay service. In *Wikipedia, The Free Encyclopedia*. Retrieved in January 2019 from https://en.wikipedia.org/w/index.php?title=Video_relay_service&oldid=905944879

Xu, J. (2017). *Sign Language Translation Using Kinect And Dynamic Time Warping*. Available online at https://web.stanford.edu/class/cs231m/projects/final-report-xu.pdf

Yoshida, M. (2008). *Barriers for telecommunication accessibility and needs assessment of video relay services (VRS): Utilization of VRS for the deaf community* (Thesis). Rochester Institute of Technology.

Chapter 7
Strategic Interaction Among Content Providers in Sponsored Content Markets

Hamid Garmani
https://orcid.org/0000-0002-7294-6722
Information Processing and Decision Support Laboratory Sultan Moulay Slimane University, Morocco

Mohamed El Amrani
Information Processing and Decision Support Laboratory, Sultan Moulay Slimane University, Morocco

Driss Ait Omar
https://orcid.org/0000-0002-4096-8273
Information Processing and Decision Support Laboratory Sultan Moulay Slimane University, Morocco

Mohamed Baslam
Information Processing and Decision Support Laboratory Sultan Moulay Slimane University, Morocco

Mostafa Jourhmane
Information Processing and Decision Support Laboratory Sultan Moulay Slimane University, Morocco

ABSTRACT

The data-sponsored scheme allows the content provider to cover parts of the network subscription fees for end-users. As sponsored data gains traction in the industry, it is important to understand its implications. This work considers content providers (CP) choice of how much content to sponsor and the implications for end-users and CPs. In particular, the authors model the interactions of CPs as a non-cooperative game in terms of pricing, the credibility of content, and the amount of sponsored content. Additionally, they have proved through a detailed analysis of the existence and uniqueness of the Nash equilibrium. Based on the game properties, they implement a learning scheme using best-response dynamics that allows CPs to learn their strategies in a fast, accurate, yet completely distributed fashion. Extensive simulations provide attractive insights on how the sponsoring content may increase the CPs payoff.

DOI: 10.4018/978-1-7998-3355-0.ch007

INTRODUCTION

Currently, we are witnessing the explosive growth of global mobile data traffic, which reached 2.5 exabytes per month in 2014 and is predicted to exceed 24.3 exabytes per month by 2020. Furthermore, the total quantity of mobile devices worldwide will be 75 billion, while being more and more intelligent. According to Cisco, mobile video traffic will contribute to nearly 75% of the total mobile traffic. Therefore, one of the important challenges for content providers is how to attract more mobile users to access their content, and thus achieve a higher revenue gain. For this reason, a new promising scheme has been introduced "sponsored content". The key idea of data sponsoring is to allow the CPs to subsidize the ISP access fees, and thereby the end-users can access contents from the CP through the internet service provider (ISP) for free. In 2014, AT&T launched a data sponsored plan, where the CP (e.g., Youtube, Twitter) can pay the access price of end-users to their content through ISP. Examples in the USA are Netflix or Binge-On with T-Mobile, DIRECTV and U-verse Data Free TV with AT&T, etc. ISPs are even offering services such that CPs can if they wish, apply sponsored data: examples are Verizon Wireless, AT&T Mobility and T-Mobile US, or Orange with DataMI in France, among others.

With the sponsored content Schema, the CP cooperates with the ISP, and the former will pay the latter based on the content traffic requested by end-users. Correspondingly, the end-users will be partially charged or they can consume the content for free. The sponsored content scheme creates a positive impact on the ISP, CP, and end-users, i.e., a multi-win situation. In other words, the end-users want to access more content that will not be calculated into their data caps. The CP can encourage more end-users to have a higher demand for content. Accordingly, the higher demand contributes to the more payoff of the CP and ISP. For this reason, the academia and the industry has a remarkable interest to investigate the sponsored content scheme.

In this chapter, we introduce the integration of the sponsored content schemes. In particular, we investigate the effects of their interaction and their coexistence on the end-user behavior, the CPs. The main contributions of this chapter:

- We present new models that include one important feature, the CP's revenue models with a sponsored content shema.
- We formulate the interactions among CPs as a non-cooperative game as a function of tree market parameters pricing, number of sponsored content, and the credibility of content.
- End-users' behavior is modeled as a function of CP strategies (*i.e.*, pricing, number of sponsored content, and the credibility of content).
- We formally prove the existence and the uniqueness of the Nash equilibrium of the non-cooperative game among CPs, which means that there exists a stable state where all CP do not have an incentive to change their strategies. So, our model ensures the existence of an equilibrium for keeping the economy stable and achieving economic growth. We propose an iterative algorithm to achieve the Nash equilibrium.
- We conduct thorough numeric investigation to determine the effect of sponsored content on the strategies of CPs. In addition, the convergence of the proposed algorithm to the Nash equilibrium point is explored. We believe, based on results, that the proposed economic model is general and feasible, and this is applicable to modeling the competition among CPs content market with a sponsored content plan.

The rest of this chapter is organized as follows. Section 2 discusses related work. In Section 3, we describe the system model. We prove the existence and uniqueness of a Nash equilibrium point in Section 4. Then, we present a numerical investigation in Section 5. In Section 6 conclusions and future.

RELATED WORK

Game theory has been applied to a wide range of networking problems to study the interaction of players seeking to maximize their profit (Ait Omar et al. 2017) (Amrani et al. 2018a) (Amrani et al. 2018b).

The authors (Omar et al. 2019a) study models that involve CPs, CDN providers and end-users. The authors formulate interactions among CPs and and among CDN as a non-cooperative game.

In (Outanoute et al.) (Garmani et al. 2018a) the authors study competition between the CPs as a non-cooperative game. The authors in (Garmani et al. 2019a) study competition among two CPs, one is rational and another one is bounded rationality. In (Altman 2012) the author considered that the CPs faced with the question of how to choose in what content to specialize. He showed that the problem of selecting the content type is equivalent to a congestion game. The authors in (Omar et al. 2019b) analyze a model with several ISP, where the end-users are bounded rational. In (Reiffers-Masson et al. 2015) the authors studied the impact of an exclusive contract between CP and ISP in a non-neutral network, and proved such agreements have positive effects for the end-users and negative effect on CPs. The authors in (Garmani et al. 2018b) (Garmani et al. 2019b) studied competition between ISP in Information Centric Network using mathematical tools game theory. In the authors (Ait Omar et al. 2019b) (Ait Omar et al. 2019c) modeled and studied competition between ISP where end-users are bounded rational.

The authors in (Maillé and Tuffin 2019) analyze a model with one CP several Internet Service Providers, where the CP can choose to sponsor a proportion of content and a level of advertisement. They show that sponsored content can be beneficial to the end-users and ISPs. In (Xiong et al. 2018b) the authors consider a sponsored content market with a single ISP, a single CP, and a set of end-users. The interactions among three entities are modeled as two-stage Stackelberg game. The authors have analyzed two scenarios: a scenario where the ISP and CP are in completion, and the scenario where the ISP and CP cooperate for a common goal. The authors (Xiong et al. 2019) (Xiong et al. 2018a) in investigate joint sponsored and caching content under the non-cooperative game. The interactions among ISP, CP, and end-users are modeled as a three-stage Stackelberg game. In (Joe-Wong et al. 2015), a three-stage optimization hierarchy was proposed to model the process of selecting content volume, sponsorship levels and unit data price by the end-users, the CPs and the monopolistic ISP in a pay-as-you-go market. Therein, the CPs' revenue obtained from ad-clicking was assumed to be proportional to the consumed content volume, and the MUs were allowed to consume contents from different CSPs at the same time. The interactions between users, CPs, and an ISP is studied in (Ante and Knutson 2013), but without a detailed examination of heterogeneity in CP incentives to sponsor data or of the relative benefits that sponsored data provides to different types of end-users and CPs. In (Zhang et al. 2016), the authors explored the interplay between ISP and CPs and presented a pricing mechanism for sponsored data that is truthful in CP's valuation as well as its underlying traffic. The authors in (Xiong et al. 2017) conducted economic analysis on sponsored content in the presence of social network effects, where the interactions among the SP, CP, and MUs are modeled as a three-stage game. In (Wang et al. 2019) the authors model the service selection process between end-users as an evolutionary population game, and pricing-then-sponsoring process between the CP and the ISP as a non-cooperative game. The authors show that sponsoring can help improve the profit of both provider (CP and ISP) and the end-user quality of experience.

PROBLEM FORMULATION

We consider an Internet market model with multiple CP and an arbitrary number of end-users. End-users can access the contents of the CP only through the ISP network while CP provides the content to the end-users. The Internet market depends on three effective factors sponsoring, pricing and credibility of content. Under the assumption that each CP can sponsored content, it can decide to either sponsored the entire or a portion of the requested content. Let N, the number of content that the CP sells. The sponsoring strategy adopted by each ISP is denoted by S that take values in the interval $[0, N]$. The CP pricing strategies set also consists of the content price p_c that the end-users should pay for content. Each CP advertises to the end-users credibility of content c. Figure 1 shows a sponsored content market with CP, ISP of a set of end-users.

Demand Model

We consider that demand of content from end-users to CP is a linear function affected by strategies of all CP, as for example in ((Garmani et al. 2019b) (Garmani et al. 2018b))as follows:

$$D_f = d_f - \sigma_f^f p_{c_f} + \varsigma_f^f c_f + \chi_f^f S_f + \sum_{g=1, f\neq g}^{F} (\sigma_f^g p_{c_g} - \varsigma_f^g c_g - \chi_f^g S_g)$$

The parameter d_f expresses the potential demand of end-users. σ_f^g, ς_f^g and χ_f^g they are tree positive parameters representing respectively the responsiveness of CP_f to price p_g, credibility of content c_g and number of sponsored content S_g. For CP_f, the demand D_f decreasing in the price it charges, p_{c_f}, and increase in the price charged by its opponent, p_{c_g}. The analogous relationship holds in credibility of content and number of sponsored content, in this case D_f is increasing in c_f (resp. S_f) and decreasing in c_g (resp. S_g).

Assumption 1

The sensitivity σ verifies:

$$\sigma_f^f \geq \sum_{g=1, g\neq f}^{F} \sigma_f^g.$$

The sensitivity ς verifies:

Figure 1. A schematic example of the sponsored content markets, where all the payments are made to the ISP

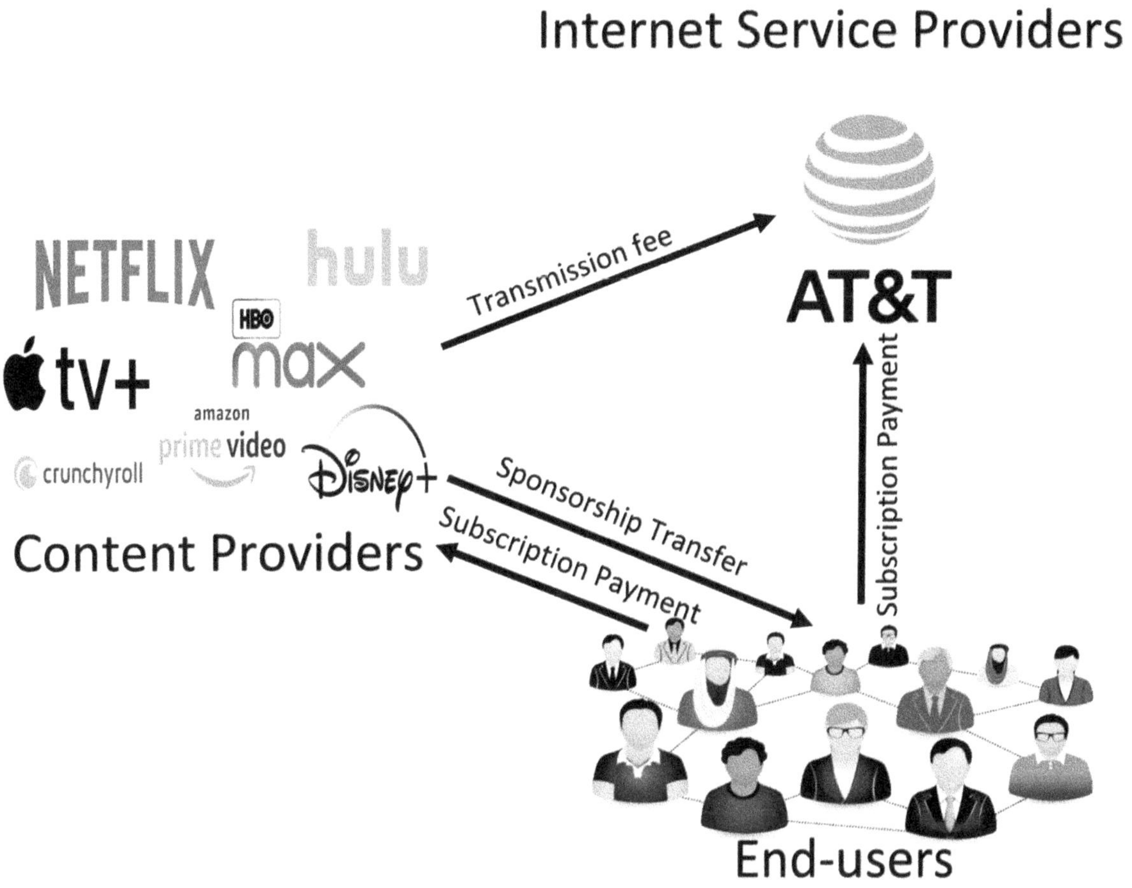

$$\varsigma_f^f \ge \sum_{g=1, g \neq f}^{F} \varsigma_f^g$$

The sensitivity χ verifies:

$$\chi_f^f \ge \sum_{g=1, g \neq f}^{F} \chi_f^g.$$

Assumption 1 implies that the influence of CP strategies on its observed demand is greater than the influence of the strategies of its opponent on their demand.

Utility Function

The utility of CP_f can be modeled as:

$$U_f = p_{c_f} D_f - p_u S_f D_f - p_t c_f D_f - \theta_f c_f$$

θ_f the cost to produce a unit of credibility of content c_f. p_t represents the transmission fee that the CPs pay per unit of credibility of content c_f. p_u is sponsoring cost of CP per unit data and paid to the ISP. In the utility function, the first term is the revenue of CP_f, the second term denotes the cost due to sponsorship, the third term is the cost to produce the credibility of content c_f, and the fourth term is transmission fee results when the CP_f forwards to the end-users demand with credibility of content c_f.

Credibility of content c_i of CP_f is a function of the quality of content q_{c_f} and the quality service q_{s_f}, which is written as follows (Garmani et al. 2018a) (Outanoute et al. 2019):

$$c_f = \lambda q_{s_f} + \mu q_{c_f}$$

where μ and λ are two positive constants. The quality of content QoC for example, quality of video streaming (Ait Omar et al. 2019a). q_{s_f} is quality of service QoS measure (e.g. delay, see (Baslam et al. 2013a) (Baslam et al. 2011)).

Then, the utility of CP_f is given by the following formula:

$$U_f = p_{c_f} D_f - p_u S_f D_f - p_t(\lambda q_{s_f} + \mu q_{c_f}) D_f - \theta_f(\lambda q_{s_f} + \mu q_{c_f})$$

GAME ANALYSIS

Let $G = \left[\mathcal{F}, \left\{\mathcal{P}_{c_f}, \mathcal{S}_f, \mathcal{Q}_{s_f}, \mathcal{Q}_{c_f}\right\}, \{U_f(.)\}\right]$ denote the non-cooperative price QoC sponsoring QoS game (NPQSQG), where $\mathcal{F}$ ={1, ..., F} is the index set identifying the CPs, $\mathcal{P}_{c_f}$ is the price strategy set of CP_f, $\mathcal{Q}_{s_f}$ is the QoS strategy set of CP_f, $\mathcal{S}_f$ is the sponsoring strategy set of CP_f, $\mathcal{Q}_{c_f}$ is the QoC strategy set of CP_f and $U_f(.)$ is the utility function of CP_f defined in equation . We assume that the strategy spaces $\mathcal{P}_{c_f}$, $\mathcal{Q}_{s_f}$, $\mathcal{S}_f$ and $\mathcal{Q}_{c_f}$ of each CP_f are compact and convex sets with maximum and minimum constraints. Thus, for each CP_f we consider as respective strategy spaces the closed intervals: $\mathcal{P}_{c_f} = \left[\underline{p}_{c_f}, \overline{p}_{c_f}\right]$, $\mathcal{Q}_{s_f} = \left[\underline{q}_{s_f}, \overline{q}_{s_f}\right]$, $\mathcal{S}_f = \left[\underline{S}_f, \overline{S}_f\right]$ and $\mathcal{Q}_{c_f} = \left[\underline{q}_{c_f}, \overline{q}_{c_f}\right]$. Let the price vector
$p_c = (p_{c_1}, ..., p_{c_f})^T \in \mathcal{P}_c^F = \mathcal{P}_{c_1} \times \mathcal{P}_{c_2} \times ... \times \mathcal{P}_{c_F}$, QoS vector
$q_s = (q_{s_1}, ..., q_{s_F})^T \in Q_s^F = \mathcal{Q}_{s_1} \times \mathcal{Q}_{s_2} \times ... \times \mathcal{Q}_{s_F}$, sponsoring vector
$S = (S_1, ..., S_F)^T \in \mathcal{S}^F = \mathcal{S}_1 \times \mathcal{S}_2 \times ... \times \mathcal{S}_F$, QoC vector
$q_c = (q_{c_1}, ..., q_{c_F})^T \in Q_c^F = \mathcal{Q}_{c_1} \times \mathcal{Q}_{c_2} \times ... \times \mathcal{Q}_{c_F}$.

Price Game

A NPQSQG in price p_c is defined for fixed $\mathbf{q}_s \in \mathcal{Q}_s$, $\mathbf{S} \in \mathcal{S}$, $\mathbf{q}_c \in \mathcal{Q}_c$ as $G(\mathbf{q}_s, \mathbf{S}, \mathbf{q}_c) = \mathcal{F}$, $\{\mathcal{P}_{c_i}\}, \{U_f(., \mathbf{q}_s, \mathbf{S}, \mathbf{q}_c)\}]$.

Definition 1: *A price vector* $\mathbf{p}_c^* = (p_{c_1}^*, \ldots, p_{c_F}^*)$ *is a Nash equilibrium of the NPQSQG* $G(\mathbf{q}_s, \mathbf{S}, \mathbf{q}_c)$ *if:*

$$\forall (f, p_{c_f}) \in (\mathcal{F}, \mathcal{P}_{c_f}),$$

$$U_f(p_{c_f}^*, \mathbf{p}_{f_{-i}}^*, \mathbf{q}_s, \mathbf{S}, \mathbf{q}_c) \geq U_f(p_{c_f}, \mathbf{p}_{c_{-f}}^*, \mathbf{q}_s, \mathbf{S}, \mathbf{q}_c)$$

Theorem 1: *For each* $\mathbf{q}_s \in \mathcal{Q}_s$, $\mathbf{S} \in \mathcal{S}$, $\mathbf{q}_c \in \mathcal{Q}_c$, *the game* $[\mathcal{F}, \{\mathcal{P}_{c_f}\}, \{U_f(., \mathbf{q}_{ss}, \mathbf{S}, \mathbf{q}_c)\}]$ *admit a unique Nash Equilibrium.*

Proof

To prove existence, we note that the strategy space $\mathcal{P}_{c_f}$ is defined in the closed interval bounded by the minimum and maximum prices. Thus, the strategy space $\mathcal{P}_{c_f}$ is nonempty, convex, and compact subset of the Euclidean space R^N. In addition, the second order derivative of the utility with respect to the prices is negative as below:

$$\frac{\partial^2 U_f}{\partial p_{c_f}^2} = -2\sigma_f^f \leq 0$$

The second derivative of the utility function is negative, then the utility function is thus concave, which ensures existence of a Nash equilibrium point in the game $G(\mathbf{q}_s, \mathbf{S}, \mathbf{q}_c)$.

We use the following proposition that holds for a concave game (Lasaulce et al. 2009): If a concave game satisfies the dominance solvability condition:

$$-\frac{\partial^2 U_f}{\partial p_{c_f}^2} \geq \sum_{g, g \neq f} \left| \frac{\partial^2 U_f}{\partial p_{c_f} \partial p_{c_g}} \right|$$

then the game $G(\mathbf{q}_s, \mathbf{S}, \mathbf{q}_c)$ admits a unique Nash equilibrium point.

The mixed partial is written as:

$$\frac{\partial^2 U_f}{\partial p_{c_f} \partial p_{c_g}} = \sigma_f^g \geq 0$$

Then,

$$-\frac{\partial^2 U_f}{\partial p_{c_f}^2} - \sum_{g,g\neq f} \left| \frac{\partial^2 U_f}{\partial p_{c_f} \partial p_{c_g}} \right| = 2\sigma_f^f - \sum_{g,g\neq f} \sigma_f^g \geq 0$$

Thus, the game $G(\mathbf{q}_s, \mathbf{S}, \mathbf{q}_c)$ admit a unique Nash equilibrium point.

Number of Sponsored Content Game

A NPQSQG in number of sponsored content is defined for a fixed $\mathbf{p}_c \in \mathcal{P}_c$, $\mathbf{q}_s \in \mathcal{Q}_s$, $\mathbf{q}_c \in \mathcal{Q}_c$ as $G(\mathbf{p}_c, \mathbf{q}_s, \mathbf{q}_c) = [\mathcal{F}, \{\mathcal{S}_f\}, \{U_f(\mathbf{p}_c, \mathbf{q}_s, \cdot, \mathbf{q}_c)\}]$.

Definition 2: *A sponsoring vector* $\mathbf{S}^* = (S_1^*, \ldots, S_F^*)$ *is a Nash equilibrium of the NPQSQG* $G(\mathbf{p}_c, \mathbf{q}_s, \mathbf{q}_c)$ *if:*

$$\forall (f, S_f) \in (\mathcal{F}, \mathcal{S}_f),$$

$$U_f(\mathbf{p}_c, \mathbf{q}_s, S_f^*, \mathbf{S}_{-f}^*, \mathbf{q}_c) \geq U_f(\mathbf{p}_c, \mathbf{q}_s, S_f, \mathbf{S}_{-f}^*, \mathbf{q}_c)$$

Theorem 2: *For each* $\mathbf{p}_c \in \mathcal{P}_c$, $\mathbf{q}_s \in \mathcal{Q}_s$, $\mathbf{q}_c \in \mathcal{Q}_c$, *the game* $[\mathcal{F}, \{\mathcal{S}_f\}, \{U_f(\mathbf{p}_c, \mathbf{q}_s, \cdot, \mathbf{q}_c)\}]$ *admit a unique Nash Equilibrium.*

Proof

To prove existence, we note that the strategy space $\mathcal{S}_f$ is defined in the closed interval bounded by the minimum and maximum number of sponsored content. Thus, the strategy space $\mathcal{S}_f$ is nonempty, convex, and compact subset of the Euclidean space R^N. In addition, the second order derivative of the utility with respect to the number of sponsored content is negative as below:

$$\frac{\partial^2 U_f}{\partial S_f^2} = -2\chi_f^f p_u \leq 0$$

The second derivative of the utility function is negative, then the utility function is thus concave, which ensures existence of a Nash equilibrium point in the game $G(\mathbf{p}_c, \mathbf{q}_s, \mathbf{q}_c)$.

We use the following proposition that holds for a concave game (Lasaulce et al. 2009): If a concave game satisfies the dominance solvability condition:

$$-\frac{\partial^2 U_f}{\partial S_f^2} \geq \sum_{g, g \neq f} \left| \frac{\partial^2 U_f}{\partial S_f \partial S_g} \right|$$

then the game $G(\mathbf{p}_c, \mathbf{q}_s, \mathbf{q}_c)$ admits a unique Nash equilibrium point.

The mixed partial is written as:

$$\frac{\partial^2 U_f}{\partial S_f \partial S_g} = \chi_f^g p_u \geq 0$$

Then,

$$-\frac{\partial^2 U_f}{\partial S_f^2} - \sum_{g, g \neq f} \left| \frac{\partial^2 U_f}{\partial S_f \partial S_g} \right| = (2\chi_f^f - \sum_{g, g \neq f} \chi_f^g) p_u \geq 0$$

Thus, the game $G(\mathbf{p}_c, \mathbf{q}_s, \mathbf{q}_c)$ admit a unique Nash equilibrium point.

QoC Game

A NPQSQG in QoC is defined for a fixed $\mathbf{p}_c \in \mathcal{P}_c$, $\mathbf{q}_s \in \mathcal{Q}_s$, $\mathbf{S} \in \mathcal{S}$ as $G(\mathbf{p}_c, \mathbf{q}_s, \mathbf{S}) = [,\ \{\mathcal{Q}_{c_f}\}, \{U_f(\mathbf{p}_c, \mathbf{q}_s, \mathbf{S}, .)\}]$.

Definition 3: *A QoC vector* $\mathbf{q}_c^* = (q_{c_1}^*, \ldots, q_{c_F}^*)$ *is a Nash equilibrium of the NPQSQG* $G(\mathbf{p}_c, \mathbf{q}_s, \mathbf{S})$ *if:*

$$\forall (f, q_{c_f}) \in (\mathcal{F}, \mathcal{Q}_{c_f}),$$

$$U_f(\mathbf{p}_c, \mathbf{q}_s, \mathbf{S}, q_{c_f}^*, \mathbf{q}_{c_{-f}}^*) \geq U_f(\mathbf{p}_c, \mathbf{q}_s, \mathbf{S}, q_{c_f}, \mathbf{q}_{c_{-f}}^*)$$

Theorem 3: *For each* $\mathbf{p}_c \in \mathcal{P}_c$, $\mathbf{q}_s \in Q_s$, *the game* $[\mathcal{F},\ \{\mathcal{Q}_{c_f}\}, \{U_f(\mathbf{p}_c, \mathbf{q}_s, \mathbf{S}, .)\}]$ *admit a unique Nash Equilibrium.*

Proof

To prove existence, we note that the strategy space ∂_{c_f} is defined in the closed interval bounded by the minimum and maximum QoCs. Thus, the joint strategy space ∂_{c_f} is nonempty, convex, and compact subset of the Euclidean space R^N. In addition, the second order derivative of the utility with respect to the QoC is negative as below:

$$\frac{\partial^2 U_f}{\partial q_{c_f}^2} = -2\mu\chi_f^f p_t \leq 0$$

The second derivative of the utility function is negative, then the utility function is thus concave, which ensures existence of a Nash equilibrium point in the game $G(\mathbf{p}_c, \mathbf{q}_s, \mathbf{S})$.

We use the following proposition that holds for a concave game (Lasaulce et al. 2009): If a concave game satisfies the dominance solvability condition:

$$-\frac{\partial^2 U_f}{\partial q_{c_f}^2} \geq \sum_{g,g\neq f} \left| \frac{\partial^2 U_f}{\partial q_{c_f} \partial q_{c_g}} \right|$$

then the game $G(\mathbf{p}_c, \mathbf{q}_s, \mathbf{S})$ admits a unique Nash equilibrium point.

The mixed partial is written as:

$$\frac{\partial^2 U_f}{\partial q_{c_f} \partial q_{c_g}} = \mu\chi_f^j p_t$$

Then,

$$-\frac{\partial^2 U_f}{\partial q_{c_f}^2} - \sum_{g,g\neq f} \left| \frac{\partial^2 U_f}{\partial q_{c_f} \partial q_{c_g}} \right| = \mu(\chi_f^f - \sum_{g,g\neq f} \chi_f^g) p_t$$

Thus, the game $G(\mathbf{p}_c, \mathbf{q}_s, \mathbf{S})$ admit a unique Nash equilibrium point.

QoS game

A NPQSQG in QoS is defined for a fixed $\mathbf{p}_c \in \rho_c$, $\mathbf{S} \in S$, $\mathbf{q}_c \in \partial_c$ as $G(\mathbf{p}_c, \mathbf{q}_c, \mathbf{S}) = [F, \{\partial_{s_f}\}, \{U_f(\mathbf{p}_c, ., \mathbf{S}, \mathbf{q}_c)\}]$.

Definition 4: *A QoS vector* $\mathbf{q}_s^* = (q_{s_1}^*, ..., q_{s_F}^*)$ is a Nash equilibrium of the NPQSQG $G(\mathbf{p}_c, \mathbf{q}_c, \mathbf{S})$ if:

$$\forall (f, q_{s_f}) \in (F, \partial_{s_f}),$$

$$U_f(\mathbf{p}_c, q_{s_f}^*, \mathbf{q}_{s_{-f}}^*, \mathbf{S}, \mathbf{q}_c) \geq U_f(\mathbf{p}_c, q_{s_f}, \mathbf{q}_{s_{-f}}^*, \mathbf{S}, \mathbf{q}_c)$$

Theorem 4: *For each* $\mathbf{p}_c \in P_c$, $\mathbf{S} \in S$, $\mathbf{q}_c \in \partial_c$, the game $[F, \{\partial_{s_f}\}, \{U_f(p_c,.,S,q_c)\}]$ admit a unique Nash Equilibrium.

The QoS game has the same analysis as QoC game, just we replace μ by λ.

LEARNING NASH EQUILIBRIUM

The section mentioned above shows clearly that the Nash equilibrium point is unique. Now, we devise an algorithm that converges to the Nash equilibrium in a distributed manner based on the best-response dynamic. A best response dynamics scheme contains a series of rounds, wherein each round, each CP observe the policy taken by its competitors in previous rounds and input them in its decision process to update its policy. In the first round, the choice of each CP is the best response based on its arbitrary belief about what the other players will choose.

Algorithm 1 summarizes the best response learning steps that each CP has to perform in order to find the Nash equilibrium point.

Algorithm 1. *Best response algorithm dynamics*

Initialize vectors $\mathbf{x}(0) = [x_1(0), \ldots, x_F(0)]$ randomly;
For each CP_f, $f \in K$ at time instant t compute:
- $x_f(t+1) = \underset{x_f \in X_f}{\operatorname{argmax}}\left(U_f(\mathbf{x}(t))\right)$.

If $\forall f \in, |x_f(t+1) - x_f(t)| < \varepsilon$, then STOP. Else, $t = t+1$ and go to the previous step

Such as:

x refers to the vector price p_c, vector of sponsored content S, vector q_s and vector q_c.

X_k refers to the policy profile price, number of sponsored content, QoS or QoC.

Price of Anarchy

The concept of Price of Anarchy (PoA) has been widely used in economics and in game theory and mechanism design theory, to measure the performance degradation of a system due to the selfish decisions made by independent players (Koutsoupias and Papadimitriou 1999). The Price of Anarchy (PoA) measuring how the efficiency of a system degrades due to selfish behavior of its players as the ratio of the social utility at a Nash equilibrium divided by optimal social utility (the social utility being the sum of utilities over all players), for example in network economics (Baslam et al. 2013b) (Amrani et al. 2018a) (Ait Omar et al. 2019c).

PoA is defined as the ratio between the performance measures of the worst equilibrium and the optimal outcome. A PoA close to 1 indicates that the equilibrium is approximately socially optimal, and thus the consequences of selfish behavior are relatively benign.

In (Guijarro et al. 2011) we measure the loss of efficiency due to actors' selfishness as the quotient between the social welfare obtained at the Nash equilibrium and the maximum value of the social welfare:

$$PoA = \frac{min_{\mathbf{p}_c,\mathbf{q}_c,\mathbf{S}} W_{NE}(\mathbf{p}_c,\mathbf{q}_c,\mathbf{S})}{max_{\mathbf{p}_c,\mathbf{q}_c,\mathbf{S}} W(\mathbf{p}_c,\mathbf{q}_c,\mathbf{S})}$$

Where $W(\mathbf{p}_c,\mathbf{q}_c,\mathbf{S}) = \sum_{f=1}^{F} U_f(\mathbf{p}_c,\mathbf{q}_c,\mathbf{S})$ a social welfare function and $W_{NE}(\mathbf{p}_c^*,\mathbf{q}_c^*,\mathbf{S}^*)$ $= \sum_{f=1}^{F} U_f(\mathbf{p}_c,\mathbf{q}_c,\mathbf{S})$ is a sum of utilities of all CP at Nash equilibrium.

NUMERICAL INVESTIGATION

In this section, we perform the simulations to evaluate the strategy of CPs under competitive cases, where each CP sponsored a number of content. We consider a scenario with two homogeneous CP seeking to maximize their payoff. The parameter values used in the numerical investigation are summarized in Table 1.

Table 1. Setting used for numerical simulations.

$\sigma_1^1 = \sigma_2^2$	$\sigma_f^g, f \neq g$	$\varsigma_1^1 = \varsigma_2^2$	$\varsigma_f^g, f \neq g$	F
7	0.3	0.7	0.3	2
λ	μ	p_u	$\chi_1^1 = \chi_2^2$	$\chi_f^g, f \neq g$
5	0.5	10	0.7	0.3
d_2	d_1	p_t	$\theta_1 = \theta_2$	N
250	250	10	0.5	100

Figures 2, 3, 4 and 5 illustrate the convergence to equilibrium price, number of sponsored content, QoS and QoC. This figure shows the number of iterations needed for convergence to the Nash equilibrium point, it is clear that the speed of convergence is relatively high. The figures 1, 2, 3 and 4 demonstrates the existence and uniqueness of a Nash equilibrium point at which no CPs can profitably deviate given the strategies of another CP.

Figure 2. Price game: Convergence to the Price at Nash equilibrium.

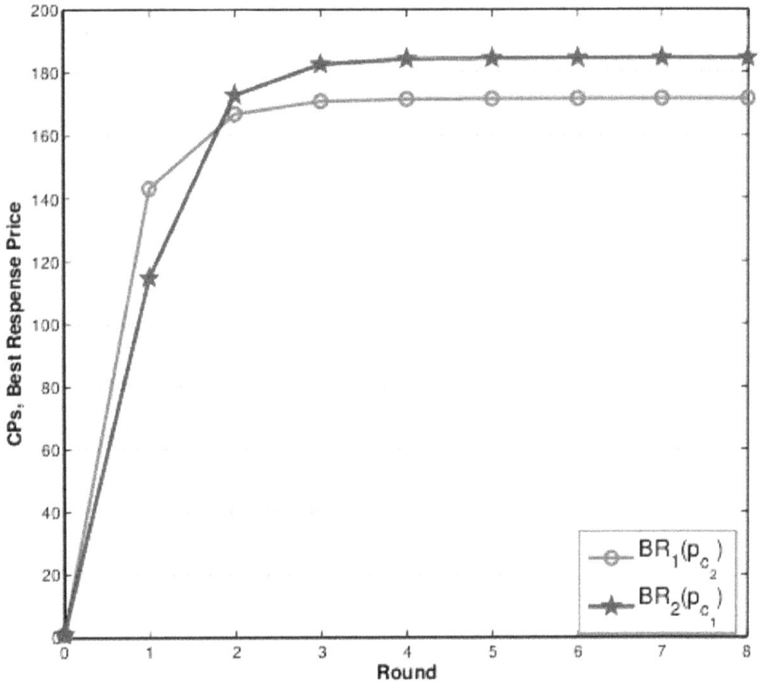

Figure 3. Price game: Convergence to the QoC at Nash equilibrium.

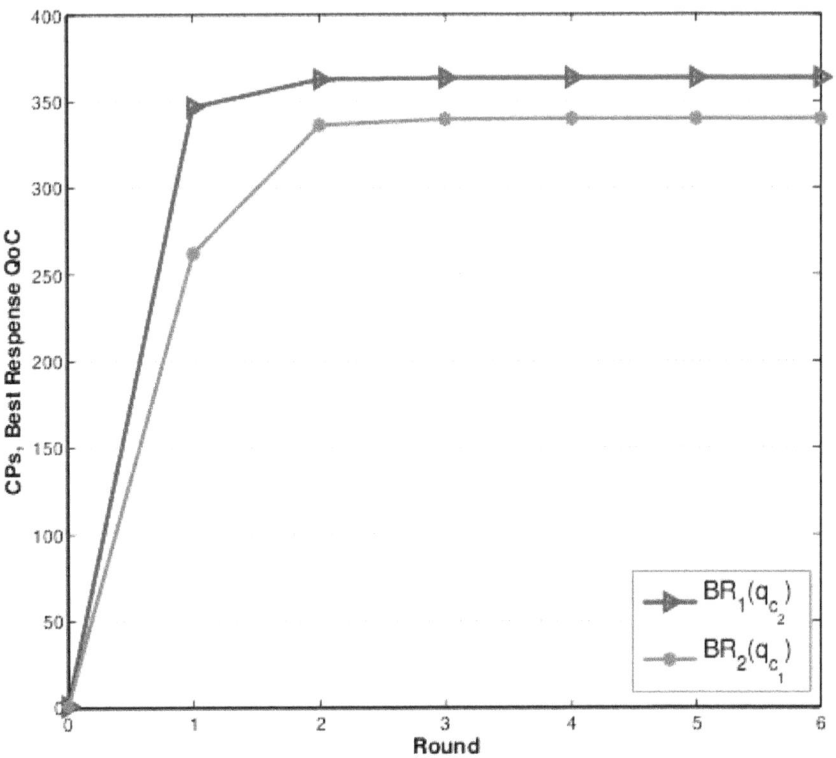

Figure 4. Price game: Convergence to the QoS at Nash equilibrium.

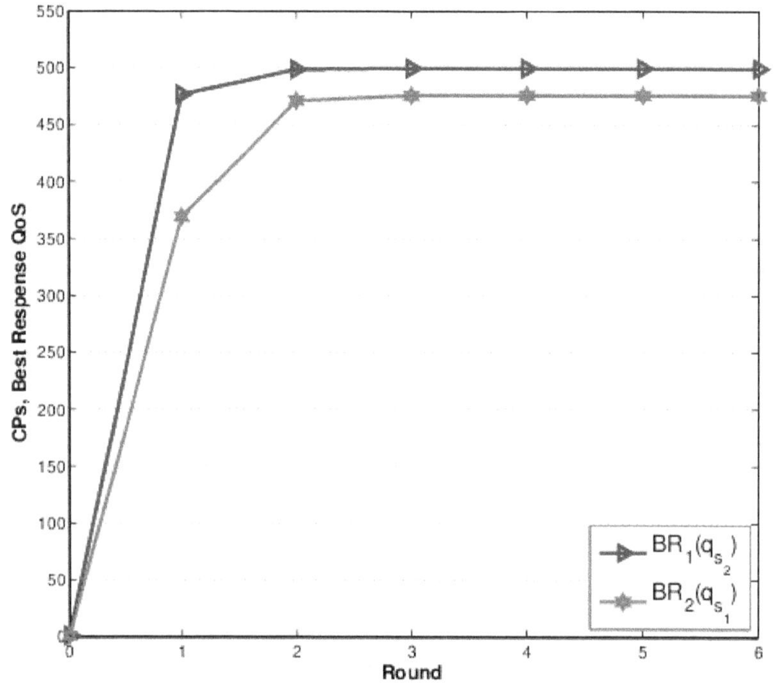

Figure 5. Price game: Convergence to the number of sponsored content at Nash equilibrium

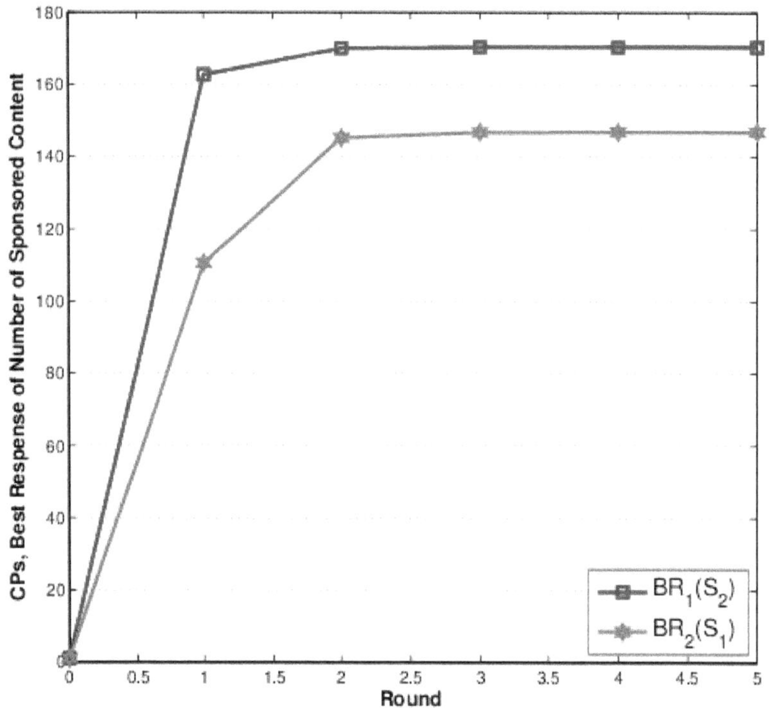

Figure 6. Equilibrium price p_c as a function of number of sponsored content S

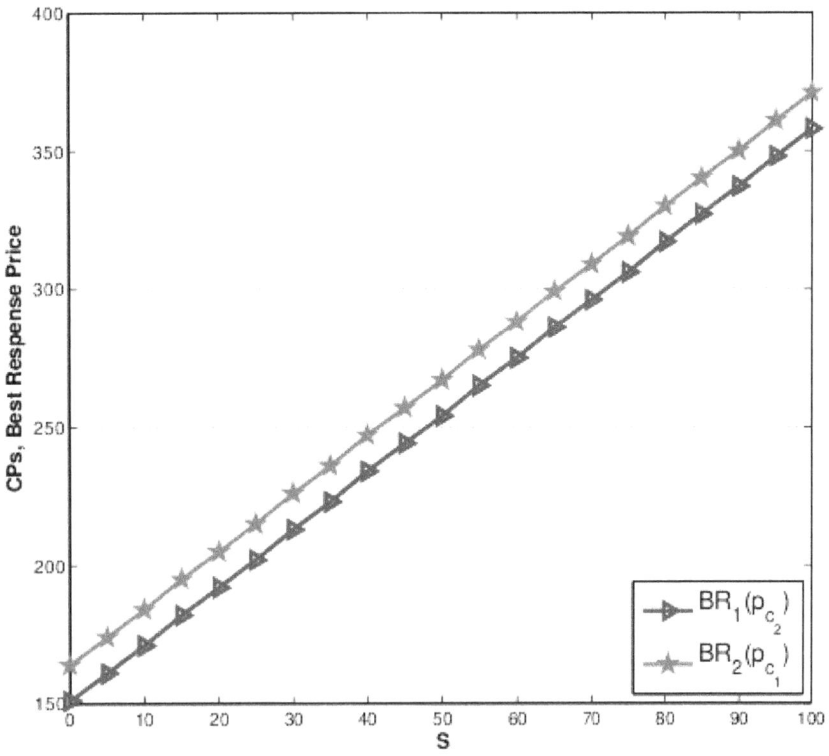

We investigate the impact of the number of sponsored content S on the price p_c, as illustrated in figure 6. When the number of sponsored content S increase the price p_c increase. The intuition is that as the number of sponsored content S increase, the sponsoring cost increase. Therefore, the CP need to increase their price p_c in order to compensate the increase in the sponsoring cost. The CP needs to choose a lower price to attract more end-users to consume more sponsored content.

Figures 7 and 8 shows the QoS and QoC as function of number of sponsored content S. The QoS and the QoC decrease as the number of sponsored content get increases S. The reason is that as the number of sponsored content S increases, the sponsored content demand of the end-users increases. Therefore, the transmission fee increases. Thus, the CP needs to slightly decrease its QoS and QoC to compensate the increase in the transit cost.

We plot the expected utility of CPs as a function of the number of sponsored content S in Figure 9. Utility increase with respect to the number of sponsored content S. As S increases, the sponsored content demand of the end-users increases. Moreover, the increase of the sponsored content demand leads to the profit improvement of both CP since they can sell more content to the incoming end-users.

Figure 10 shows the PoA variation curve as a function of transmission fee p_t. In that figure, the PoA increases with respect to p_t. When the transmission fee p_t is lower, the PoA is lower, and then the Nash equilibrium is not socially efficient, CPs are selfish and each one seeks to maximize its profit individually. But when the transmission fee p_t increases, the equilibrium becomes more and more socially efficient. This increase finds the simple intuition that when the transmission fee increases; CPs cooperate with each other for optimizing the Nash equilibrium.

Figure 7. Equilibrium QoC q_c as a function of number of sponsored content S

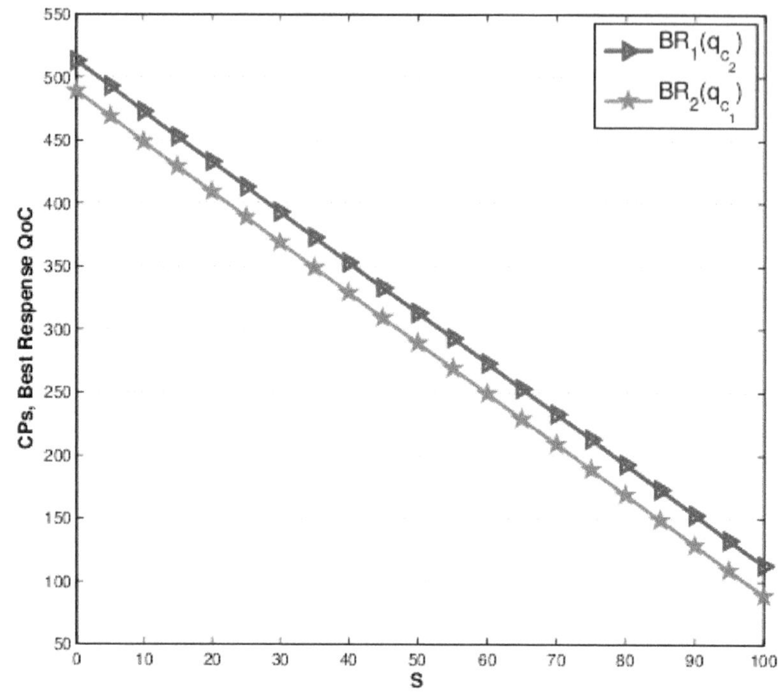

Figure 8. Equilibrium QoS q_s as a function of number of sponsored content S

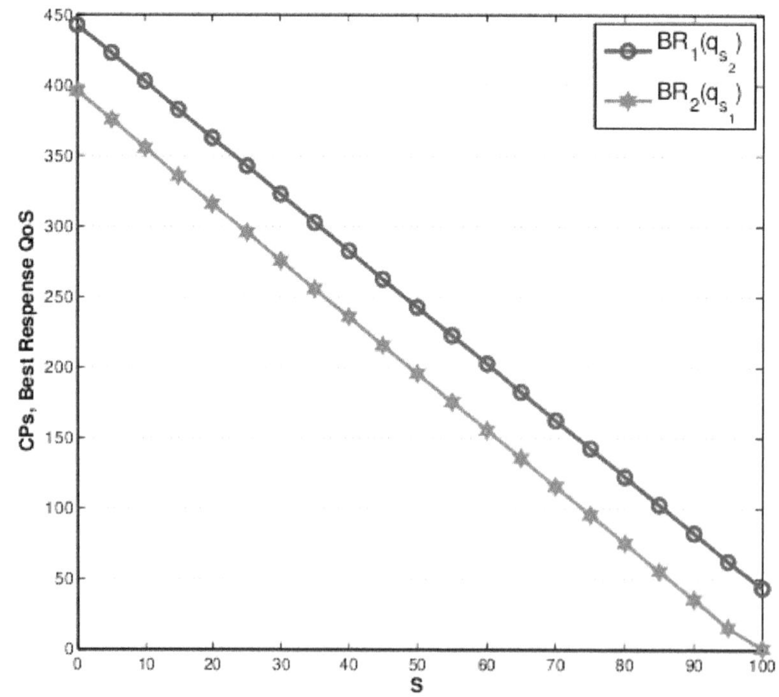

Figure 9. Utility as a function of number of sponsored content S

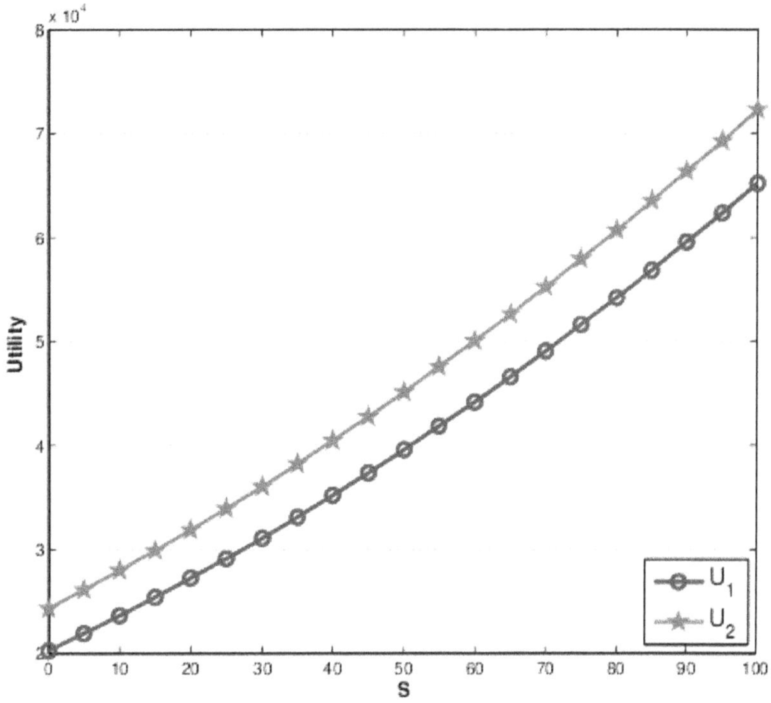

Figure 10. Price of Anarchy as a function of transmission fee p_t.

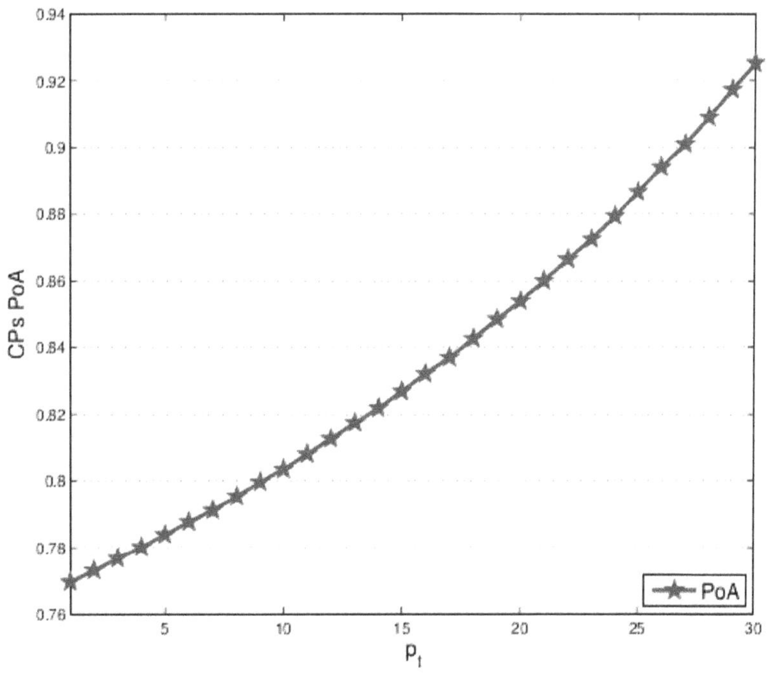

Figure 11. Price of Anarchy as a function of cost θ.

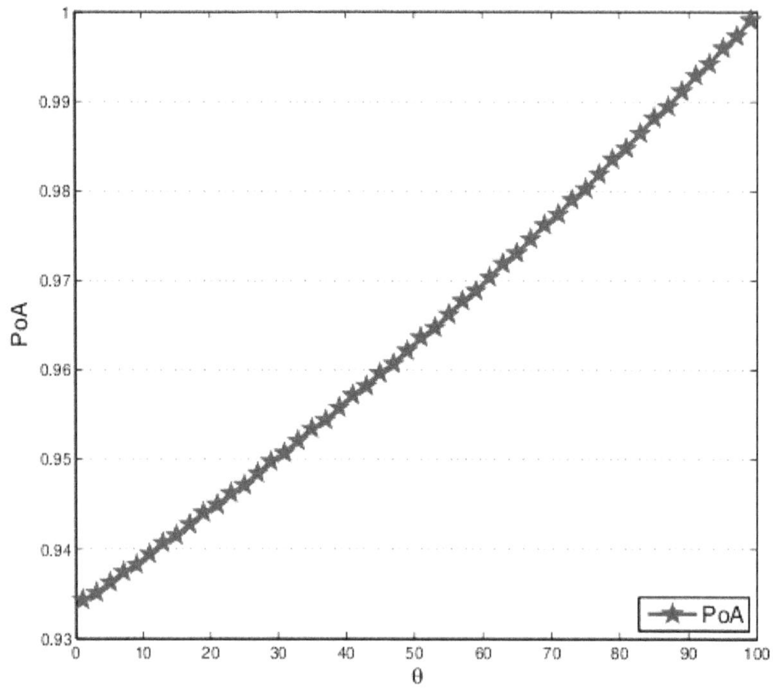

Figure 11 shows the PoA variation curve as a function of cost θ. In that figure, we first notice that the PoA increases when θ increases. A special feature is that the Nash equilibrium performs well, and the loss of efficiency is only a round 8%. This result indicates that the Nash equilibrium of this game is fair and socially efficient.

CONCLUSION

In this chapter, we derive the optimal sponsored content behaviors for end-users and CPs and then consider their implications for CPs and end-users. We investigate the interactions among the CPs as a non-cooperative game. Then, we have analytically analyzed the game. Furthermore, we proved the existence and uniqueness of the Nash equilibrium point in a competitive market under our proposed model. This result is significant because it implies that a stable solution with suitable economic incentives in sponsoring content is feasible in the internet paradigm. Moreover, we have adopted an iterative best response-based algorithm that enables CPs to learn their Nash equilibrium strategies in a fully distributed fashion. Through the extensive simulation analysis, it is revealed that sponsoring content is the best choice of CP. As ongoing work, we are working on generalizing our proposal for a large number of CPs and ISPs.

REFERENCES

Ait Omar, D., El Amrani, M., Baslam, M., & Fakir, M. (2019a). A Game-Theoretic Approach for the Internet Content Distribution Chain. In A. Podelski & F. Taïani (Eds.), *Networked Systems* (pp. 270–285). Cham: Springer International Publishing. doi:10.1007/978-3-030-05529-5_18

Ait Omar, D., Garmani, H., El Amrani, M., Baslam, M., & Fakir, M. (2019b). A Customer Confusion Environment in Telecommunication Networks: Analysis and Policy Impact. *International Journal of Cooperative Information Systems, 28*(02), 1930002. doi:10.1142/S021884301930002X

Ait Omar, D., Outanoute, M., Baslam, M., Fakir, M., & Bouikhalene, B. (2019c). On understanding price-QoS war for competitive market and confused consumers. *Computing, 101*(9), 1327–1348. doi:10.100700607-018-0642-5

Ait Omar, D., Outanoute, M., Baslam, M., Fakir, M., & Bouikhalne, B. (2017). Joint Price and QoS Competition with Bounded Rational Customers. In *Networked Systems: 5th International Conference, NETYS 2017, Marrakech, Morocco, May 17-19, 2017, Proceedings*. Springer International Publishing. 10.1007/978-3-319-59647-1_33

Altman, E. (2012). In which content to specialize? a game theoretic analysis. In *International Conference on Research in Networking*. Springer.

Amrani, M. E., Garmani, H., Baslam, M., Minaoui, B., & Ayachi, R. E. (2018a). Cournot Game of ISP under Net Neutrality. *International Journal of Control and Automation, 11*(7), 16. doi:10.14257/ijca.2018.11.7.07

Amrani, M. E., Omar, D. A., Baslam, M., & Minaoui, B. (2018b). A Game Theoretic Approach to Optimize Transmission Strategies for Mobile Applications. In *2018 6th International Conference on Wireless Networks and Mobile Communications (WINCOM)*. IEEE. 10.1109/WINCOM.2018.8629618

Ante, S. E., & Knutson, R. (2013, July). Good luck leaving your wireless phone plan. *Wall Street Journal*.

Baslam, M., El-Azouzi, R., Sabir, E., & Echabbi, L. (2011). Market share game with adversarial access providers: A neutral and a non-neutral network analysis. In *Network Games, Control and Optimization (NetGCooP), 2011 5th International Conference on*. IEEE.

Baslam, M., El-Azouzi, R., Sabir, E., Echabbi, L., & Bouyakhf, E.-H. (2013a). A Game Theoretic Analysis of Price-QoS Market Share in Presence of Adversarial Service Providers. In H. Hanappi (Ed.), *Game Theory Relaunched*. InTech. doi:10.5772/54380

Baslam, M., El-Azouzi, R., Sabir, E., Echabbi, L., & Bouyakhf, E.-H. (2013b). A Game Theoretic Analysis of Price-QoS Market Share in Presence of Adversarial Service Providers. In H. Hanappi (Ed.), *Game Theory Relaunched*. InTech. doi:10.5772/54380

Garmani, H., Ait Omar, D., El Amrani, M., Baslam, M., & Jourhmane, M. (2019a). Analysis of a dynamics duopoly game with two content providers. *Chaos, Solitons, and Fractals, 109466*. doi:10.1016/j.chaos.2019.109466

Garmani, H., Baslam, M., & Jourhmane, M. (2018a). Analysis of Competition between CPs Fronting Advertisers. *International Journal of Advanced Science and Technology*, *117*, 53–66. doi:10.14257/ijast.2018.117.05

Garmani, H., Baslam, M., & Jourhmane, M. (2018b). Caching Games between ISP in Information Centric Network. *International Journal of Control and Automation*, *11*(4), 125–142. doi:10.14257/ijca.2018.11.4.12

Garmani, H., Outanoute, M., Baslam, M., & Jourhmane, M. (2019b). New Competition-Based Approach for Caching Popular Content in ICN. In A. Podelski & F. Taïani (Eds.), *Networked Systems* (pp. 286–300). Cham: Springer International Publishing. doi:10.1007/978-3-030-05529-5_19

Guijarro, L., Pla, V., Vidal, J. R., & Martinez-Bauset, J. (2011). *Analysis of price competition under peering and transit agreements in Internet Service provision to peer-to-peer users. In 2011 IEEE Consumer Communications and Networking Conference* (pp. 1145–1149). CCNC. doi:10.1109/CCNC.2011.5766356

Joe-Wong, C., Ha, S., & Chiang, M. (2015). *Sponsoring mobile data: An economic analysis of the impact on users and content providers. In 2015 IEEE conference on computer communications (infocom)* (pp. 1499–1507). IEEE.

Koutsoupias, E., & Papadimitriou, C. (1999). Worst-case equilibria. In *Annual Symposium on Theoretical Aspects of Computer Science*. Springer.

Lasaulce, S., Debbah, M., & Altman, E. (2009). Methodologies for analyzing equilibria in wireless games. *IEEE Signal Processing Magazine*, *26*(5), 41–52. doi:10.1109/MSP.2009.933496

Maillé, P., & Tuffin, B. (2019). Analysis of Sponsored Data Practices in the Case of Competing Wireless Service Providers. In H. Peng, C. Deng, Z. Wu, & Y. Liu (Eds.), *Computational Intelligence and Intelligent Systems* (pp. 55–70). Springer.

Omar, D. A., Garmani, H., El Amrani, M., Baslam, M., & Fakir, M. (2019a). Analysis of Bargaining Game Policy in the Internet Content Distribution Chain. *International Journal of Mobile Computing and Multimedia Communications*, *10*(3), 47–73. doi:10.4018/IJMCMC.2019070103

Omar, D. A., Garmani, H., El Amrani, M., Baslam, M., & Fakir, M. (2019b). Analyzing the Customers' Dynamic Confusion in Telecommunication Networks Share Game. *International Journal of Business Data Communications and Networking*, *15*(2), 15–34. doi:10.4018/IJBDCN.2019070102

Outanoute, M., Garmani, H., Baslam, M., Ayachi, R. E., & Bouikhalene, B. (2019). A Non-Cooperative Game Analysis of Competition between Content Providers in the Internet Market. *International Journal of Business Data Communications and Networking*, *15*(1), 88–104. doi:10.4018/IJBDCN.2019010106

Reiffers-Masson, A., Hayel, Y., & Altman, E. (2015). *Pricing Agreement between Service and Content Providers: A Net Neutrality Issue*. arXiv preprint arXiv:150503555

Wang, W., Xiong, Z., Niyato, D., Wang, P., & Han, Z. (2019). A Hierarchical Game With Strategy Evolution for Mobile Sponsored Content and Service Markets. *IEEE Transactions on Communications*, *67*(1), 472–488. doi:10.1109/TCOMM.2018.2869820

Xiong, Z., Feng, S., Niyato, D., Wang, P., Leshem, A., & Han, Z. (2019). Joint Sponsored and Edge Caching Content Service Market: A Game-Theoretic Approach. *IEEE Transactions on Wireless Communications, 18*(2), 1166–1181. doi:10.1109/TWC.2018.2890469

Xiong, Z., Feng, S., Niyato, D., Wang, P., Leshem, A., & Zhang, Y. (2018a). Game Theoretic Analysis for Joint Sponsored and Edge Caching Content Service Market. In *2018 IEEE Global Communications Conference (GLOBECOM)*. IEEE. 10.1109/GLOCOM.2018.8647436

Xiong, Z., Feng, S., Niyato, D., Wang, P., & Zhang, Y. (2017). Economic analysis of network effects on sponsored content: a hierarchical game theoretic approach. In *GLOBECOM 2017-2017 IEEE Global Communications Conference*. IEEE. 10.1109/GLOCOM.2017.8254738

Xiong, Z., Feng, S., Niyato, D., Wang, P., & Zhang, Y. (2018b). *Competition and cooperation analysis for data sponsored market: A network effects model. In 2018 IEEE Wireless Communications and Networking Conference (WCNC)* (pp. 1–6). Barcelona: IEEE.

Zhang, L., Wu, W., & Wang, D. (2016). TDS: Time-dependent sponsored data plan for wireless data traffic market. In *IEEE INFOCOM 2016-The 35th Annual IEEE International Conference on Computer Communications*. IEEE.

Chapter 8
A Pricing Model to Optimize Transmission Strategies for Mobile Devices

Mohamed El Amrani
Information Processing and Decision Support Laboratory, Sultan Moulay Slimane University, Morocco

Hamid Garmani
https://orcid.org/0000-0002-7294-6722
Information Processing and Decision Support Laboratory Sultan Moulay Slimane University, Morocco

Driss Ait Omar
https://orcid.org/0000-0002-4096-8273
Information Processing and Decision Support Laboratory Sultan Moulay Slimane University, Morocco

Mohamed Baslam
Information Processing and Decision Support Laboratory Sultan Moulay Slimane University, Morocco

Brahim Minaoui
Information Processing and Decision Support Laboratory Sultan Moulay Slimane University, Morocco

ABSTRACT

The simultaneous multiple data transmission can improve the use of the network. Unlike existing solutions in the literature, in this chapter, the authors propose a solution to the network resource allocation problem under the selfish behavior of mobile device with multiple connections to several available network interfaces simultaneously, to resolve the conflict of interest in network. They analyze the impact of interactions between users based on two conflicting factors (i.e., throughput and monetary cost). Also, a diverse set of user service types is taken into consideration, which makes the proposed approach suitable for an integrated service network. Analytical and numerical results demonstrate the validity of the proposed approach, which show that the non-cooperative game has an equilibrium point that depends on all parameters of the system, and they show that this situation between mobile devices is much more beneficial in terms of the performance of mobiles, cost, and the data transfer rate.

DOI: 10.4018/978-1-7998-3355-0.ch008

INTRODUCTION

Normally, a single connection does not allow full use of the network, especially when the quality of the network is not good. To accommodate the explosive growth in mobile data traffic, multihoming networks where users with multiple network interfaces can simultaneously transmit data traffic over this interfaces. Traditional modeling does not specifically consider this situation according to the needs of users (e.g., voice and video traffic). But today, mobile devices need more and more transmission speed and often when these latter are authorized to use multiple connections (sessions) and they share, simultaneously, multiple network interfaces (Wifi, Bluetooth, 3G,4G, ...), by ensuring user's satisfaction to offer multi-farious services in the most resource- and cost-efficient manners . In this case mobile devices, interact with each other, which means that the pay-off (maximizing net data transfer and minimizing cost) of each device (user) depends not only on the number of concurrent connections established by it, but also on the number of concurrent established connections by the other devices using the same network interfaces as it. These common uses between devices create an atmosphere of interaction that translates by the number of concurrent established connections by each mobile device. The choice of this number must be taken in such a way that each device maximizes its gain while maintaining the stability of the systems (Mobile devices in equilibrium).

So, the objective of this work is to establish a complete and a realistic model, which takes into consideration the interaction between mobile devices and the state of the system. After we show existing and uniqueness of equilibrium, and based on this model, we will show the effectiveness of the equilibrium, we analyze the impact of the model parameters on the state of the system.

The rest of this paper is as follows: in the section 2 we present related works and in the section 3 we develop our model. Section 4 provides theorems for the existence and uniqueness of equilibria, the section 5 presents a numerical study to validate our claims. Finally, we conclude the paper is section 6.

Related Works

The resource management of devices that have multiple interface is not obvious. Several works have begun this problem, and the researchers who deal with the notion of bandwidth aggregation, this last is an important technical that allows devices to use multiple network interfaces to improve throughput, which can be studied at different network layers, Application, transport, network and link (Ramaboli et al. 2012) (Chebrolu et al. 2005) (Chebrolu and Rao 2006) (Habak et al. 2015). Various methods focus on the transport layer. In (Nguyen et al. 2011), authors use a Multi-TCP protocol on mobile devices and provide an analysis of the overall throughput of the network. In (Liu et al. 2001), authors study the impact of concurrent downloading on the fairness and system's transient behavior.

The contributions that are most related to ours are (Lu et al. 2013) and (Zhang et al. 2005); (Lu et al. 2013) presents a concurrent transmission method for mobile device in multihoming devices, based on the use of strategies: maximum throughput strategy, minimum delay strategy and minimum consumption strategy. In (Zhang et al. 2005), authors study a selfish behavior of TCP users in which users are allowed to open multiple concurrent connections to maximize their individual goodputs.

Problem Modeling

We first define the transmission of mobile as an interaction game. The definition of the utility function will be presented in the next subsection.

Problem Defining

We consider a set of mobiles $= \{1,...,J\}$, each mobile j is in interaction on the services of a set of network interface i $\forall i \in= \{1,...,I\}$.

The mobile are in interaction for the data transfer rate of available network interfaces. Individual mobiles are treated as players. A strategy n_j^i available to a mobile j is a feasible number of concurrent connections it can open concurrently to a network interface i. Let N_j denote the feasible strategy set of player j, then $\mathbf{n}_j \in N_j$. And the feasible strategy space of this game is $= N_1 \times N_2 ... \times N_J$. Then, a feasible strategy $\mathbf{n}_j$ is an I-dimension vector, $\mathbf{n}_j = (n_j^1, n_j^2, ..., n_j^I) \in N_J$. The objective of each player is to maximize its utility U_j by adjusting $\mathbf{n}_j \in \mathbf{n} = (\mathbf{n}_1, \mathbf{n}_2, ..., \mathbf{n}_J)$.

We define, for a specific network interface, some factors, such as V_{max}^i the maximum data transfer rate between a client and a server, and V_a^i, the average data transfer rate for a single connection, as: $N_{max}^i = V_{\max}^i / V_a^i$, N_{max}^i is defined as the maximum number of concurrent data transmission connections provided that the data transfer rate reaches V_{max}^i and that an average speed of Single connection reaches V_a^i. For a network interface i, T^i is the number of all opened connections by all connected devices such as:

$$T^i = \sum_{j\in} n_j^i$$

We now seek to see the relationships between the various factors of the problem. We define variable Υ_e^i to represent the proportion of data transmission time of the total time, defined as:

$$\Upsilon_e^i = \frac{S^i / V_s^i}{\delta^i + S^i / V_s^i}$$

From the view point of probability analysis is also a function of T^i and S^i. By the above expression of the function, as T^i increases or S^i increases it is possible to increase the number of data transmission connections. With V_s^i, S^i and δ^i represent, respectively: a single data transfer speed, the data size of each request and the transmission delay. Then, we define a variable N_d^i as the number of data transmission connections at a particular time and after replacement Υ_e^i by its value we will have.

$$N_d^i = T^i \Upsilon_e^i = \frac{T^i S^i}{V_s^i \delta^i + S^i}$$

We differentiate between two data transfer rates, V_s^i defined as the experienced data transfer rates and V_u^i defined as the net data transfer rate (i.e. the data transfer rate without considering redundancy ($\delta^i = 0$)), these last are defined by:

$$V_u^i = \frac{V_{\max}^i}{T^i} \quad , \quad V_s^i = \frac{V_{\max}^i}{N_d^i}$$

When the number of data transmission connections is less than N_{max}^i, each connection maintains a rate V_a^i, indicating that the network interface is not yet fully used. On the other hand, when the number of data transmission connections exceeds N_{max}^i, the single connection rate V_s^i is limited by $V_{\max}^i$. Then we simplify and get the formulas of V_u^i and V_s^i;

$$V_u^i = \begin{cases} V_a^i & , N_d^i < N_{\max}^i \\ \dfrac{V_a^i N_{max}^i}{T^i} & , N_d^i \geq N_{\max}^i \end{cases}$$

$$V_s^i = \begin{cases} V_a^i & , N_d^i < N_{\max}^i \\ \dfrac{V_a^i * N_{max}^i S^i}{T^i S^i - \delta^i V_a^i N_{\max}^i} & , N_d^i \geq N_{\max}^i \end{cases}$$

Utility Model

The natural object of each user, in such communication system, is to maximize its utility function. This utility function is composed of two terms, the total net data transmission (gain: $G_j(\mathbf{n})$) and the cost ($C_j(\mathbf{n})$) (the total price paid to have internet service access).

From above, the total net data transmission ($G_j(\mathbf{n})$) of a mobile j, is the product of the net data transfer rate and the number of established connections by this device from all network interfaces;

$$G_j(\mathbf{n}) = \alpha_j \sum_{i \in \blacklozenge} n_j^i V_u^i$$

Also, the total price ($C_j(\mathbf{n})$) of a mobile j, is the product of the price per byte, the net data transfer rate and the number of established connections by this device from each network interfaces;

$$C_j(\mathbf{n}) = (1-\alpha_j)\sum_{i\in\blacklozenge} P_i n_j^i V_s^i$$

where P_i is the price per byte paid by the mobile device to the interface network i (to the network access)and α_j is a parameter allows to determine the mobile user priority.

Thus, we can consider a more complete utility function, including this total cost.Then the final utility function is given as follows:

$$U_j(\mathbf{n}) = G_j(\mathbf{n}) - C_j(\mathbf{n})$$

$$= \sum_{i\in} \left(\alpha_j n_j^i V_u^i - (1-\alpha_j) P_i n_j^i V_s^i\right)$$

$$= \begin{cases} \sum_{i\in\blacklozenge} \left(\alpha_j - (1-\alpha_j)P_i\right) n_j^i V_a^i & , N_d^i < N_{\max}^i \\ \sum_{i\in\blacklozenge} n_j^i \left(\dfrac{\alpha_j V_a^i N_{max}^i}{T^i} - \dfrac{(1-\alpha_j) V_a^i N_{max}^i S^i}{(T^i S^i - \delta^i V_a^i N_{\max}^i)}\right) & , N_d^i \geq N_{\max}^i \end{cases}$$

Game Analysis

In the rest of this work, we will analyze a non cooperative game. The players are the mobile devices that must define their strategies (number of concurrent connections $\mathbf{n}_j$) for each network interface i. The concept of non-cooperation (competition) is considered by assuming that all mobile devices just looking to maximize their utility functions.

Definition of Nash equilibrium of the game of number of concurrent connections: We consider a game of strategic form of J-players

$$\Gamma = \{N_1, ..., N_J, U_1, ..., U_J\},$$

where N_j is the set of number of concurrent connection strategies of player j and U_j its utility function.

Definition 1: ***Nash equilibrium*** *specifies a strategy* $\mathbf{n}_j^* \in N_j$ for each player j (with $j = 1,..,J$) in such a way that:

$$U_j(\mathbf{n}^*) = \max_{n_j^i \in N_i} U_j(\mathbf{n}_1^*, ..., \mathbf{n}_{j-1}^*, \mathbf{n}_j, \mathbf{n}_{j+1}^*, ..., \mathbf{n}_J^*),$$

Below, we analyze non-cooperative number of concurrent connections for J -devices that maximize their utilities. To do so, we demonstrate the existence and uniqueness of the game equilibrium between J -devices, after we calculate the equilibrium point. To analyze equilibrium of the game, we need to find properties of the utility function.

Learning Nash Equilibrium

In many papers, the main concern was the search for the equilibrium situations (namely Nash) (Ait Omar et al. 2019b) (Amrani et al. 2018b) (Ait Omar et al. 2019a) (Outanoute et al. 2019) (Garmani et al. 2019a) (Omar et al. 2019) (Garmani et al. 2018a) (Garmani et al. 2018b) (Garmani et al. 2019b) (Garmani et al. 2018c) (Amrani et al. 2018a) (Ait Omar et al. 2017) (Omar et al. 2018). Lately, researchers are questioning the need for convergence of a learning algorithm to a Nash equilibrium, there are several reasons for this. First, there may the multiple equilibria in a game, and it cannot any method for coordinating choice of agent.

Learning algorithms resemble the behavior of competitors in many naturally arising games, and thus results, on the convergence or non-convergence properties of such dynamics may inform our understanding of the applicability of Nash equilibria as a plausible solution concept in some settings. In the reality, when every mobile device tries to maximize its revenue, it is the most natural to accept Nash equilibrium as the optimal solution concept. In Nash equilibrium, each mobile device's strategy is the best response to the other devices' strategies. Thus, no mobile device can gain from unilateral deviation.

Best Response Algorithm

Initialization of number of concurrent connections vectors; For each mobile device $j \in$� at iteration t:

- $\mathbf{n}_j^{t+1} = \underset{\mathbf{n}_j \in N_j}{\operatorname{argmax}} \left(U_j(\mathbf{n}^t) \right)$

Non-Cooperative Devices

We consider a non cooperative game of strategic form of J-mobile devices

$$\Gamma_1 = \{\mathbf{n}_1, ..., \mathbf{n}_j, ..., \mathbf{n}_J ..., U_1, ..., U_j, ..., U_J\},$$

where $\mathbf{n}_j = [1, ..., n_j^{max}]$ is the set of connection strategies of device j and U_j its utility function.

The question is, under general assumptions, when can we guarantee the existence and uniqueness of the equilibrium due through the devices? We consider that the number of concurrent connections is the only parameter of the game . Thus, we have the following general result on the uniqueness of the number of concurrent connections based on Nash equilibrium for all J mobile devices.

Theorem 1: *(**existence and uniqueness**) Consider the game of number of concurrent connections (), there is only one number of concurrent connections based Nash equilibrium* $\mathbf{n}^*$ of the mobile devices' game.

Numerical Results

We turn now to discuss how to take advantage of our analytic results. We propose to study numerically the two cases of interactions between mobile devices. Taking account of previous expressions of utilities of mobile devices. To illustrate, we consider 2 scenarios, for each scenario, mobile devices are looking to maximize their respective payoffs. In particular, we discuss the influence of the increase of the numbers of interacting mobile devices and the maximum data transfer rate V^i_{max}.

Scenario 1: *For this Scenario, we consider a case when the number of mobile devices increases (from 1 to 45) of a mobile* ($\alpha_j = 0.75$, $p = 0.01$) and all mobile devices are connected to the same network interface ($V^i_{max} = 1000000$, $V^i_a = 100000$, $N^i_{max} = 10$, $S^i = 10000000$, $\delta^i = 150$).

Figure 1. Scenario 1: Impact of the number of mobile devices on the number of concurrent connections

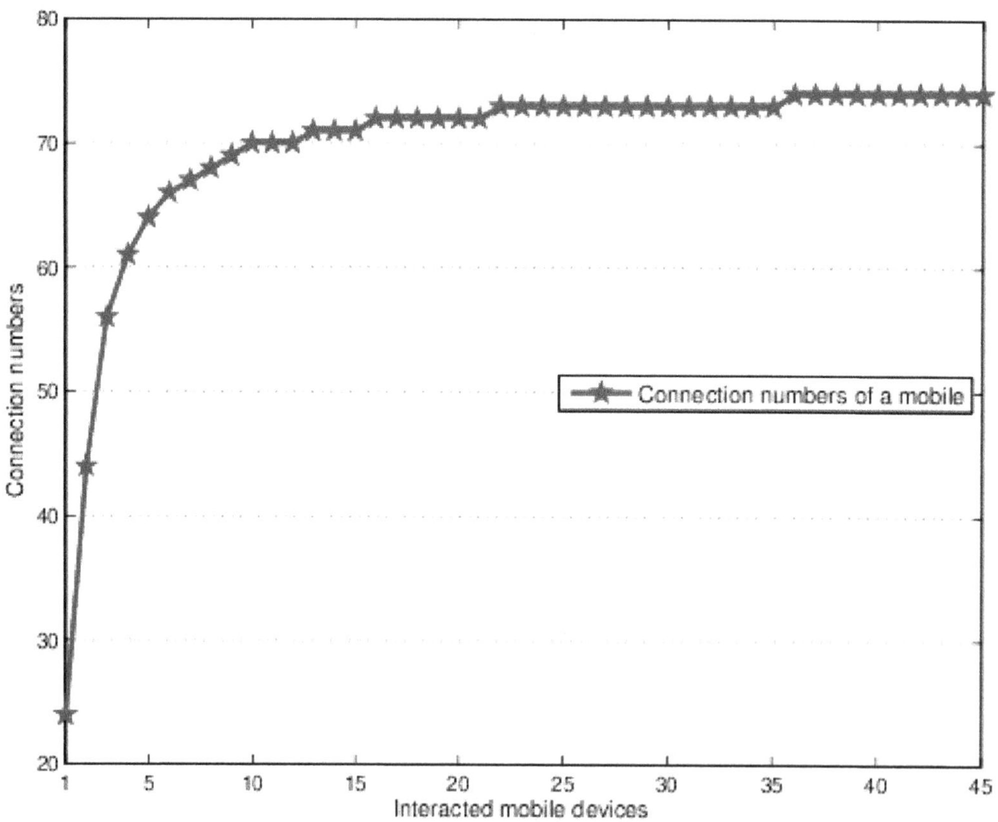

Figure 2. Scenario 1: Impact of the number of mobile devices on the computational cost

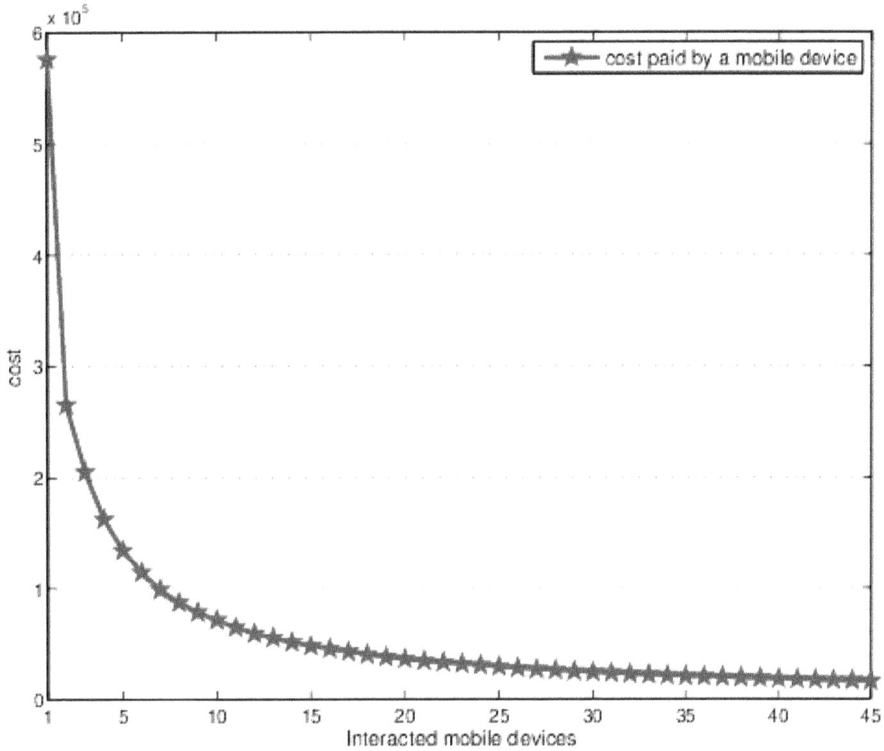

Figure 3. Scenario 1: Impact of the number of mobile devices on the net data transfer rate

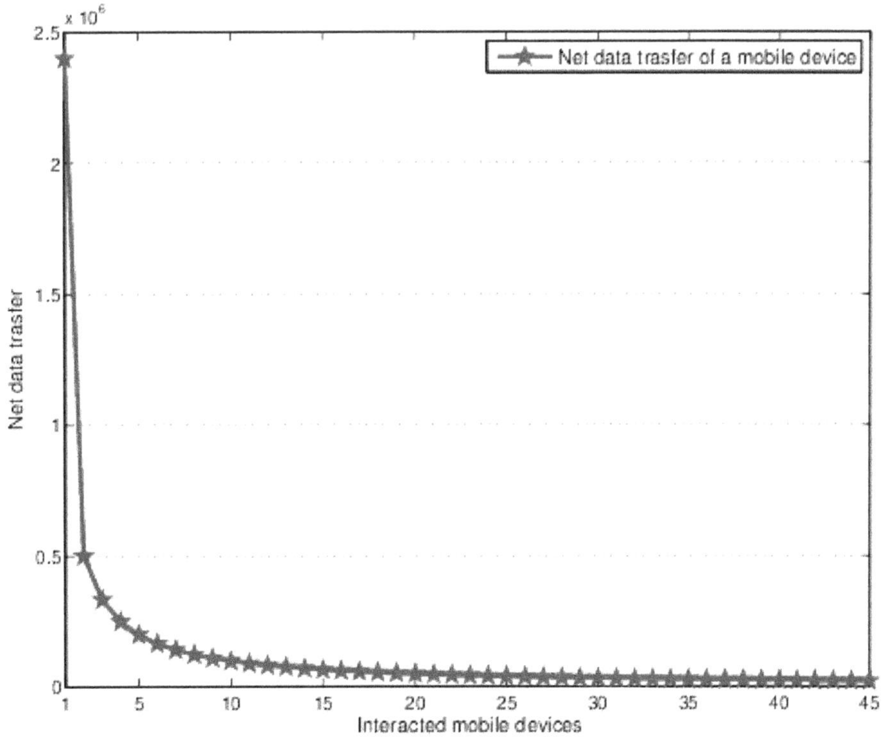

Figure 4. Scenario 1: Impact of the number of mobile devices on the device gain

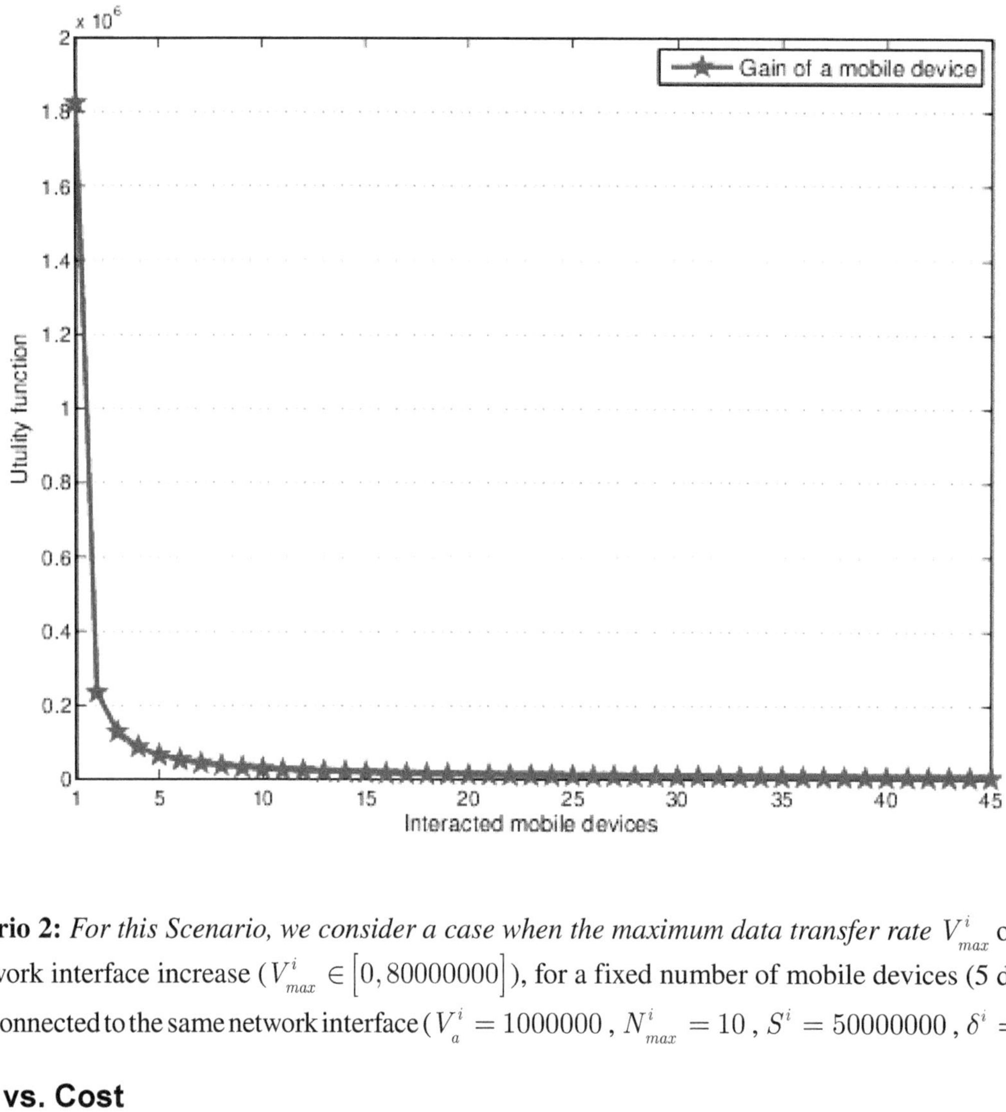

Scenario 2: *For this Scenario, we consider a case when the maximum data transfer rate* V_{max}^{i} of a network interface increase ($V_{max}^{i} \in [0, 80000000]$), for a fixed number of mobile devices (5 devices) connected to the same network interface ($V_{a}^{i} = 1000000$, $N_{max}^{i} = 10$, $S^{i} = 50000000$, $\delta^{i} = 120$).

Gain vs. Cost

To increase their gains, mobile devices choose optimal strategies(Nash equilibrium). The scenarios 1 and 2 show, respectively, the variation of the strategies taken, the cost, net data transfer rate and of course the gain obtained, by varying numbers of mobile devices and the maximum data transfer rate V_{max}^{i}.

For scenario 1, we note that the Nash equilibrium strategies behaves differently. For a mobile device, the number of concurrent connections increases when the number of mobile devices in interaction increases also, because the mobile devices tride to maximaise ther payoff. This solution results allow to use the maximum of the capacity of the network interface (Figure 1). despite this increase of the strategies (Figure 1), this does not allow mobile devices to minimize costs (Figure 1), which explains the low payoff (Figure 1).

Figure 5. Scenario 2: Impact of the maximum data transfer rate on the number of concurrent connections

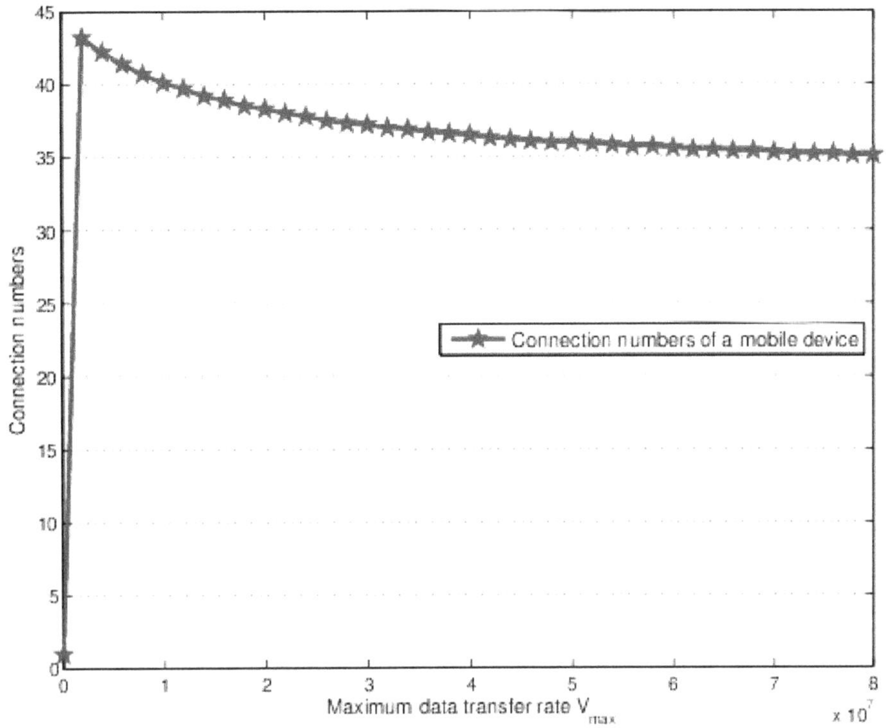

Figure 6. Scenario 2: Impact of the maximum data transfer rate on the computational cost

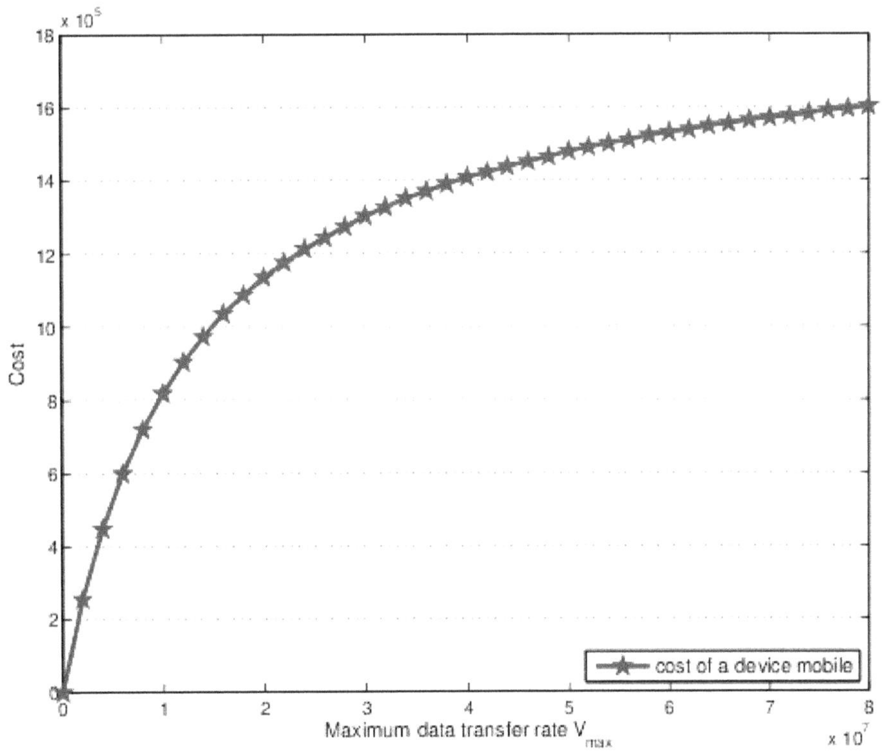

Figure 7. Scenario 2: Impact of the maximum data transfer rate on the net data transfer rate

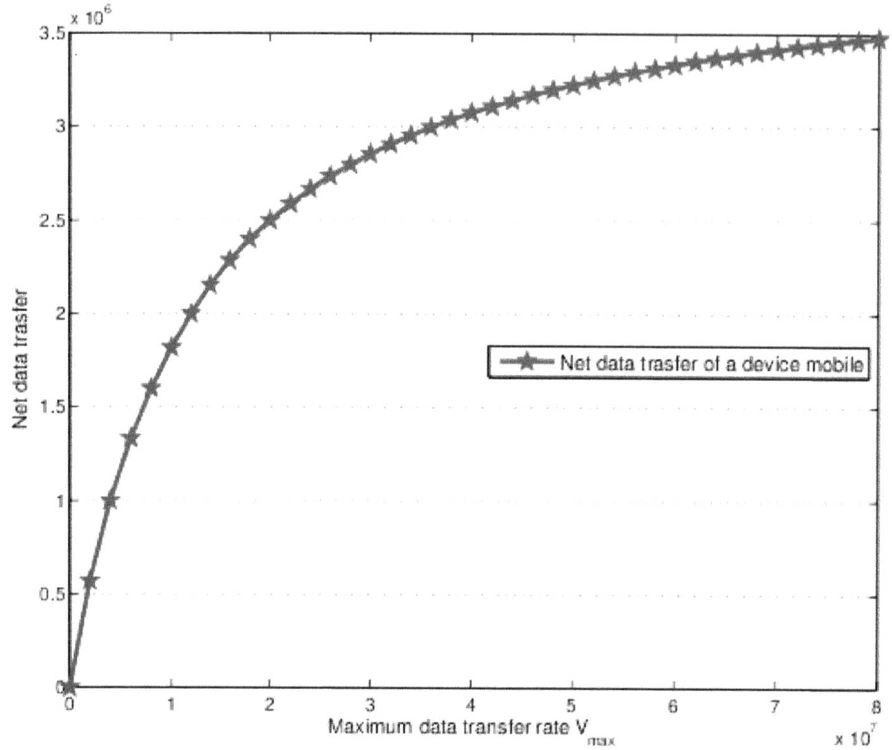

Figure 8. Scenario 2: Impact of the maximum data transfer rate on the devices gain

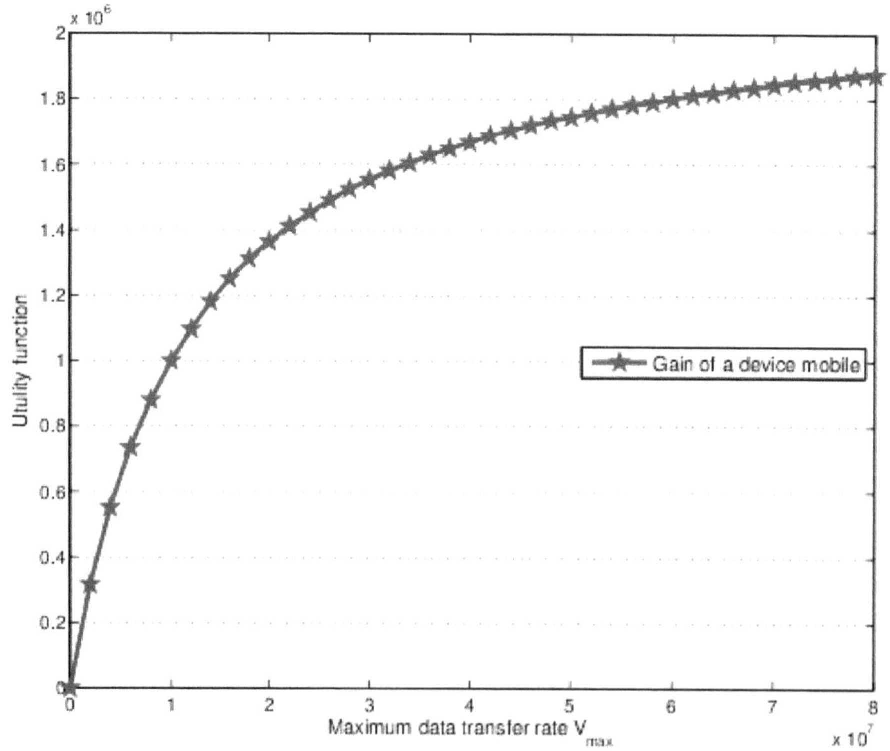

For scenario 2, it's clear that Nash equilibrium strategies become less important, for a mobile device, when the number of mobile devices in interaction increases, Nash equilibrium strategies decrease (Figure 2), but this result allows the mobile devices to maximize the net data transfer rate (Figure 2) and maximize the cost (Figure 2), which explains the high payoff of mobile devices (Figure 2).

CONCLUSION

In this paper, presented in this work a game for bandwidth aggregation for evaluating user's satisfaction level in terms of resource allocation and cost-efficient manners.

An interacting concurrent transmission game for mobile in multihoming devices. Each mobile devices tries to maximize its payoff by adjusting its strategy (i.e. the number of concurrent established connections). In addition, based on the Rosen supermodularity condition, we proved the existence and uniqueness of a Nash equilibrium for the game. We have shown numerically that is beneficial for mobiles, because it involves higher gains with the best price depending on the user priority.

For the perspectives, we can differentiate between the heterogeneous network interfaces, its means that we will specificity the cost of maintenance of connection according to the type of network interface. Additionally, we will provide an implementation on Android-based devices.

REFERENCES

Ait Omar, D., El Amrani, M., Baslam, M., & Fakir, M. (2019a). A Game-Theoretic Approach for the Internet Content Distribution Chain. In A. Podelski & F. Taïani (Eds.), *Networked Systems* (pp. 270–285). Cham: Springer International Publishing; doi:10.1007/978-3-030-05529-5_18

Ait Omar, D., Garmani, H., El Amrani, M., Baslam, M., & Fakir, M. (2019b). A Customer Confusion Environment in Telecommunication Networks: Analysis and Policy Impact. *International Journal of Cooperative Information Systems*, *28*(02), 1930002. doi:10.1142/S021884301930002X

Ait Omar, D., Outanoute, M., Baslam, M., Fakir, M., & Bouikhalne, B. (2017). Joint Price and QoS Competition with Bounded Rational Customers. In A. El Abbadi & B. Garbinato (Eds.), *Networked Systems* (pp. 457–471). Cham: Springer International Publishing; doi:10.1007/978-3-319-59647-1_33

Amrani, M. E., Garmani, H., Baslam, M., Minaoui, B., & Ayachi, R. E. (2018a). Cournot Game of ISP under Net Neutrality. *International Journal of Control and Automation*, *11*(7), 16. doi:10.14257/ijca.2018.11.7.07

Amrani, M. E., Omar, D. A., Baslam, M., & Minaoui, B. (2018b). A Game Theoretic Approach to Optimize Transmission Strategies for Mobile Applications. In *2018 6th International Conference on Wireless Networks and Mobile Communications (WINCOM)*. IEEE. doi:10.1109/WINCOM.2018.8629618

Chebrolu, K., Raman, B., & Rao, R. R. (2005). A network layer approach to enable TCP over multiple interfaces. *Wireless Networks*, *11*(5), 637–650. doi:10.100711276-005-3518-5

Chebrolu, K., & Rao, R. R. (2006). Bandwidth aggregation for real-time applications in heterogeneous wireless networks. *IEEE Transactions on Mobile Computing*, *5*(4), 388–403. doi:10.1109/TMC.2006.1599407

Gabay, D., & Moulin, H. (1978). *On the uniqueness and stability of Nash's equilibrium in non cooperative games*. Université Paris IX-Dauphine, Centre de recherche de mathématiques de la decision.

Garmani, H., Ait Omar, D., El Amrani, M., Baslam, M., & Jourhmane, M. (2019a). Analysis of a dynamics duopoly game with two content providers. *Chaos, Solitons, and Fractals*, *109466*. doi:10.1016/j.chaos.2019.109466

Garmani, H., Baslam, M., & Jourhmane, M. (2018a). Analysis of Competition between CPs Fronting Advertisers. *International Journal of Advanced Science and Technology*, *117*, 53–66. doi:10.14257/ijast.2018.117.05

Garmani, H., Baslam, M., & Jourhmane, M. (2018c). Caching Games between ISP in Information Centric Network. *International Journal of Control and Automation*, *11*(4), 125–142. doi:10.14257/ijca.2018.11.4.12

Garmani, H., Omar, D. A., Amrani, M. E., Baslam, M., & Jourhmane, M. (2019b). Analysis of sponsored content in the case of competing content providers. *International Journal of Control and Automation*, *12*(10), 14. doi:10.33832/ijca.2019.12.10.04

Habak, K., Harras, K. A., & Youssef, M. (2015). Bandwidth aggregation techniques in heterogeneous multi-homed devices: A survey. *Computer Networks*, *92*, 168–188. doi:10.1016/j.comnet.2015.08.039

Liu, Y., Gong, W., & Shenoy, P. (2001). On the impact of concurrent downloads. In *Simulation Conference, 2001. Proceedings of the Winter*. IEEE.

Lu, F., Wang, Z., Liao, X., & Jin, H. (2013). Concurrent Transmission Method for Mobile Applications using Multihoming Devices. *2013 IEEE 9th International Conference on Mobile Ad-hoc and Sensor Networks,* 240–247 doi:10.1109/MSN.2013.114

Nguyen, S. C., Zhang, X., Nguyen, T. M. T., & Pujolle, G. (2011). Evaluation of throughput optimization and load sharing of multipath TCP in heterogeneous networks. In *Wireless and optical communications networks (WOCN), 2011 eighth international conference on*. IEEE. doi:10.1109/WOCN.2011.5872966

Omar, D. A., Baslam, M., Nachaoui, M., & Fakir, M. (2018). Towards a New Approach to Rationalize the Operators' Choice in Telecommunication Networks. *International Journal of Control and Automation*, *11*(8), 39–50. doi:10.14257/ijca.2018.11.8.04

Omar, D. A., Garmani, H., El Amrani, M., Baslam, M., & Fakir, M. (2019). Analysis of Bargaining Game Policy in the Internet Content Distribution Chain. *International Journal of Mobile Computing and Multimedia Communications*, *10*(3), 47–73. doi:10.4018/IJMCMC.2019070103

Outanoute, M., Garmani, H., Baslam, M., El Ayachi, R., & Bouikhalene, B. (2019). A Non-Cooperative Game Analysis of Competition between Content Providers in the Internet Market. *International Journal of Business Data Communications and Networking*, *15*(1), 88–104. doi:10.4018/IJBDCN.2019010106

Ramaboli, A. L., Falowo, O. E., & Chan, A. H. (2012). Bandwidth aggregation in heterogeneous wireless networks: A survey of current approaches and issues. *Journal of Network and Computer Applications*, *35*(6), 1674–1690. doi:10.1016/j.jnca.2012.05.015

Rosen, J. B. (1965). Existence and uniqueness of equilibrium points for concave n-person games. *Econometrica*, *33*(3), 520–534. doi:10.2307/1911749

Zhang, H., Towsley, D., & Gong, W. (2005). Tcp connection game: A study on the selfish behavior of tcp users. In *Network Protocols, 2005. ICNP 2005. 13th IEEE International Conference on*. IEEE.

APPENDIX

Proof 1 (Existence): *Equation () represent the second derivative of the utility function (6) w.r.t. the number of concurrent connections:*

$$\frac{\partial^2}{\partial^2 n_J^i} U_J(\mathbf{n}) =$$

$$\begin{cases} \left(\alpha_j - (1-\alpha_j)P_i\right)V_a^i & , N_d^i < N_{\max}^i \\ \left[-2\frac{(\alpha_j - 1)P_i V_a^i N_{max}^i S^i \left(T^i S^i - p^i V_a^i N_{max}^i - n_j^i S^i\right)^2}{\left(T^i S^i - p^i V_a^i N_{max}^i\right)^3} \atop -2\frac{\alpha_j V_a^i N_{max}^i \left(T^i - n_j^i\right)}{\left(T^i\right)^3} \right] & , N_d^i \geq N_{\max}^i \end{cases}$$

and we have, $V_s^i < V_a^i$, then, $0 < \dfrac{N_{max}^i S^i}{T^i S^i - p^i V_a^i N_{\max}^i} < 1$, then we have,

$T^i S^i - p^i V_a^i N_{\max}^i > T^i S^i - p^i V_a^i N_{\max}^i - N_{max}^i S^i > 0$.

So,

$$\frac{\partial^2}{\partial^2 n_j^i} U_j(\mathbf{n}) < 0$$

The second derivative of the utility function is negative, then the utility function is concave, hence the existence of the Nash equilibrium follows, (Rosen 1965).

Proof 2 (Uniqueness): *The uniqueness of the equilibrium point is guaranteed if the utility function satisfies the conditions of Rosen* (Rosen 1965). *Moulin* (Gabay and Moulin 1978), *derived from the supermodularity condition, which is another alternative to satisfy the conditions of Rosen: The Nash equilibrium point is unique if:*

$$-\frac{\partial^2}{\partial^2 n_j^i} U_j(\mathbf{n}) - \sum_{l,l \neq j} \left| \frac{\partial^2}{\partial n_j^i \partial n_l^i} U_j(\mathbf{n}) \right| > 0$$

the mixed partial derivative is given by:

$$\frac{\partial^2}{\partial n_j^i \partial n_l^i} U_j(\mathbf{n}) =$$

$$\begin{cases} 0 & , N_d^i < N_{\max}^i \\ \dfrac{(1-\alpha_j)P_i V_a^i N_{max}^i (S^i)^2 n_j^i \left(T^i S^i - p^i V_a^i N_{max}^i - 2S^i n_j^i\right)}{\left(T^i S^i - p^i V_a^i N_{max}^i\right)^3} - \dfrac{\alpha_j V_a^i N_{max}^i \left(T^i - 2n_j^i\right)}{\left(T^i\right)^3} & , N_d^i \geq N_{\max}^i \end{cases}$$

In the worst case scenario, $p = 0$ and we have $\alpha_j > n_k^i$ (i.e. the power of computing of a mobile j is higher then the number of connection can an application choose)

$$\frac{\partial^2}{\partial n_j^i \partial n_l^i} U_j(\mathbf{n}) \geq 0$$

From the progression, and after replacing (10) and (13) in, we have:

$$\begin{cases} \left((1-\alpha_j)P_i - \alpha_j\right)V_a^i & , N_d^i < N_{\max}^i \\ \dfrac{(\alpha_j - 1)P_i V_a^i N_{max}^i \left(\left(T^i - \dfrac{p^i V_a^i N_{max}}{S^i} - n_j^i\right)\left(3 - J\right) + J - 1\right)}{\left(T^i - \dfrac{p^i V_a^i N_{max}}{S^i}\right)^3} + \dfrac{(\alpha_j)P_i V_a^i N_{max}^i \left(\left(T^i - n_j^i\right)\left(3-J\right) + \left(J-1\right)n_j^i\right)}{\left(T^i\right)^3} & , N_d^i \geq N_{\max}^i \end{cases}$$

and we have, $V_s^i < V_a^i$, then, $0 < \dfrac{N_{max}^i S^i}{T^i S^i - p^i V_a^i N_{\max}^i} < 1$ then we have,
$T^i S^i - p^i V_a^i N_{\max}^i > T^i S^i - p^i V_a^i N_{\max}^i - N_{max}^i S^i > 0$, and finally we have:

$$-\frac{\partial^2}{\partial^2 n_k^i} U_k(\mathbf{n}) - \sum_{l,l \neq k} \left| \frac{\partial^2}{\partial n_k^i \partial n_l^i} U_k(\mathbf{n}) \right| > 0$$

The positivity of the last expression means that the supermodularity condition of Moulin holds, then this game satisfies the conditions of Rosen, where the uniqueness of the equilibrium is verified following, (Gabay and Moulin 1978).

Chapter 9
Decision Choice Optimization With Genetic Algorithm in Communication Networks

Driss Ait Omar
https://orcid.org/0000-0002-4096-8273
Information Processing and Decision Support Laboratory Sultan Moulay Slimane University, Morocco

Hamid Garmani
https://orcid.org/0000-0002-7294-6722
Information Processing and Decision Support Laboratory Sultan Moulay Slimane University, Morocco

Mohamed El Amrani
Information Processing and Decision Support Laboratory, Sultan Moulay Slimane University, Morocco

Mohamed Baslam
Information Processing and Decision Support Laboratory Sultan Moulay Slimane University, Morocco

Mohamed Fakir
Information Processing and Decision Support Laboratory Sultan Moulay Slimane University, Morocco

ABSTRACT

Optimization is an essential tool in the field of decision support. In this chapter, the authors study an inverse problem applied in the telecommunication networks. Indeed, in the telecommunication networks, service providers have subscription offers to customers. Since competition is strong in this sector, most of these advertising offerings, totally or partially ambiguous, are prepared to attract the attention of consumers. For this reason, customers face problems in making decisions about the choice of the operators that gives them a better report price/QoS. Mathematical modeling of this decision support problem led to the resolution of an inverse problem. More precisely, the inverse problem is to find the function of the QoS real knowing the QoS theoretical or advertising. This model will help customers who seek to know the degree of sincerity of their operators, and it is an opportunity for operators who want to maintain their resources so that they gain the trust of customers.

DOI: 10.4018/978-1-7998-3355-0.ch009

INTRODUCTION

In the recent years, the privatization and liberation of services in telecommunication networks sectors lead to a diversification of services providers. This diversity leads to strong competition between them, each of which tries to attract and / or retain customers. Since do not give true information about their systems (client confusion), customers do not have complete information to make a good decision. This confusion presents an obstacle to customers to have all the information on the operator's offer. For it, consumer confusion has been studied as an important in many markets due to dynamics, such as over choice, excessive marketing communications, and similar tariffs and promotions. Hence, in the presence of confusion, the customer's choice is often uncertain (bounded rationality).

Telecommunication networks also have been considered in the economic literature as one of the sectors that causes confusion among consumers (Ait Omar, Outanoute, Baslam, Fakir, & Bouikhalne, 2017) Complex fee structures introduced by service providers, number portability, loyalty programs and complicated service variations make it difficult for consumers to understand the type of transaction and get the best report price /QoS.

The problems related to the choice of an operator is on several parameters including real quality of service, theoretical quality of service, bandwidth, price... Operators decide a price and QoS for services offered to its customers. QoS proposed remains a parameter that depends on other variables namely bandwidth, the share of this market operator. In these circumstances, an operator can never guarantee the quality of service it promises to customers. We call it then theoretical QoS, while the QoS perceived by customers is the real QoS. In the telecommunications networks, the credibility of each operator is measured by the difference between its theoretical QoS and the real QoS. A customer is interested in the recognition of with good credibility (which has a real QoS close to the theoretical QoS). In addition, as there is no real QoS to operators cannot solve this problem in a direct way. Hence, this kind of problem can be modeled within the meaning of the inverse problem. The inverse problem is generally ill-posed problem, on the contrary live with the problem said that the solution exists, is unique and depends on data. For example, if it is to reconstruct the past state of a system knowing its current state, we are dealing with an inverse problem; but the fact of predicting the future state given the current state is a direct problem. Similarly, in the case of a determination of parameters of a system knowing a part of the stage (a part of the set of parameters); we speak of parameter identification problems.

In related work, many studies have contributed to the study of the problem of consumer confusion and their bounded rationality. However, to my knowledge there is no work that is interested in finding the form of real QoS to know the confidence level of service providers in telecommunication networks. The authors of the paper (Ait Omar et al., 2017) they proposed a model that is based on the economic weighing of and the Luce probabilistic choice model. They concluded that the client is confused or not by the SPs, if its degree of irrationality increases, the earns more; this model of Luce's choice is often more complicated than the discrete choice model implied by the authors of the paper ((Coucheney, Maille, & Tuffin, 2013),(Ait Omar, Garmani, El Amrani, Baslam, & Fakir, 2019)). In addition, the authors of the book (Lorkowski & Kreinovich, 2018) debate the confusion concept, display that irrational decision-making can be explained if one takes into account that human capacities to process information are limited. Thus, they used and improved heuristic techniques to predict the quality of the decision by formulating and solving optimization problems. The use of heuristic methods for the optimization of the proposed models is due to their efficiency in the search for the global optimum.

The instead of this chapter is organized as follows: in section 2, we present modeling problematic using the inverse problem theory. We present in section 3 the genetic algorithm as optimization method used to solve the models proposed in our work. In section 4, we present the different numerical results obtained. We conclude this study with perspectives in section 5.

FORMULATION AND MODELING OF THE PROBLEM

In this section, we model the profit of a customer if he chooses to subscribe in the SP_i and we use the Luce model to mathematise the discrete choice of clients by exploiting the softmax function or the normalized exponential function (Bishop, 2006), as in the article (Morgan, 1974).

Model of Utility Customers

The benefit of a consumer is often calculated based on the strategies of its operator, which are the QoS and the price it offers. The profit u_i of a consumer registers with the operator's services SP_i is as follows:

$$u_i\left(q_i, p_i\right) = \alpha.q_i - p_i \tag{1}$$

with αq_i is a client's revenue if he chooses the SP_i and $\alpha > 0$.

The Luce Model

The Luce model is a first probabilistic choice model that incorporates boundedly rational choice of customers(Kim & Yoon, 2004)-(Qi, Zhang, Zhang, & Shi, 2006)-(Shapiro, 2001). With this model, customers can choose the operator that will maximize profits by choosing one that has the maximum probability; but forcing this choice is inadequate in this area given the existence of hidden information that has not been represented in this model. The following equation represents the probability that the customer chooses SP_i:

$$\rho_i(p,q) = \frac{e^{\frac{u_i}{\lambda}}}{\sum_{j=1}^{N} e^{\frac{u_j}{\lambda}}} \tag{2}$$

with $\lambda \in [0,1]$ is the degree of customer irrationality, N is the number of SPs and p and q are respectively the vector of price and QoS.

When λ tends to 0, means that customer behavior is rational, i.e., they have all the information and rules that allow them to maximize their profit, while customers are irrational when λ approaches 1.

Demand Model D_i

We consider a market size n (the total number of customers), the function of the application to the operator's services i, D_i is the probability that a customer selects the operator multiplied by the size, n, of the market. It is expressed by:

$$D_i(p,q) = n.\rho_i(p,q) \tag{3}$$

Theoretical Quality of Service

We consider $Delay_i^u$ the time required for data transmission to a SP_i user u. This time is expressed in telecommunications (M. Baslam, El-Azouzi, Sabir, & Echabbi, 2011) as a function of bandwidth available at I SP_i and the demand D_i:

$$Delay_i^u = \frac{1}{\Phi_i - D_i} \tag{4}$$

That means more demand is greater than the time increases, and vice versa, over the bandwidth increases the time becomes less important. This proportionality is logical since:

- More as demand increases, the number of customers connected to the SP_i becomes large and thus the time becomes more important.
- More than the bandwidth increases, the SP largely has capacity to cover all customers and therefore the time became smaller.

In the model of L. Kleinrock (Kleinrock, 1975)with queues, Quality of service QoS is the inverse of the total response time when the user wants to access the service. Let c_i the deadlines for transmission of data between the service provider and the provider services l' SP_i, the total time of the answer is accumulates between c_i and $Delay_i^u$. Thus the quality of service is expressed by the following equation (as (Mohamed Baslam, Echabbi, El-Azouzi, & Sabir, 2012)):

$$q_i = \frac{1}{Delay_i^u + c_i} \tag{5}$$

From the two equations (4) et (5), we show the existence of the relationship between the quality of service, demand $D_i(p,q) = n\rho_i(p,q)$ and bandwidth Φ_i by the following equation (M. Baslam et al., 2011):

$$q_i = \frac{\Phi_i - D_i}{1 + c_i(\Phi_i - D_i)} \tag{6}$$

or by the following equation:

$$\Phi_i = D_i + \frac{q_i}{1 - q_i c_i} \tag{7}$$

From the equations (6), we can deduce that when demand of SP_i approach covering all the bandwidth; QoS becomes less.

Problem Formulation

From equation (1), we find that the theoretical benefit of a user is:

$$u_i(q_i, p_i) = \alpha.q_i - p_i, \tag{8}$$

then the real benifit is:

$$u_i^R(q_i^R, p_i^R) = \alpha.q_i^R - p_i^R, \tag{9}$$

with q_i^R is a real QoS and p_i^R is a real price.

Assumption 1: *In telecommunications, there is no general difference between promotional price (theoretical) and the real price that the user paid when the invoice settlement. However, in our study we assume that* $p_i = p_i^R$.

Taking into consideration this assumption 1, the difference between the real and theoretical benefit becomes:

$$u_i - u_i^R = \alpha.q_i - p_i - (\alpha.q_i^R - p_i^R) = \alpha(q_i - q_i^R)$$

Resource Management Model

Used to help operators in the telecommunications field to maintain their resources so that the difference between their offers and advertising that benefit the customer actually is optimal. It is a tool to customers who want to register with the operator that meets their needs.

$$\min \quad F_i(p, q) = \frac{1}{2}.\alpha.\rho_i \cdot [\left(q_i - q_i^R\right)^2]$$

under the constraints:

$$\begin{cases} p_i < P_{max} \\ q_{min} < q_i^R < q_i \end{cases}$$

The first constraint is related to customer purchasing power. the second constraint is a formulation in terms of customer needs in real QoS. it must meet a minimum threshold and it should not exceed the theoretical QoS.

Discrete Choice Model Customers

Lets customers know the weight (sincerity) of all operators in the telecommunications market. This problem is formulated as a this multi-objective model:

$$min \quad G(p,q) = \begin{pmatrix} F_1(p,q) \\ F_2(p,q) \\ \vdots \\ F_N(p,q) \end{pmatrix}$$

under the constraints:

$$\begin{cases} \forall\ i \in \{1,...,N\}\ p_i < P_{max} \\ \forall\ i \in \{1,...,N\}\ q_{min} < q_i^R < q_i \end{cases}$$

To solve the multi-objective problem (MOP), we must transform it into a single-objective problem weighted. For this, we applied the aggregation method and the result of the transformation is:

$$minG(p,q) = \sum_{i=1}^{N} \gamma_i . F_i(p,q)$$

under the constraints:

$$\begin{cases} \forall\ i \in \{1,...,N\}\ 0 < \gamma_i < 1\ et \sum_{i=1}^{N} \gamma_i = 1 \\ \forall\ i \in \{1,...,N\}\ p_i < P_{max} \\ \forall\ i \in \{1,...,N\}\ q_{min} < q_i^R < q_i \end{cases}$$

and

$$F_i(p,q) = \frac{1}{2}.\alpha.\rho_i \cdot [\left(q_i - q_i^R\right)^2],\ i = 1,2,...,N$$

with γ is considered vector weight of the operators in the telecommunications market.

GENETIC ALGORITHM

Genetic Algorithms (GAs) developed by Holland (Holland, 1992) and his student Goldberg (Shapiro, 2001), are based on the mechanics of natural evolution and natural genetics(Michalewicz, 1996)-(Deep & Thakur, 2007). GAs differ from usual inversion algorithms because they do not require a starting value. The GAs use a survival-of-the-fittest scheme with a random organized search to find the best solution to a problem. Solve an optimization problem is find the optimum of a function from a finite number of choices, often very large. The practical applications are numerous, whether in the field of industrial production, transport or economics - wherever there is need to minimize or maximize digital functions in systems simultaneously operate a large number of parameters. Algorithm (1) represents the genetic algorithm used to optimize the models proposed in this work.

Algorithm 1: Genetic Algorithm

1. Initialize the initial population **P.**
2. Evaluate $P(t)$.
3. **While** No convergence **do**
4. a. $P(t+1)$= Selection of Parents in $P(t)$.
    b. $P(t+1)$ = Apply Crossing Operator on $P(t+1)$
    c. $P(t+1)$ = Apply on Mutation Operator $P(t+1)$
    d. $P(t)$ = Replace elders of $P(t)$ Descendants of their $P(t+1)$
    e. Evaluate $P(t)$
5. **End while**

NUMERICAL RESULTS

In this section, we present the numerical results obtained by assuming that we have SP s in this telecommunications market. We use the genetic algorithm with the parameters that will allow us to obtain the optimal solution for our proposed models.

The Real Quality of the Function Study

Study of a Limited Case

In the telecommunications market, the real quality of service is a function that depends on the bandwidth Φ_i and the demand D_i of SP_i. In reality, we know that when the bandwidth increases, the real quality of service (QoS) increases and vice versa; also when demand increases, real service quality decreases and increases when demand decreases. In this context, we observed that the real quality can be expressed as a polynomial of degree 2, the variable x_i is the ratio between Φ_i and D_i, as following:

$$q_i^R(\Phi_i, D_i) = \alpha_i^1 * (\frac{\Phi_i}{D_i})^2 + \alpha_i^2 * (\frac{\Phi_i}{D_i}) + \alpha_i^3$$

We used the genetic algorithm (with the table settings 1) to find these coefficients for different values of bandwidth Φ_i and of demand D_i. The figures 1 and 2 show the influence of respectively Φ_i and D_i on QoS (theorical and real).

Table 1. Genetic algorithm parameters to the figures of the results 1 and 2

Population size N	16
Type selection	at roulette
Type of crossover	Single-points
Probability of crossover P_c	0.7
Type of mutation	uniform
Probability of mutation	0.05
Maximum number of generations	100

From Figures 1 and 2 we note that with the change of bandwidth Φ_i and the demand D_i, the genetic algorithm was able to find the good coefficients of the polynomial to minimize the gap between what is theoretical and what is real. In the next part, we will not restrict ourselves to the case presented above, but we are expanding the study of the variation of the actual quality using the technique of discretization dominates definition of theoretical quality seeking at each point the value of the actual quality by solving the model for managing resources.

Study of a General Case by Discretization

The quality theoretical of a SP_i may vary within a range delimited by a minimum and maximum value: $q_i^t \in [q_{min}, q_{max}]$. To make a digital resolution, we will discretize the interval 3, that is to say, turn it into an approximate problem (discrete) to find the values of the actual quality q_i^R at each point of the discrete domain.

Figure 1. Variation of q_i^t *and* q_i^R *in terms of* Φ_i

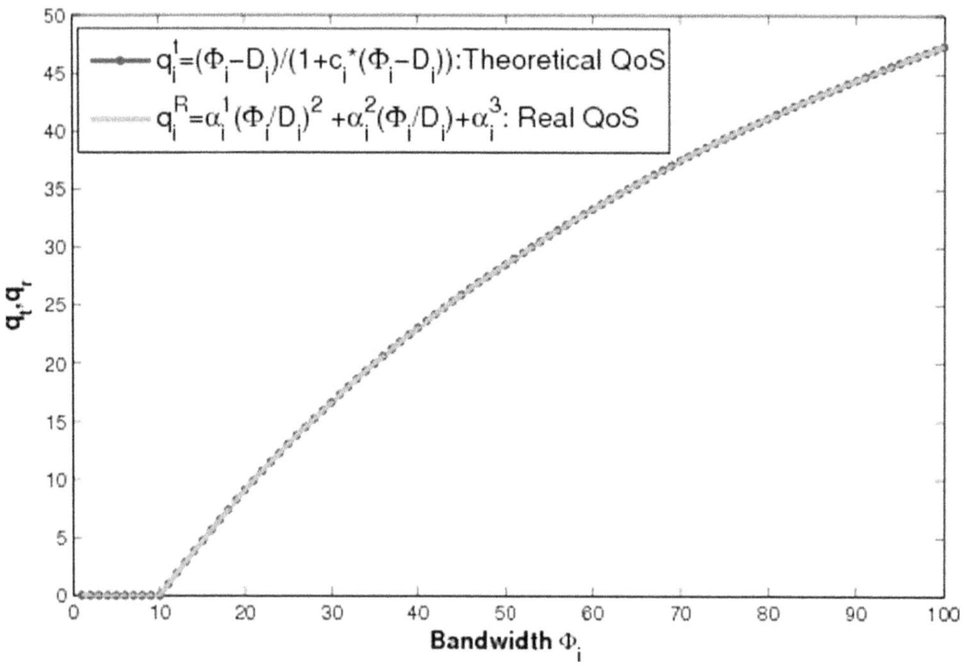

Figure 2. Variation of q_i^t *and* q_i^R *in terms of* D_i

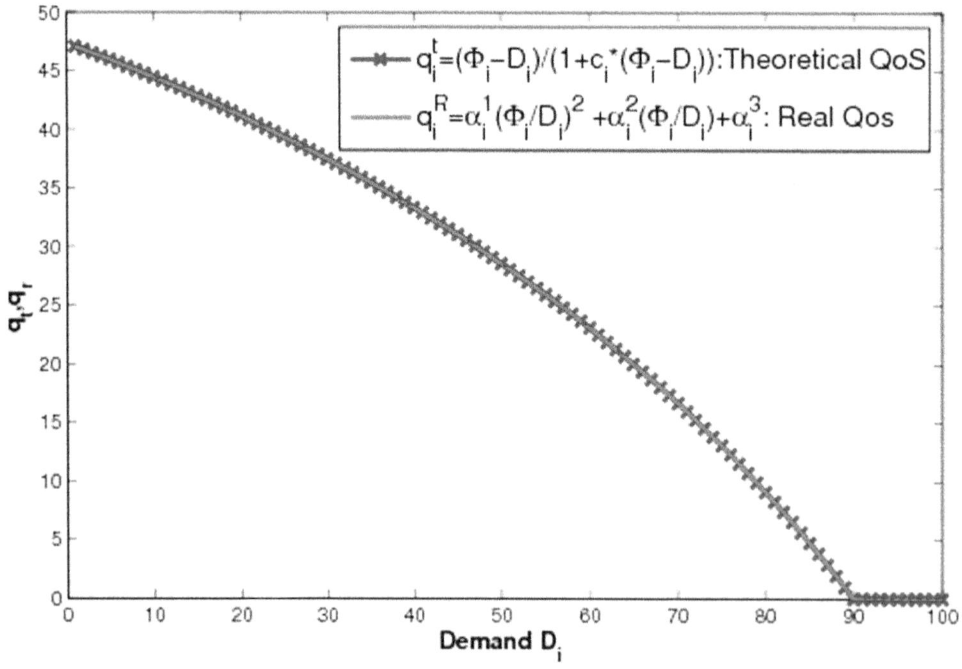

Figure 3. Discretization of an interval
or h is a positive regular pitch.

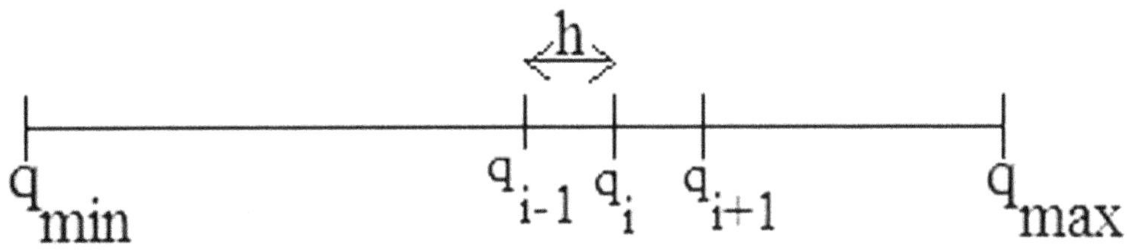

A) Study The Impact of Bandwidth on q_t and q_r

We launched the genetic algorithm; Matlab programmed with the parameters listed in Table 2; on the model of resource management with variation of bandwidth Φ_i, and we obtain the results shown in Figures 4, 5 and table 3.

Figure 4 shows the decrease of the function $Fitness$ relative to iterations (generations) the genetic algorithm. The objective function value begins 10^3, in the first generation, to reach the value 10^{-4}, in the 253ème generation.This result shows that the decrease is Igue (remarkable). From Figure 5, we note that to achieve the same minimum value, the algorithm needs to go to the 253 generation.

Table 2. Genetic Algorithm Parameters to the Figures of the results 4 and 5

Population size N	20
Type selection	at roulette
Type of crossover	Multi-point
Probability of crossover P_c	0.65
Type of mutation	non uniform
Probability of mutation	0.05
Maximal Number of generation	300

B) Impact on Demand Study q_t and q_r

We launched the genetic algorithm; with the parameters listed in Table 4;on the model of resource management with variation of the demand D_i, and we obtain the results shown in the following figures 6, 7 and table 5.

Figure 6 shows the decrease of the function $Fitness$ relative to iterations (generations) the genetic algorithm. the objective function value begins with 10^2 in the first generation, reaching the values $5,694 * 10^{-4}$, in the 167th generation. This result shows that the decrease is Igue (remarkable). From the figure 7, We note that to achieve the same minimum value, the algorithm needs to go to the 167 generation.

Figure 4. Decrease in the Objective function according generations

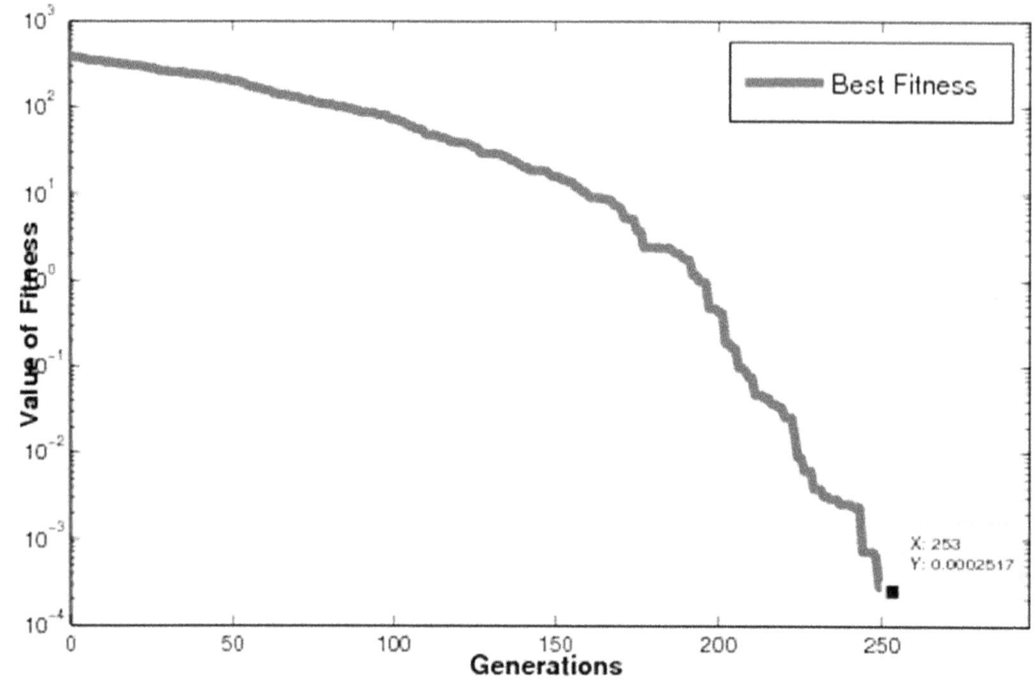

Figure 5. Changes in the real quality of service over the generations

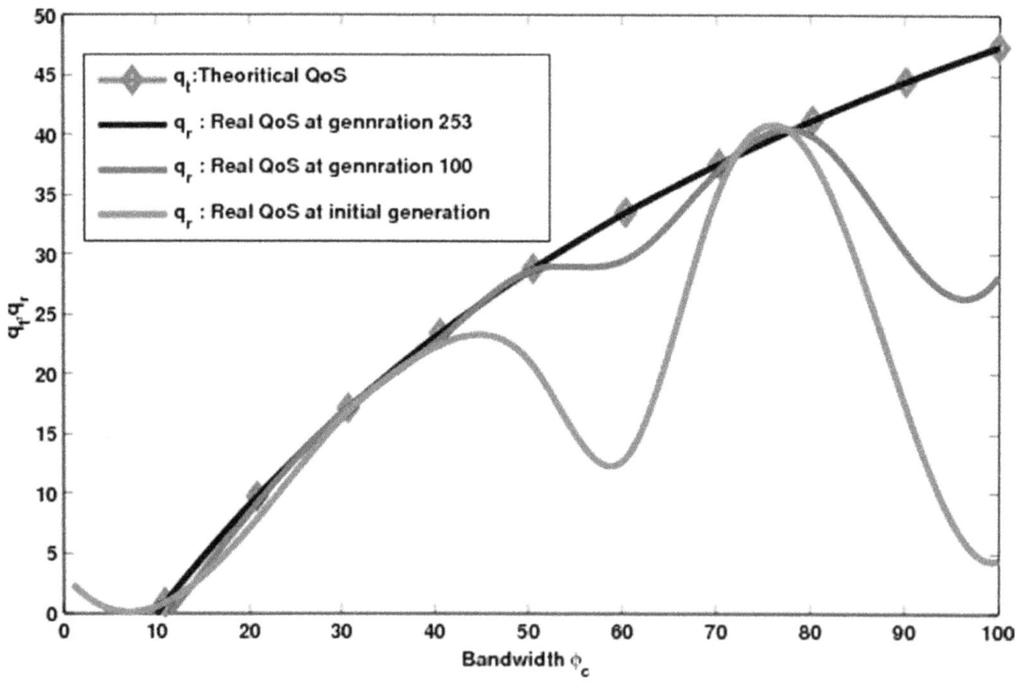

Table 3. Convergence results of the genetic algorithm (variation of Φ_i)

Number of generation	253
minimum cost	$2.517 * 10^{-4}$
step of discretization h	$\frac{1}{10}$

Table 4. Genetic algorithm parameters to the figures of the results 6 and 7

Population size N	10
Type of selection	at roulette
Type of crossover	Multi-point
Probability of crossover P_c	0.60
Type of mutation	not uniform
Probability of mutation	0.05
Maximum number of generation	200

Figure 6. Decrease in the Objective function according generations

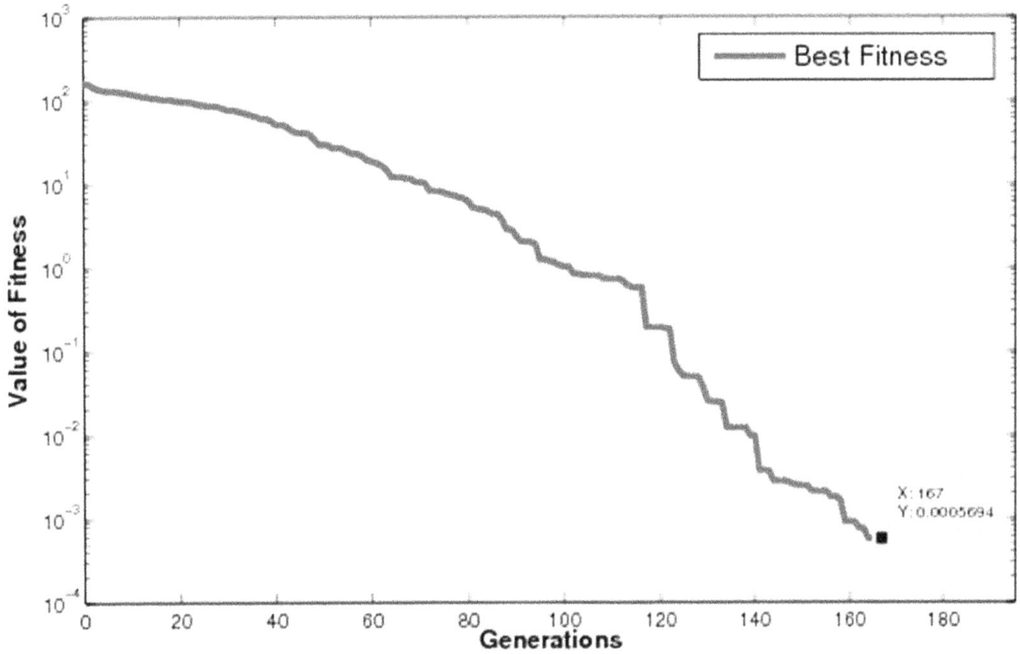

Figure 7. Changes in the real quality of service over the generations

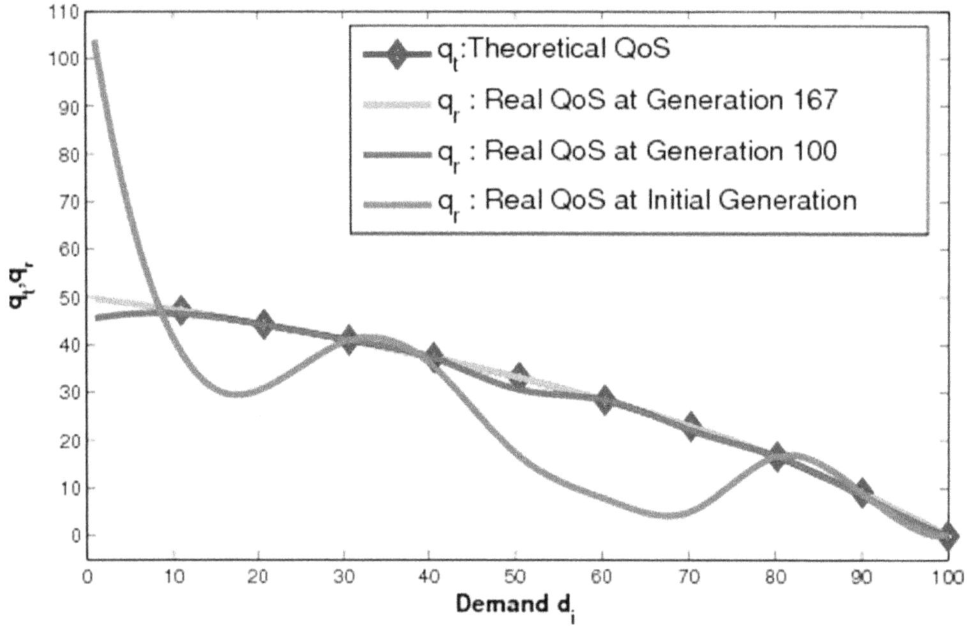

Table 5. Convergence results of the genetic algorithm (variation of D_i)

Number of generation	167
Minimum cost	$5.694 * 10^{-4}$
Step of discretization h	$\frac{1}{20}$

Model Resolution of the Weight Calculation

In this part, we consider a telecommunication network system that we have two services providers. We will use the model of discrete choice of customers to find their weight in this market ranging Φ_i and D_i an operator. The calculation of these weights is a kind of decision support for customers seeking to register with the services of the most sincere operator (who has more confidence in the sense of the difference between q^t and q^R).

Impact of Bandwidth Φ on the Weight of Operators

We vary the bandwidth of the operator 1 and observe the influence on weight and that of the adversary. Figure 8 show that the weight of the operator is an increasing function compared to bandwidth φ_1, then the weight of its adversary is a decreasing function with respect to φ_1. This result is real, since the increase in bandwidth φ_1 causes the improvement of the real QoS q_1^R and therefore the operator 1 must have a good reputation and a good weight for his adversary.

Impact of Demand D on the Weight of Operators

We will vary the request of the operator 1 and observe the influence on weight and that of the adversary. Figure 9 shows that the wight of operator 1 is decreasing function compared to demand D_1, then the wight of its adversary is a increasing function compared to D_1. This result is real, because the increase in demand D_1 causes degradation of the real QoS q_1^R and therefore the operator 1 must not have a good reputation and a good weight for his adversary.

Figure 8. Change in weight of SPs increasing Φ_1

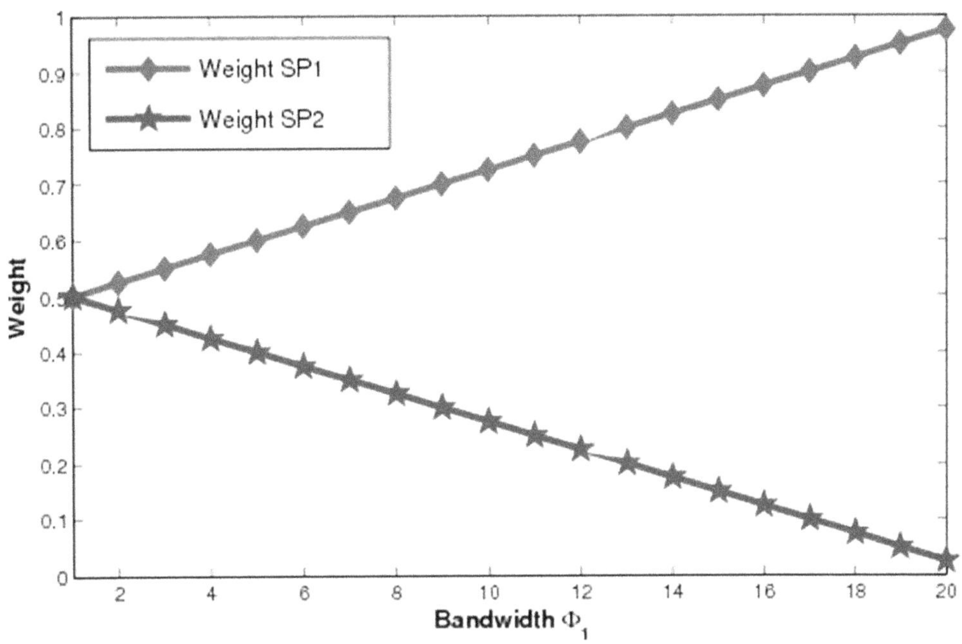

CONCLUSION

In this chapter, we have studied the problem of real quality, which is the key to rational customer decision-making in telecommunications networks. We were able to propose an inverse problem reformulated in the form of optimization models. These models will allow customers to know the service provider that best meets their requirements by knowing the demand and the bandwidth they use. Also, it is a modeling that allows service providers to adapt their bandwidth to user demand so that the difference between what is theoretical and what is perceived is optimal. However, we used the genetic algorithm for the numerical solution of optimization models. This choice is due to the efficiency it shows in the field of global optimization.

Figure 9. Change in weight SPs increasing D_1

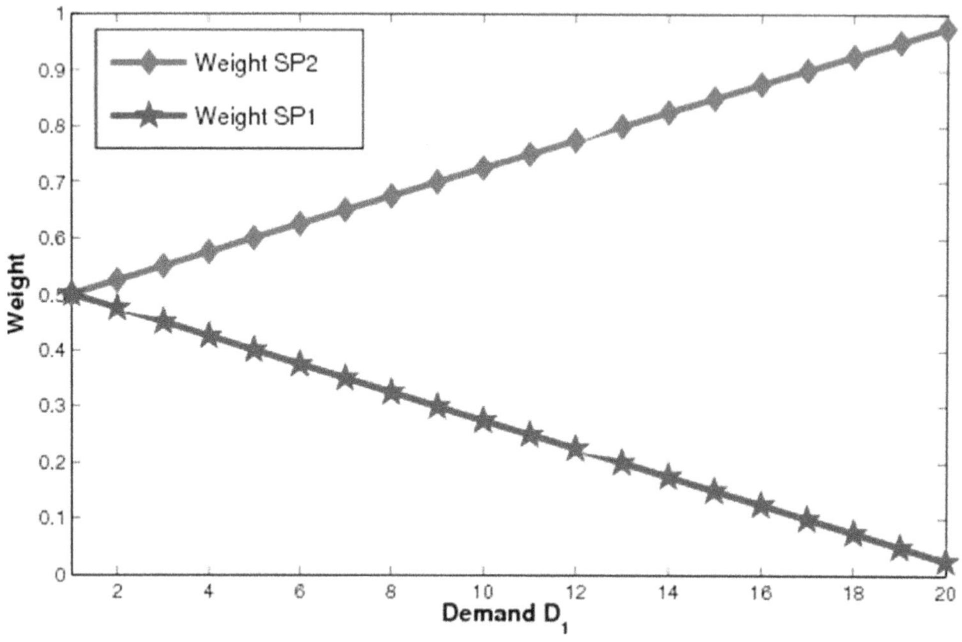

In future work, we aim to merge our model with the study of cache in information centric networks (Garmani, Baslam, & Jourhmane, 2018) to understand the impact of caching on the quality of service perceived by customers.

REFERENCES

Ait Omar, D., Garmani, H., El Amrani, M., Baslam, M., & Fakir, M. (2019). A Customer Confusion Environment in Telecommunication Networks : Analysis and Policy Impact. *International Journal of Cooperative Information Systems*, *28*(02), 1930002. doi:10.1142/S021884301930002X

Ait Omar, D., Outanoute, M., Baslam, M., Fakir, M., & Bouikhalne, B. (2017). Joint Price and QoS Competition with Bounded Rational Customers. In A. El Abbadi & B. Garbinato (Eds.), *Networked Systems* (pp. 457–471). Springer International Publishing; doi:10.1007/978-3-319-59647-1_33

Baslam, M., Echabbi, L., El-Azouzi, R., & Sabir, E. (2012). Joint Price and QoS Market Share Game with Adversarial Service Providers and Migrating Customers. In R. Jain & R. Kannan (Eds.), *Game Theory for Networks* (pp. 642–657). Springer Berlin Heidelberg.

Baslam, M., El-Azouzi, R., Sabir, E., & Echabbi, L. (2011). Market share game with adversarial Access providers : A neutral and a non-neutral network analysis. *International Conference on NETwork Games, Control and Optimization (NetGCooP 2011)*, 1-6.

Bishop, C. M. (2006). *Pattern recognition and machine learning*. New York: Springer.

Coucheney, P., Maille, P., & Tuffin, B. (2013). Impact of Competition Between ISPs on the Net Neutrality Debate. *IEEE eTransactions on Network and Service Management*, *10*(4), 425–433. doi:10.1109/TNSM.2013.090313.120326

Deep, K., & Thakur, M. (2007). A new mutation operator for real coded genetic algorithms. *Applied Mathematics and Computation*, *193*(1), 211–230. doi:10.1016/j.amc.2007.03.046

Garmani, H., Baslam, M., & Jourhmane, M. (2018). Caching Games between ISP in Information Centric Network. *International Journal of Control and Automation*, *11*(4), 125–142. doi:10.14257/ijca.2018.11.4.12

Holland, J. H. (1992). *Adaptation in Natural and Artificial Systems : An Introductory Analysis with Applications to Biology, Control and Artificial Intelligence*. Cambridge, MA: MIT Press; doi:10.7551/mitpress/1090.001.0001

Kim, H.-S., & Yoon, C.-H. (2004). Determinants of subscriber churn and customer loyalty in the Korean mobile telephony market. *Telecommunications Policy*, *28*(9), 751–765. doi:10.1016/j.telpol.2004.05.013

Kleinrock, L. (1975). *Queueing Systems*. New York, NY: Wiley-Interscience.

Lorkowski, J., & Kreinovich, V. (2018). *Bounded Rationality in Decision Making Under Uncertainty : Towards Optimal Granularity*. Consulté à l'adresse https://www.springer.com/gp/book/9783319622132

Michalewicz, Z. (1996). *Genetic Algorithms + Data Structures = Evolution Programs* (3e éd.). Consulté à l'adresse https://www.springer.com/gp/book/9783540606765

Morgan, B. J. T. (1974). On Luce's choice axiom. *Journal of Mathematical Psychology*, *11*(2), 107–123. doi:10.1016/0022-2496(74)90002-9

Qi, J., Zhang, Y., Zhang, Y., & Shi, S. (2006). TreeLogit Model for Customer Churn Prediction. *2006 IEEE Asia-Pacific Conference on Services Computing (APSCC'06)*, 70-75. doi:10.1109/APSCC.2006.111

Shapiro, J. (2001). Genetic Algorithms in Machine Learning. In G. Paliouras, V. Karkaletsis, & C. D. Spyropoulos (Eds.), *Machine Learning and Its Applications : Advanced Lectures* (pp. 146–168)., doi:10.1007/3-540-44673-7_7

Chapter 10
Online Self-Presentation Strategies Among Social Networking Sites' Users in the United Arab Emirates

Azza Abdel-Azim M. Ahmed
Zayed University, UAE & Cairo University, Egypt

ABSTRACT

This study explored the strategies of self-presentation (ingratiation, supplication, and enhancement) among United Arab Emirates users (n=230) of popular social networking sites (SNS). The size of social networks, degree of network connectivity, and perceptions of self-presentation success were examined. The results indicated a significant positive correlation between the frequency of SNS use and ingratiation and enhancement strategies. Greater diversity of online friends among the respondents was positively associated with the perception of online self-presentation success. Males and females differed in the size of the online social network they interacted with, diversity of online friends, and preferred self-presentation strategies. However, no significant gender differences were found in the levels of network connectivity and perceptions of self-presentation success.

INTRODUCTION

Internet-based social networking platforms such as Facebook, YouTube, Instagram, and Twitter are often considered innovative. In contrast with traditional media such as radio, film, and television, they allow for information sharing and social interaction (Pempek, Yermolayeva & Calvert, 2009). Social networking sites (SNS) offer users a degree of connection with others, relational satisfaction, and a way to learn about the surrounding culture (Croucher, 2011). Unlike traditional media, they support many-to-many communication modes, where information presented by each participant reaches many recipients at a time. Rather than viewing mass-produced content, with social networking sites, users become the creators of their own content. They also become the "stars" of their own productions (Pempek et al.,

DOI: 10.4018/978-1-7998-3355-0.ch010

2009). Moreover, social media offer the potential to promote those opportunities, communicate with youth, and eventually facilitate positive youth development (Lee & Horsley, 2017).

New opportunities for self-presentation and impression management offered by SNS allow users to create custom pages and report personal information strategically. Users can interact in a variety of communication modes using plain text, posting status updates, writing comments on friends' pages, and sharing images (Rui & Stefanone, 2013). As noted by Boyd (2007), SNS are based around a "profile," a form of individual (or, less frequently group) home page, which offers a description of each member. In addition to text, images, and video created by a member, a social network site profile contains comments from other members, and a public list of people that one identifies as "friends" within the network. Member profiles are usually identified by participants' real names and often include photographs; thus, the network of connections is displayed as an integral piece of one's online self-presentation (Donath & Boyd, 2004).

SNS in the Gulf Cooperation Countries

In the past decade, social networking sites have established their place as an integral and interdependent actor in society in the Gulf area, specifically in the United Arab Emirates (UAE). The most popular types of SNS in the UAE include Facebook, video sharing sites like YouTube, and microblogging platforms such as Twitter (Al Jenaibi, 2011). In 2019, WhatsApp comes on the top of social media used by 96% of GCC youth daily, then Facebook 81% and Instagram 79%; while nine in ten young Arabs use at least one of the major social media channels daily (Arab Youth Survey, 2019).

In the Arab region, over 125 million residents are using the Internet, with a near 30% annual growth rate and more than 53 million active SNS users. In 2012, countries like the UAE, Bahrain, Qatar, and Kuwait achieved Internet connectivity rates above 50%, while on average regional rates stood around 28%. Although countries like Morocco, Sudan, and Yemen had some of the lowest Internet penetration rates in the region, they also had the region's highest growth rates (Dubai School of Government, 2013). In a recent Arab Youth Survey report (2019), it was reported that 99% of UAE residents are active users of internet and social media, with a 1.5 growth compared to 2018; then Kuwait 92%, Bahrain 81%, Saudi Arabi (67%) and Oman (50%) (p. 75).

A study of the Arab social networks (Ahmed, 2010) found that Facebook was the most popular online social network (65.2%) among a sample of 325 respondents from Egypt and UAE. Similar results were presented by the 2013 Arab World Online report, which demonstrated that Facebook continues to be the most popular social network among Arabs, followed by Google+, and Twitter (Dubai School of Government, 2013). The report showed that 54% of respondents indicated using Facebook more than once a day, while 30% used Google+ at the same frequency. Only 14% of the sample used Twitter more than once a day.

Donath & Boyd (2004) noted that within SNS, members can find information about one another before a connection is made by looking at profile pages. Thus, common ground can be established, and new connections can be formed. Al Jenaibi (2011) concluded that social media have a strong presence in the lives of residents of the United Arab Emirates, using a sample of 556 participants from the UAE. The study found that social media are a vital source of news among respondents, and recognized their day-to-day reliance on social media in acquiring new friends and receiving news updates. In a recent survey, 71% among 3,373 respondents from 22 Arab countries agreed that online communication often replaced traditional communication (Dubai School of Government, 2013). However, to date there has been little attention from Arab media scholars given to online self-presentation strategies among SNS users in the Arab region. Most empirical research on self-presentation has been conducted in the U.S., Western Europe, and China. The current study aimed to examine self-presentation strategies of UAE users on the most popular social networking sites.

THEORETICAL FRAMEWORK AND LITERATURE REVIEW

Self-Presentation and Impression Management

Interpersonal self-presentation has been studied extensively by psychologists and sociologists. This concept was developed by Goffman (1959) who described how people negotiate and validate their identities in face-to-face communication and introduced "frames" within which to evaluate the meaning of interpersonal encounters. Later, electronic communication has established a new range of frames of interaction. Although mediated communications are more limited in nature and less rich than those in which participants are physically present, social interaction via electronic means provides new opportunities for self-presentation. For example, there have been discussions about the possible nature of "electronic selves" (e.g., see Miller, 1995).

Self-presentation has sometimes been distinguished from impression management, which has been defined as an attempt to control the images presented to others, usually to increase the power of the individual. However, the tactics used to engage in both impression management and self-presentation are the same (Lee, Quigley, Nesler, Corbett, & Tedeschi, 1999).

At a conceptual level, Dominick (1999) suggested that impression management involves two processes: impression motivation and impression construction:

Some common motives are a desire to maximize material rewards (making your boss think you are competent to get a promotion), maintaining self-esteem (presenting yourself in a favorable way so that other people will like you), and to create a desired self-identity (a new lawyer can solidify his image as a member of the legal profession by dressing and acting like a lawyer should). Impression construction concerns how a person creates the desired impression through his or her choice of various self-presentation strategies (Dominick, 1999: 647).

According to Miller (1995), Goffman saw embarrassment as an important indicator of a person's failure to present an acceptable self, and as an important motivator to project an improved self-image. Thus, people will seek to present themselves more effectively in order to minimize the embarrassment of a failed self-presentation. Likewise, they will be motivated to improve their performance in order to avoid the embarrassment they feel at its failure. Therefore, most often people will interact through positive self-presentations in which they attempt to match the self-presentations of others.

Online Self-Presentation

It refers to a process by which individuals engage in impression management and information control in everyday life (Schlenker & Pontari, 2000). In social networking sites, personal information is stated explicitly, and can be managed by the author of the information. Papacharissi's analysis (2002) showed that personal homepages on the Internet fit Goffman's (1959) concept of self because people can control what they present about themselves online for others to convey a particular image. Boyd (2007) argued that people have more control online with respect to self-presentation than they have in everyday interactions. They can carefully choose what information to put forward, thereby eliminating visceral reactions that might have seeped out in everyday communication.

Birnbaum, 2008 found that individuals were careful to shape the impression they gave off to others on Facebook, the most popular social media platform. He concluded that the aim of self-presentation was to communicate and interact with others in beneficial ways likely to engender support through Facebook. Gonzales and Hancock (2011) found that people took time when posting information about themselves, and carefully selected what aspects they would like to emphasize. Evidence of selective self-presentation was found in a variety of Internet spaces, including e-mails, discussion boards, and online dating websites. Similarly, Tice, Butler, Muraven, and Stillwell (1995) found that people alter their self-presentation to be more favorable with strangers (those who possess no base-rate information); and more modest with friends who possess base-rate information.

Self-presentation has been viewed as the presentation of self that an individual tends to perform intentionally and desires to be seen by others (Wong, 2012). Given that Facebook profiles are visible to large audience of friends, family, and acquaintances, the motivation to carefully control one's self-presentation should be high. Not only should Facebook self-presenters be highly motivated to control their images, but they should also be able to exercise this high degree of control (Toma & Carlson, 2012). Individuals make a series of conscious decisions regarding how to self-present (Vitak, 2012) based on the people with whom they are interacting at any given time. Facebook and MySpace profiles serve as a stage on which users can make public or semipublic presentations of themselves, and most often users will strive to portray themselves in a positive light. As this self-image is publicly displayed to a peer audience and subject to constant sanctioning via public feedback, it is even more important to the SNS user to be perceived as role model or as compliant to peer norms. It is therefore plausible to assume that SNS users use strategies that assist in presenting (and promoting) themselves as attractive to the audience (Loss, Lindacher, & Curbach, 2013). Lee and Horsley (2017) studied the impact of adolescents' Facebook use on the six "Cs" (competence, confidence, connection, character, compassion/caring, and contribution) of the positive youth development (PYD) framework. They concluded that the participants could use Facebook as an effective tool to organize leisure activities which in turn influenced adolescents' social competence and social connections positively over time.

DeAndrea and Walther (2011) conducted an experimental study to investigate how inconsistent online information affects interpersonal impressions. The authors concluded that subjects rated the inconsistencies of acquaintances as more intentionally misleading, more hypocritical and less trustworthy relative to the inconsistencies of friends, that is, people they knew and interacted with more closely. On the other hand, Rui and Stefanone (2013) introduced a distinction useful in the analysis of self-presentation in SNS: self-provided information (SPI) versus other-provided information (OPI), that is, information about someone that is provided by others. Rui and Stefanone noted that while previously self-presentation was governed by self-provided information, recent developments in social networking technologies allowed for OPI from SNS user contacts. This type of information may involve identifying people in shared photographs as well as comments from other network members. The advent of OPI via SNS increases the possibility of a profile owner's reduced control over his or her self-presentation. Similarly, Loss, Lindacher, and Curbach (2013) stated that SNS enable persons to "inspect, edit and revise" their self-presentation on the Internet before making it available to others. These authors argued that arena of "editable self-presentation" (p. 3) might increase the motivation to engage in tactics of impression management intentionally, and might also lead to a more intense use of proactive, assertive self-presentation.

Self-Presentation Classifications

Scholars differ in their overall classification of self-presentation strategies. Leary (1996) discussed self-presentation strategies in everyday life through various direct and subtle self-presentation tactics such as self-description, attitude statements, nonverbal behavior, social associations, conformity and compliance, aggression and risk-taking. Lee et al (1999) introduced a two-component model of self-presentation: defensive self-presentation that includes the use of excuse, justification, disclaimer, self-handicapping and apology, as well as assertive self-presentation that includes the use of ingratiation, intimidation, supplication, entitlement, enhancement, blasting and exemplification. The study concluded that these tactics have measured a general tendency to desire or to avoid the social approval of people in their daily lives. Avia et al. (1998) explained that defensive self-presentation stems from looking for social approval and avoiding social rejection, whereas assertive self-presentation emerges from actively seeking power and social status.

Jones (1990) introduced five strategies of self-presentation construction drawn from research in interpersonal communication. They were ingratiation, competence, intimidation, exemplification, and supplication. According to Jones, a person using ingratiation strategy has a goal of being liked by others. Among some common tactics of ingratiation are saying positive things about others or saying mildly negative things about yourself, as well as statements of modesty, familiarity, and humor. With a supplication strategy, the goal is to appear helpless so that others would come to your aid. Characteristics of this approach include entreaties for help and self-deprecation.

Following Jones' (1990) classification, Chua and Choib (2010) focused on three major strategies for self-presentation on social networking sites: competence, ingratiation, and supplication. Wong (2012) similarly examined the elements of self-presentation by using three types of assessments—ingratiation assessment, supplication assessment, and enhancement assessment—in an analysis of self-presentation and social support on Facebook among university students. The study reported that students were consistent in shaping their behaviors and presenting themselves according to their desired images. These results showed that the three types of self-presentation (ingratiation, supplication, and enhancement) were related to social support sought or received by students on Facebook. Likewise, a study by Kim and Lee (2011) found that Facebook users had the tendency to provide support to others when they saw others were in need of help. Mehdizadeh (2010) also stated that Facebook users tend to take actions to promote themselves on Facebook in order to receive positive feedback from the public, an enhancement strategy of online self-presentation.

Little is known about the social media users' strategies in presenting themselves via social media and the implications of this presentations. Therefore, the current study examined self-presentation strategies used by a sample of Emiratis in social media networks, focusing on three strategies ingratiation, supplication and enhancement. The level of social networking connectivity and the size of the network were also analyzed. Differences between genders were investigated in terms of the preferred SNS, social networking connectivity, and diversity of users with whom they interact.

RESEARCH HYPOTHESES

Wong (2012) found that frequency of Facebook usage was statistically significant and positively correlated with the three types of self-presentation on Facebook. In addition, among the three types of self-presentation, ingratiation had the strongest positive correlation with frequency of use of an application on Facebook among university students. Therefore, it can be hypothesized that:

H1: There will be a correlation between the level of network connectivity and self-presentation strategies (ingratiation, supplication, and enhancement).

Metzler & Scheithauer (2017) found that adolescents' positive self-presentation and number of friends were related to a higher frequency of receiving positive feedback, which in turn was negatively associated with self-esteem. Rui and Stefanone (2013) indicated that size is one of the most important characteristics of a social networking site's audience that affects online self-presentation. As online social networks increase in size, so does the need to maintain online relationships. Smock (2010) concluded that one possible outcome associated with large audience is more interactions between profile owners and their network. Therefore, it can be predicted that:

H2: There will be a correlation between network size and self-presentation strategies (ingratiation, supplication, and enhancement).

Rui and Stefanone (2013) noted that relationship maintenance is one of the main motives of using SNS. For example, one may self-present in a significantly different way when in a business meeting versus when on a date. SNS, for example, can place employers and romantic partners on the same communication plane, making it more difficult for users to segment audiences and present varied versions of the self (Vitak, 2012). Also, Tice et al. (1995) concluded that self-presentation differs when interacting with friends versus with strangers. Having multiple types of audiences, that is, the diversity of social categories of online friends, can affect strategies and practices used in self–presentation. Therefore, it can be predicted that:

H3: There will be a correlation between diversity (relationships and national origin) and self-presentation strategies (ingratiation, supplication, and enhancement).

Gibbs, Ellison, and Heino, (2006) examined the relationship between the strategic success of online dating and self-presentation. They found that those with more positive online self-disclosure felt they had greater self-presentation success. Also, it was found that the ability to learn about people and the number of people met in online dating led to greater self-presentation success. Therefore, it can be hypothesized that:

H4: There will be a correlation between perception of self-presentation success and self-presentation strategies (ingratiation, supplication, and enhancement).

The literature on self-presentation has emphasized differences in self-presentation strategies between males and females. Hilsen and Helvik (2014) investigated two generations of social network users in their online self-presentation: Those who were introduced to the Internet and social media as adults (40 years old) and those who have grown up with the technology (under 25 years old, called "the net-generation"). They found a difference in how these two groups related to social media. However, it was shown that both groups developed a "new" or revised sense of self and sociability through social media. Boyd (2007) also concluded that gender influences participation on social network sites. Younger boys are more likely to participate in SNS than younger girls (46% vs. 44%), but older girls are far more likely to participate than older boys (70% vs. 57%). Also, older girls are far more likely to use these sites to communicate with friends they see in person than with younger people or boys of their age. Therefore, the last hypothesis states that:

H5: There will be differences between males and females in their online self-presentation strategies.

METHODOLOGY

Sampling and Data Collection

The study sample consisted of 230 Emiratis from two Emirates, Dubai, the largest and most populous city in the United Arab Emirates, and Abu Dhabi, UAE capital. Due to the conservative nature of the population, snowball sampling was used to collect the research data. Five students from the Mass Communication Program at Abu Dhabi University helped in collecting data from the two Emirates according to their residency. The sample composed of 48.3% males and 51.7% females. The age ranged from 16 to 35 years old. The mean age was 25.5 years. The education levels of the respondents were: university level 67.4%, high school level 17.8%, post graduate level 10.9%, and preparatory or less 3.9%. A 14-question questionnaire was used to collect the data. The questions were written in Arabic, the mother-tongue of the respondents. Item reliability was tested using Cronbach's alpha and was above 0.8 for all variables. The age and the gender of respondents were recorded for sampling demographics.

Measurements

Types of Social Networks and Frequency of their Use:

Respondents were provided a list of major social networking sites (Facebook, Instagram, Kik, WhatsApp, BBM, and Twitter) and were asked to indicate how frequently (always, sometimes, rarely, and never) they used each of them.

Online Network Size

To measure online network size, respondents were asked how many friends they had on social networking sites.

Level of Social Network Connectivity

Three questions were used to measure the level of social network connectivity, adapted from Lee (2009). The respondents were asked how long they had been using the social networking sites. The score ranged from 1-3. The highest score (3) was for "6 years and above" and the lowest was (1) for "1 year to less than 3 years" category. A second question asked about how often they use social media in an average week. The highest score (3) was for "almost every day" and the lowest (1) was for "once or twice a week." A third question asked about the number of hours the respondents use social networking sites daily. The highest score was for the "4 hours or more a day" category and the lowest was for "less than 1 hours a day," $\alpha = 0.823$.

Diversity Categories

Diversity was measured in terms of relationships and nationality. Two questions were asked to measure this variable. The first question asked how frequently the participants communicated with the following social categories (family, colleagues, friends, relatives, work partners, and strangers). The second question asked about the national origin of their social network friends and followers (Emiratis, non-Emirati Arabs, Americans and European, and Asians), $\alpha = 0.851$.

Perception of Online Self-Presentation Success

This variable refers to the extent to which the respondents consider themselves successful in achieving their objectives for using SNS. A 5-point Likert scale ranging from "strongly agree" (5) to "strongly disagree" (1) was used to measure this variable. It consisted of four items adopted from Walther, Slovacek, and Tidwell (2001): "Online social networks sites allow me to present myself in a favorable way"; " I think I have made a good impression on others through online social networks"; "I feel I can communicate with my friends through social networks effectively"; and "I feel I am able to achieve my online communication goals effectively". The overall scores ranged from 4-20 points: very successful 16-20 (68.7%), successful 1015 (29.1%), and not successful 4-9 (2.2%), $\alpha = 0.837$.

Self-Presentation Strategies via SNS

Adopted from Chua and Choib (2010) and Wong (2012), this variable included the following items measured on a 5-point Likert scale ranged from "strongly agree" (5) to "strongly disagree" (1), $\alpha = 0.868$.

(1) Ingratiation: "I regularly choose and upload photos that make me look attractive"; " I always express the same attitudes as others to gain acceptance"; " I regularly present myself as helpful to others"; " I always post interesting news, articles, or photos to attract my friends to read them"; and " I always comment on friends' posts to express caring". (2) Supplication: " I always seek help from my friends in online social networking sites most of the time"; "I tend to appear weak or helpless to get care or concern from others on SNS"; and " I always show an inability to complete work or get help from others on SNS" . (3) Enhancement: " I always put up posts to show that I am knowledgeable"; "I always put up posts with the intent to show intelligence"; " I used to tell people when I complete tasks which others find difficult"; and " I emphasize to others the importance of a task when I am successful".

Statistical Techniques

The SPSS statistical program was used for analyzing the data. Frequencies, Cronbach's alpha, t-tests, and correlation analysis were used to test the research hypotheses and to answer the research questions.

RESULTS

Usages of Social Networking Sites: The results indicated that the highest percentage of respondents (81.3%) use WhatsApp. This was followed by Instagram (72.2%), Twitter (50.9%), Facebook (35.7%), Kik (24.3%), and MySpace (11.7%). This finding is supported by a recent survey (Arab Youth Survey, 2019) on social media usages in Gulf countries. It showed that WhatsApp was on the top among the most frequent social media used (96%), followed by Facebook (81%) and Instagram (79%).

In terms of network size, 46.5% of the sample had more than 100 friends on SNS while 53.5% had less than 50 to 100 friends. For level of network connectivity, the results showed that 33.9% of the sample was highly connected to SNS, 60.4% was moderately connected, and 5.7% were weakly connected. The Digital 2019 report supported this result as it showed that people in UAE tend to spend almost three hours daily on social media sites.

Hypotheses Test

The first research hypothesis stated that there would be a correlation between level of social network connectivity and self-presentation strategies (ingratiation, supplication, and enhancement). The self-presentation strategies via SNS varied among the respondents:

1. **Ingratiation:** The findings showed that 8.3% of the respondents were low, 52.2% were moderate, and 39.6% were high in ingratiation;
2. **Supplication**: 53% of the respondents were low, 30.4% were moderate, and 16.5% were high in supplication;
3. **Enhancement:** 24.3% of the respondents were low, 49.1% were moderate, and 26.5% of the respondents were high in enhancement.

The Pearson correlation coefficient indicates that there was a significant positive correlation between the level of network connectivity and both ingratiation and enhancement (see Table 1); while the relationship between network connectivity and supplication was non-significant. This means that the higher the level of network connectivity, the higher the ingratiation and enhancement among the respondents. Enhancement ($r = .180$) was stronger than ingratiation ($r = 0.159$) in its positive correlation with the level of social network connections among Emiratis. Therefore, the hypothesis was supported for ingratiation and enhancement, but not for the supplication strategy.

Table 1. Correlation between level of network connectivity and self-presentation strategies (n=230)

Variables	Self-presentation Strategies via Online SN		
	Ingratiation	**Supplication**	**Enhancement**
Level of Social Networking Connection	.159 (*)	0.111(NS)	.180(**)

Note: *P≤ 0.01; **P≤ 0.000 (2-tailed)

The second hypothesis stated that there would be a correlation between online social network size and self-presentation strategies.

The results indicated that there was a positive and significant correlation between the size of social network and all three self-presentation strategies: ingratiation, supplication, and enhancement (see Table 2). Therefore, the bigger the size of online SNS, the more respondents used various self-presentation strategies. The hypothesis was retained.

Table 2. Correlation between online network size and self-presentation strategies (n=230)

Variables	Self-presentation Strategies via Online SN		
	Ingratiation	**Supplication**	**Enhancement**
Online Social Networking Size	.262(**)	0.129(***)	.167(*)

Note: * P ≤ 0.01; **P <= 0.000; *** P≤ 0.05 (2-tailed)

The third hypothesis stated that there would be a correlation between diversity of social categories and self-presentation strategies. The association between diversity of social categories and self-presentation strategies was tested using the Pearson correlation procedure (see Table 3).

Table 3. Correlation between diversity and self-presentation strategies (n=230)

Variables		Self-presentation Strategies via Online SNs		
		Ingratiation	**Supplication**	**Enhancement**
Diversity of social categories in SNSs	Relationships	.178(**)	.130(*)	.142(*)
	Nationalities	.153(*)	.073 (NS)	.060 (NS)

Note: *P ≤ 0.05; **P≤ 0.005 (2-tailed)

The Pearson correlation coefficient showed that there was a significant positive association between the diversity of the types of relationships in SNS and the three self-presentation strategies. It was strongest in ingratiation (r = .178) and weakest in supplication (r = .130). The diversity in the nationalities of SNS friends was found to have no correlation with the supplication and enhancement strategies, while the ingratiation had a weak positive correlation with the diversity in nationalities (r = .153). Therefore, the hypothesis was supported for diversity of relationships and partially supported for diversity of national origin.

Most of the respondents (77.4%) are "always" connected with their friends, followed by those who are "always" connected with family members (64.2%), relatives (52.2%), colleagues (42.6%), and partners at work (31.3%). The least percentage was for respondents who "always" connect with strangers (10.5%). The results also showed that most of the participants (81.7%) "always" connect with locals (Emiratis), 38.7% "always" connect with other Arab nationalities, and the least connect with Americans and Europeans (10% for each).

The fourth hypothesis stated that there would be a correlation between perception of self-presentation success and self-presentation strategies (ingratiation, supplication, and enhancement).

There was a significant positive correlation between respondents' perception of self-presentation success and the three strategies of online self-presentation (see Table 4); the hypothesis was accepted. The correlation was strongest with enhancement strategy (r = .398), and ingratiation (r = .384). It was weaker with the supplication strategy (r = .212). On average, the respondents considered themselves successful in creating good and positive impressions among their friends in the SNS (M = 4.32), communicating efficiently with friends online (M = 4.26), achieving the communication objectives successfully through SNS (M = 4.10), and presenting themselves to others in a favorable way (M = 4.07). The results showed that 68.7% of respondents perceived their online self-presentation as highly successful, while 29.1% found it moderate, and only 2.2% believed it was unsuccessful.

Table 4. Correlation between perceptions of self-presentation success and self-presentation strategies (n=230)

Variables	Self-presentation Strategies via Online SN		
	Ingratiation	Supplication	Enhancement
Perception of Self-presentation success	.384(*)	.212(*)	.398(*)

Note: * P < 0.01

The correlation between self-presentation success and diversity (relationships and national origin) and the level of network connection was also examined (Table 5).

The Pearson correlation coefficient showed a significant positive relationship between the perception of self-presentation success and level of network connection (r = .149). The higher the level of network connectivity, the more the respondents tend to perceive that they succeed in their online self-presentation. Also, there was a positive correlation between perceptions of self-presentation success and the diversity

Table 5. Correlation between perception of self-presentation success, diversity, and level of social network connectivity (n=230)

Variables	Diversity of Social categories		Level of network connection
	Relationships	Nationalities	
Perception of Self-presentation success	.199(**)	.138(*)	.146 (*)

Note: * P < 0.05; ** P < 0.01

of social categories of the respondents' friends in terms of relationship diversity (r = .199) and diversity of national origin of their SNS friends (r = .138). The correlation was stronger with

the presence of relationship diversity. This means that the more diverse friends the respondents have, the more they perceive their online presentation as successful.

The last hypothesis stated that there would be differences between genders in their online self-presentation strategies. A t-test was used to examine the differences in self-presentation strategies between male and female participants (see Table 6).

Table 6. Gender differences across variables (n=230)

The variables	Gender		Mean	Std. D.	t
Level of Social networking Connection	Males		2.27	0.571	-0.320 (NS)
	Females		2.29	0.558	
Online social networking size	Males		1.53	0.501	1.955 (**)
	Females		1.40	0.493	
Diversity of social categories (Family, friends, colleagues….etc.)	Males		2.37	0.538	0.717(NS)
	Females		2.32	0.520	
Diversity of nationalities (Arabs, Americans, Europeans …etc.)	Males		1.78	0.594	2.423(***)
	Females		1.61	0.524	
Perception of online self-presentation success	Males		2.65	0.550	-0.469(NS)
	Females		2.68	0.486	
Self-presentation Strategies via Online SNSs	Ingratiation	Males	2.46	0.600	3.557(*)
		Females	2.18	0.606	
	Supplication	Males	1.89	0.755	5.299(*)
		Females	1.39	0.667	
	Enhancement	Males	2.17	0.686	3.121(*)
		Females	1.88	0.715	

Note: * P≤ 0.000; **P ≤ 0.05; *** P≤ 0.01 (2-tailed)

The t-test showed a significant difference between males and females in the following variables:

- **Online Social Network Size**: Males interacted with SNS with a larger size (M = 1.53) relative to females (M = 1.40);
- **Diversity:** Males were associated with a wider diversity of nationalities (M = 1.78). Additionally, the results showed that the males were higher than females in their connections to colleagues (t = 4.172, M = 2.47, P < 0.000), and work partners (t = 3.097, M = 1.97, P < 0.000). The males were also higher than females in their connection to various Arab nationalities (t = 2.871, M = 2.29, P < 0.001);
- **Self-Presentation Strategies via SNS**: Males and females differed in their online self-presentation strategies. Males were found to be higher than females in all strategies: ingratiation (M = 2.46), supplication (M = 1.89), and enhancement (M = 2.17).

There were no significant differences between males and females in their levels of social network connectivity, diversity of relationships (family, friends, colleagues, etc.), and perception of online self-presentation success.

CONCLUSION

This study investigated strategies of self-presentation used in SNS by a sample of United Arab Emirates social media users. The results showed that Males interacted with SNS with a larger size (M = 1.53) relative to females (M = 1.40). This is supported by the recent Digital 2019 UAE report that showed that males are higher than females from all age groups in social media usages.

Ingratiation and enhancement were found to be the most common self-presentation strategies online. The study found a significant positive correlation between the level of social network connectivity and two of the three categories of self-presentation. The correlation was strongest for enhancement and ingratiation, but non-significant for supplication. The participants generally did not display weakness or a dependency to solicit favorable responses from others on SNS; however, this does not mean that they would not support those who might seek their help.

The results go in line with several previous findings. For example, ingratiation was the most common type of self-presentation strategy used online (Dominick, 1999; Trammell & Keshelashvili, 2005). An analysis of self-presentation tactics in A-list blogs by Trammell and Keshelashvili (2005) similarly indicated that ingratiation and enhancement were among the most commonly used forms of self-presentation. The same conclusion was supported in a study of self-presentation in teenage girls' weblogs (Bortree, 2005). In a recent analysis of Hong Kong high school students' behavior on Facebook, the strongest correlation was between the frequency of Facebook use and ingratiation (Wong, 2012). However, although Facebook profiles tend to present positive and flattering self-portrayals, they are not always sufficient to convey an idealized version of self (Toma & Carlson, 2012).

Additionally, the results of the current study showed that the diversity of relationships was associated with the type of strategies used in self-presentation and with perceptions of self-presentation success. The analysis revealed that the higher the level of connectivity, the more the participants tended to perceive that they succeeded in their online self-presentation. The higher the use of various self-presentation strategies, the higher the respondents rated their success. This finding supports the conclusion made by Gibbs et al. (2006) that those with a positive attitude toward online self-disclosure believed they had greater success in Internet dating. Finally, in the sample, male respondents interacted with larger networks as well as had a higher diversity in the national origin of network members they were connected to. Males tended to use all three self-presentation strategies more often than females. However, no differences between genders were found in the levels of network connectivity and perceptions of self-presentation success.

Given the limitations of a non-random sample, the findings are not generalizable to the entire population of users of SNS in the Emirates. Future analyses may determine how these variables interact among other populations of users. Future studies may also include cross-cultural examinations on online self-presentation. Researchers may employ alternate research methods (e.g., experiments, longitudinal data) to understand self-presentation strategies and associations between other variables, such as social capital and social support. Finally, throughout the study, the focus was on most popular social networking sites; future research might extend these findings by focusing on a particular site or another popular social network that was not included in this study.

REFERENCES

Ahmed, A. (2010). Online privacy concerns among social networks' users. *Cross-Cultural Communication*, *6*(4), 74–89.

Al Jenaibi, B. N. A. (2011). Use of social media in the United Arab Emirates: An initial study. *Global Media Journal*, *1*(2), 3–27. Retrieved from http://www.gmjme.com/gmj_custom_files/volume1_issue2/articles_in_english/volume1-issue2-article-3-27.pdf

Arab Youth Survey. (2019). *A Call for Reform*. A White Paper on the findings of the 11th annual ASDA'A BCW Arab Youth Survey 2019. Retrieved from: file:///C:/Users/Z10509/Downloads/2019%20Arab%20Youth%20Survey.pdf

Avia, M. D., Sánchez-Bernardos, M. L., Sanz, J., Carrillo, J., & Rojo, N. (1998). Self-presentation strategies and the five-factor model. *Journal of Research in Personality*, *32*(1), 108–114. doi:10.1006/jrpe.1997.2205

Birnbaum, M. G. (2008). Taking Goffman on a tour of Facebook: College students and the presentations of self in a mediated digital environment. *Dissertation Abstracts International*, Section A, 69.

Bortree, D. (2005). Presentation of self on the Web: An ethnographic study of teenage girls' weblogs. *Education Communication and Information*, *5*(1), 25–39. doi:10.1080/14636310500061102

Boyd, D. (2007). Why youth (heart) social network sites: The role of networked publics in teenage social life. In D. Buckingham (Ed.), *Youth, identity, and digital media* (pp. 119–142). Cambridge, MA: MIT Press.

Croucher, S. M. (2011). Social networking and cultural adaptation: A theoretical model. *Journal of International and Intercultural Communication*, *4*(4), 259–264. doi:10.1080/17513057.2011.598046

DeAndrea, D. C., & Walther, J. B. (2011). Attribution for inconsistencies between online and offline self-presentations. *Communication Research*, *36*(6), 805–825. doi:10.1177/0093650210385340

Digital. (2019). *Report: The United Arab Emirates*. Retrieved from: https://image.slidesharecdn.com/datareportal20190131gd100digital2019unitedarabemiratesv02-190131132748/95/digital-2019-united-arab-emirates-january-2019-v02-2-638.jpg?cb=1548941304

Dominick, J. R. (1999). Who do you think you are? Personal home pages and self-presentation on the World Wide Web. *Journalism & Mass Communication Quarterly*, *76*(4), 646–658. doi:10.1177/107769909907600403

Donath, J., & Boyd, D. (2004). Public displays of connection. *BT Technology Journal*, *22*(4), 71–82. doi:10.1023/B:BTTJ.0000047585.06264.cc

Dubai School of Government. (2013). *The Arab World Online report: Trends in Internet usage in the Arab region*. Retrieved from: http://www.dsg.ae/en/Publication/Pdf_En/424201311017185100000.pdf

Gibbs, J. L., Ellison, N. B., & Heino, R. D. (2006). Self-presentation in online personas: The role of anticipated future interaction, self-disclosure, and perceived success in Internet dating. *Communication Research*, *33*(2), 152–177. doi:10.1177/0093650205285368

Goffman, E. (1959). *The presentation of self in everyday life*. Garden City, NY: Doubleday.

Gonzales, A. L., & Hancock, J. T. (2011). Mirror, mirror on my Facebook wall: Effects of exposure to Facebook on self-esteem. *Cyberpsychology, Behavior, and Social Networking*, *14*(1-2), 79–83. doi:10.1089/cyber.2009.0411 PMID:21329447

John, D., & Catherine, T. (2010). Social capital and self-presentation on social networking sites: A comparative study of Chinese and American young generations. *Chinese Journal of Communication*, *3*(4), 402–420. doi:10.1080/17544750.2010.516575

Hilsen, A. I., & Helvik, T. (2014). The construction of self in social media, such as Facebook. *AI & Society*, *29*(1), 3–10. doi:10.100700146-012-0426-y

Jones, E. E. (1990). *Interpersonal perception*. New York: W.H. Freeman.

Kim, J., & Lee, J. E. R. (2011). The Facebook paths to happiness: Effects of the number of Facebook friends and self-presentation on subjective well-being. *Cyberpsychology, Behavior, and Social Networking*, *14*(6), 359–364. doi:10.1089/cyber.2010.0374 PMID:21117983

Leary, M. R. (1996). *Self-presentation: Impression management and interpersonal behavior*. Boulder, CO: Westview.

Lee, S. J. (2009). Online communication and adolescent social ties: Who benefits more from Internet use? *Journal of Computer-Mediated Communication*, *14*(3), 509–531. doi:10.1111/j.1083-6101.2009.01451.x

Lee, A. R., & Horsley, J. S. (2017). The role of social media on positive youth development: An analysis of 4-H Facebook page and 4-H'ers' positive development. *Children and Youth Services Review*, *77*, 127–138. doi:10.1016/j.childyouth.2017.04.014

Lee, S. J., Quigley, B. M., Nesler, M. S., Corbett, A. B., & Tedeschi, J. T. (1999). Development of a selfpresentation tactics scale. *Personality and Individual Differences*, *26*(4), 701–722. doi:10.1016/S0191-8869(98)00178-0

Loss, J., Lindacher, V., & Curbach, J. (2013). Do social networking sites enhance the attractiveness of risky health behavior? Impression management in adolescents' communication on Facebook and its ethical implications. *Public Health Ethics*, *7*(1), 1–12.

Mehdizadeh, S. (2010). Self-presentation 2.0: Narcissism and self-esteem on Facebook. *Cyberpsychology, Behavior, and Social Networking*, *13*(4), 357–364. doi:10.1089/cyber.2009.0257 PMID:20712493

Metzler, A., & Scheithauer, H. (2017). The Long-Term Benefits of Positive Self-Presentation via Profile Pictures, Number of Friends and the Initiation of Relationships on Facebook for Adolescents' Self-Esteem and the Initiation of Offline Relationships. *Frontiers in Psychology*, *8*, 1981. doi:10.3389/fpsyg.2017.01981 PMID:29187827

Miller, H. (1995). *The presentation of self in electronic life: Goffman on the Internet. In Embodied knowledge and virtual space conference* (Vol. 9). Retrieved from: http://www.dourish.com/classes/ics234cw04/ miller2.pdf

Papacharissi, Z. (2002). The presentation of self in virtual life: Characteristics of personal home pages. *Journalism & Mass Communication Quarterly*, *79*(3), 643–660. doi:10.1177/107769900207900307

Pempek, T. A., Yermolayeva, Y. A., & Calvert, S. L. (2009). College students' social networking experiences on Facebook. *Journal of Applied Developmental Psychology*, *30*(3), 227–238. doi:10.1016/j.appdev.2008.12.010

Rui, J., & Stefanone, M. A. (2013). Strategic self-presentation online: A cross-cultural study. *Computers in Human Behavior*, *29*(1), 110–118. doi:10.1016/j.chb.2012.07.022

Schlenker, B. R., & Pontari, B. A. (2000). The strategic control of information: Impression management and self-presentation in daily life. In A. Tesser, R. B. Felson, & J. M. Suls (Eds.), *Perspectives on self and identity* (pp. 199–232). Washington, DC: American Psychological Association; doi:10.1037/10357-008

Smock, A. (2010, June). *Self-presentation on Facebook: Managing content created by the users and others*. Paper presented at the annual conference of the International Communication Association, Singapore.

Tice, D. M., Butler, J. L., Muraven, M. B., & Stillwell, A. M. (1995). When modesty prevails: Differential favorability of self-presentation to friends and strangers. *Journal of Personality and Social Psychology*, *69*(6), 1120–1138. doi:10.1037/0022-3514.69.6.1120

Toma, C. L., & Carlson, C. L. (2012, January). *I'm so much cooler online: An examination of self-presentation in Facebook profiles*. Paper presented at the annual conference of the International Communication Association, Phoenix, AZ.

Trammell, K. D., & Keshelashvili, A. (2005). Examining the new influencers: A self-presentation study of Alist blog. *Journalism & Mass Communication Quarterly*, *82*(4), 968–982. doi:10.1177/107769900508200413

Vitak, J. (2012). The impact of context collapse and privacy on social network site disclosures. *Journal of Broadcasting & Electronic Media*, *56*(4), 451–470. doi:10.1080/08838151.2012.732140

Walther, J. B., Slovacek, C. L., & Tidwell, L. C. (2001). Is a picture worth a thousand words? Photographic images in long-term and short-term computer-mediated communication. *Communication Research*, *28*(1), 105–134. doi:10.1177/009365001028001004

Wong, W. K. W. (2012). Faces on Facebook: A study of self-presentation and social support on Facebook. *Discovery-SS Student E-Journal*, *1*, 184–214.

Chapter 11
Searching for Answers to Hybrid Approaches in Communication and Learning Environments

Michael Anthony Brown Sr.
Florida Institute of Technology, USA

ABSTRACT

This examination focuses on hybrid approaches and highlights two in particular: blended learning and the hybrid shift. Blended learning can support online and face-to-face (F2F) students equally by providing similar learning experiences. Incorporating synchronicity is recommended as a way to address the limitations of asynchronous learning resident in the proposed hybrid approach. Synchronicity is connecting two or more people in related educational events at the same time. Another approach, the "hybrid shift," features faculty design content and structure with a focus on being thoughtful, engaging, and intentional in promoting learning, effective collaboration, critical thinking, and communication skills. George Santayana said, "those who cannot remember the past are condemned to repeat it." This is relevant in that social media grew, in part, out of a need to overcome communication limitations of geographical distance and time. This examination looks closely at hybrid approaches and offers some answers for educational implications.

INTRODUCTION

Today's scholars are beginning to understand the importance of taking a hybrid approach to social communication that combines the best of face-to-face and online interactions. This approach considers internal and external forces on people and their communication environment. My research suggests a hybrid approach that combines the best of face-to-face and online communication. Researchers have found that even when digital communication is the better method, it was more effective when combined with face-to-face interactions (M. A. Brown, 2017, p. 209). All parties in a communication must strive to get the best possible message-medium fit tailored to the choice of media and type of use. This should result

DOI: 10.4018/978-1-7998-3355-0.ch011

in enhanced collaborations focused on message, media-mix, task, digital communication competence, interaction partners, optimal outcomes, and considerations of context (Bubaš, 2001).

This article examines the continued growth of computer-mediated communication as a starting point in working toward a hybrid approach to combining face-to-face and online activities. We then move to a discussion of the importance of trust in effective collaborations. Trust is built, in part, by overcoming and adjusting to barriers to communication. Finally, the article discusses aspects of human interactions leading to the proposed hybrid approach.

For instance, James Arthur Baldwin, American novelist, essayist, playwright, poet, and social critic, said, "Know from whence you came. If you know whence you came, there are absolutely no limitations to where you can go (2018)." Success in social networking relies on creating relationships and building strong ties.

The prevalence of digital communications might suggest that there are no issues with communication, or even that there is no need to focus on face-to-face interactions. I disagree. For instance, researchers in the United Kingdom analyzed the quality of face-to-face interactions and the way in which they are influenced by mobile communication technology (Przybylski & Weinstein, 2013). The results point to the difficulty that the mere presence of mobile phones causes difficulties in developing interpersonal trust and closeness, or even in starting a face-to-face communication at all (M. A. Brown, 2017, p. 3). The UK research also suggests that the more personal the interaction, the higher the level of difficulty (Przybylski & Weinstein, 2013).

Computer-Mediated Communication

CMC is a popular way people use information technology to interact. CMC is portable, and it removes the boundaries or limitations of geographical distance and time. Computer-mediated relating (CMR) is another term relevant for our learning journey because it addresses interpersonal dimensions of interactions, encompassing a broader spectrum of exchanges than CMC might, such as online friendships and romantic communications (Cooper and Sportolari 1997, Whitty and Gavin 2001).

The current discussion suggests tools that are available to communicators and it suggests limitations that exist in strictly online communications. This discussion addresses how the use of face-to-face skills can enhance CMC activities. For instance, the feedback cycle is crucial in communication importance because we know that messages are rarely fully "one-way" activities. The sender needs clarification that the receiver or receivers decoded the message in the way it was intended or that they received the value that the sender intended to provide in the communication. A mutual agreement between the communicating parties is vital to ensure effective interactions that are true to the original message and its meaning. This agreement sets the stage for the continuing, cyclic conversation that should follow the original engagement. A truly effective communication features all parties accepting responsibility to exhibit authenticity and accuracy. Human contact or interaction can be important to improving the engagement, but they are not normally available in digital communications.

CMC is a powerful way to connect with people. Studies on Gen Y and Millennials demonstrate their constantly growing numbers in terms of participation. These groups are known to prefer the use of instant messaging or other social media over stopping by an office for a talk (Tardanico 2012). That preference makes it a real challenge to build and nurture a team and corporate culture unless you can create real relationships and strong ties. Feedback that leads to shared meaning and value between participants is important.

Social media allows us to communicate on very serious topics like contracts, purchases, and love. The feedback cycle helps the sender ensure that the other party or parties interpreted the message correctly, or that they received the intended value from the message. In management issues, feedback is necessary because the leader needs to know how subordinates respond to directives and plans. There is a need to understand how work is progressing and to get a sense of how employees are invested, or not invested, in the work environment.

Common understanding is critical to measure effectiveness through a mutual agreement between the parties about the message and its meaning. When the receiver understands the message in the manner intended by the sender the parties can achieve effectiveness. Deeper meaning in the exchange is achieved when there is a developing level of trust between the parties that is supportive and complementary to the interaction.

Feedback should be accomplished through a continuous loop between the parties involved. It should start when the message is received and is subsequently decoded. After that, it should reengage the communicative process, generating a newly-encoded message as the receiver becomes sender. This cycle is vital to providing information that can enhance, clarify, or restate the previous message. In this way, feedback fosters common understanding and mutual agreement. To do this, digital senders must continue sending until there is a verification of receipt and understanding. When the interaction is face-to-face, senders just use verbal and visual cues to get feedback and, hopefully, make timely and important adjustments.

Senders and receivers should conduct diachronic communications, which are activities in which the parties share the "spotlight." The communication emphasis alternates between voices with each taking the lead at the appropriate time based on the requirements of message delivery. In this way, the conversation is conducted via an ongoing and dynamic process. Communication is a flexible and exponential process in which participants, contexts, and the future communication probabilities change and adapt (Dance 1970). Meaning construction happens between players in an environment that allows the discourse to shape the conversation in ways that give it meaning and allow for continued interaction (van Ruler 2015).

Theory is important in deciding the best approach to communicating. Social presence is a relevant theory that evaluates "the degree to which a person is perceived as a 'real person' in mediated communication (Gunawardena 1995, p. 147)." The theory focuses on intimacy in communication and psychological distance. Trust and relative anonymity are often listed as benefits of online communication. For instance, social distance has been increasing since the telegraph and telephone were invented. Social distances increase every day, as technology allows us to be anywhere and communicate online. Many rich and valuable communications take place despite social distance considerations and these internet conversations can create dynamic first impressions. In face-to-face communication, for instance, the first impression is derived when physical attractiveness, gender, or age are immediately revealed or implied. However, those who prefer online communication would argue that there are opportunities for more radical first impressions because of a lack of inhibition, possibly resulting in effective relationship development. Trust can help with creating and nurturing effective messages.

Importance of Trust

Trust is important in building relationships that are good for communication, allowing the participants to develop dependence, satisfaction, and commitment. Trusting relationships in organizations involve an ongoing decision to give most people the benefit of the doubt, and it can be extended even to people one does not know from direct experience (M. Brown, Sr., 2011). The changes are still evident today in personal and organizational forums and in knowledge-intensive organizations. The relevant research indicates that new ways of communicating "alter the patterns of interdependencies and the nature and extent of uncertainty (McEvily, Perrone, & Zaheer, 2003)." These new ways make people rely more on the decisions and actions of others, creating both dependency and vulnerability. These factors are crucial to building and maintaining trust, and senders must pay attention them.

It is also important to demonstrate that organizational science has made some important advances that promote understanding of the meaning of trust and how it relates to certain factors that characterize organizations and that are important for the current communication discovery. An increasing number of journal articles and special issues (Bachmann, Knights, & Sydow, 2001; Rousseau, Sitkin, Burt, & Camerer, 1998) and books (Gambetta, 1988; Kramer, 1996; Lane & Bachmann, 1998) are devoted to the topic of trust in and between organizations. The special issue published seven papers that represent a wide range of methodological approaches, a diverse set of theoretical disciplines, a variety of levels of analysis, and a blend of empirical models (McEvily et al., 2003). These views are valuable in understanding trust as a factor in both face-to-face and online interactions.

Trust grows between people as they share emotions in addressing real issues. Stephen Covey, American educator, author and keynote speaker, said, "Trust is the highest form of human motivation. It brings out the very best in people (Frank Messina, 2011, p. 219)." Trust can minimize or eliminate uncertainty for communicating parties because it allows each party to believe that the other will behave in a way that is beneficial. When the result of the communication is shared value and common understanding, a bond is created.

All good relationships are either built on trust or grow and strengthen because of it. Trust is built when people or groups who are vulnerable to each other participate in open communication. If they are willing to admit mistakes, acknowledge weaknesses applaud strengths, and lean on each other for help, they can build trust. Leaders and managers increase their own trustworthiness when they trust their team. This is not easy, because trust is a contract that requires taking risks. The risks involve putting one's faith or a pending decision in the hands of someone you have no reason to have confidence in.

The table below lists several factors that contribute to building trust which can be a complicated pursuit of enduring relationships. This is the expectation that a person will act in an honorable and mutually beneficial way.

Barriers to Communication

It is difficult to grow and maintain trust without paying attention to barriers to communication. Electronic communication has surpassed face-to-face methods leading to a new age of business communications and the lack of comfort with traditional interpersonal communication contributes to the rise of digital communication. Some experts attribute this move to the influence of Gen Y and Millennials who tend to favor the instant kind of messaging afforded by social media and digital applications over sitting down in front of another person and interacting.

There are many indicators of barriers to communication. Claude Elwood Shannon wrote an article that explained communication as an interactive process where sender and receiver are exchanging messages and dealing with noise in the channel (Sitkin, Sutcliffe, & Barrios-Choplin, 1992). Partners in communication need to feel acceptance, which can be achieved through an effective communication climate. People can feel rejected if the communication climate does not provide good social and psychological tone that open the possibilities for interaction (Atkinson & Frechette, 2009; Guzley, 1992; Muchinsky,

Table 1. Keys to building trust (M. A. Brown, Sr., 2017)

Building Trust
Receiver Adjustment: Building trust takes time, so senders must be patient as receivers adjust to messages and the communication process. Recognition is one way to help the receiver accept the interaction. Careful conversations to help with adjustments are important.
Sender/Leader Power: Collaborate and make the decision or course of action more suggestion than direct order. Provide choices and explain the rationale for any decisions made.
Security: Work to manage risk and raise comfort levels. Take the time to help people understand and adjust to the activity.
Risk Tolerance: Acknowledge and confront risk. Explain what is at stake, analyze options, and provide a safety net. If things go wrong, don't place blame. Work through the difficulty together.
Match Interests: Trust comes from shared values, shared interests, and shared responsibility. Take everyone's point of view into consideration and accommodate when possible. The overall goal should drive common actions and shape the culture of the relationship and/or group to the best outcome.
Personality Match: Find common ground in terms of values, beliefs, and even cultural background. Emphasize "we" more than "I."
Care about Others: Make sure your actions in communicating are good for all involved. You might have to sacrifice your own wants and desires at times to allow other interests to prevail for the good of the communication. Be fair in your actions and make sure the process is fair. Try to focus on a win-win outcome wherever possible and be prepared to compromise when that's not possible.
Capability: Use your skill and competence to manage the communication or complete the task. Senders sometimes must relinquish control to address issues that threaten progress.
Integrity and Predictability: Use the adage "under promise and over deliver." Be honest in your successes and failures and share information about the values that drive the communication effort and sender actions.
Communicate-Communicate-Communicate: Communication should be timely and candid. The best relationships are formed without constraints on thinking outside the box.

1977). Strategic flexibility is also an important consideration in removing barriers to the interaction. Strategic flexibility is the ability to take action in response to external environmental changes, or to drive intentional changes and adapt to environmental changes (M. A. Brown, 2017, p. 153). Strategic flexibility addresses variations in the issue at hand and makes adjustments rapidly to keep the interaction alive (Bahrami, 1992; Evans, 1991; Nadkarni & Narayanan, 2004; Roberts & Stockport, 2009; Sanchez, 1995; Slack, 1983).

One barrier that communicators must face is the need to create cultural compatibility or find common ground. Senders can foster long-standing, mutually beneficial relationships by achieving shared values in the communication. This enhances shared understanding. Another barrier is resistance to participate based on fear, disbelief, underestimation of the benefits of communication, lack of a sense of return on investment of time and effort, misunderstanding, and misperception. Other barriers are not new. Senders must deal effectively with the pace of communication to ensure that information exchange is timely and holds the receiver's interest. There are other barriers to communication, but the ones mentioned here are most relevant for this discussion.

Senders must accept responsibility for this whole process, and for continued communications. Online communicators must search for ways to request, evaluate, and respond to feedback. This is important because online communications happen so fast that the sender often has no time to even consider adjusting.

In addition to barriers, social networks may adversely affect communication by imposing superficial experiences, and grammatical and spelling erosion of the language. Current studies focus on data about the spread of social networks and their use and impact on interpersonal communication. More specifically, they look for the answer to the question: what is the nature of interpersonal communication that is found on social networking sites: personal, emotional, private or shared, informal, and public

(Gheorghiță & Pădurețu, 2014)? These are important considerations that may not get enough emphasis in organizational and personal settings.

In research on groups engaged in face-to-face and/or online discussions, a 2011 study used a national population sample to examine trust and collaboration (Jaggars, 2014). In examining four groups, some results showed that the online only group was less trusting, less capable of producing a desired result, less interested in politics, less tolerant, and less knowledgeable than the other groups. This information is not meant to show that face-to-face is the only answer to good communication; it is meant to remind collaborators of the choices that partners must make. Despite the apparent benefits of face-to-face based on the four groups students cited two reasons for preferring online methods: they appreciated the flexibility and convenience and some felt that they could use their learning time more efficiently (Jaggars, 2014).

Even when choosing online methods, many students said they learned better in the face-to-face courses, saying they simply felt they didn't learn the material as well when taking the course online. These students focused on the need or want for student-instructor and student-student interaction.

Several studies of loneliness in college students point out face-to-face communication difficulties or reluctance for in-person methods. Studies examining students' use of social networking sites to communicate with their parents created increases in emotional loneliness, anxious attachment, and conflict in the parental relationship (C. Brown, 2013; Gentzler, Oberhauser, Westerman, & Nadorff, 2011; Kim, LaRose, & Peng, 2009; Moody, 2001). Anxious attachment is about intimacy and closeness as core needs of a person.

Social skills, anxiety and technology use, along with the preferences of college students, was the subject of a published thesis (C. Brown, 2013). The work highlights the ever-growing technological communication in America and negative and positive influences resulting from that growth. Employers complain about the lack of interpersonal communication skills they experience in interviews with prospective employees, resulting in a renewed focus on communication skills as one of the top qualities employers seek (McKay, 2005). Some degradation of social skills in students who preferred technological communication over face-to-face was evident in the published thesis (C. Brown, 2013). A social skills inventory and a conversational skills rating scale demonstrated that the lower social skills belonged to students who preferred online activities.

The challenge of online interactions starts with the impersonal nature of some communications caused by not knowing who is listening or if anyone is listening. While it is true that online communicators can check what they say before sending, the reverse can be true. The sheer speed and convenience can betray the sender and the message can be sent before the thought process is complete. The digital environment does not allow the sender to determine the receiver's tone or demeanor with full confidence. Online conversations are not recommended for bad news, emotional messages, apologies, or sharing affections because the sender cannot engage to make sure the receiver understands the seriousness or gravity of the message and is relatively safe and comfortable with the information. Many people use online communication for these situations to achieve anonymity, reduce stress, alleviate accountability, etc. The sender should take those actions at their own risk.

The barriers or challenges to communication are always at work in some way. Communicators must conquer barriers to creating and/or perfecting a hybrid approach.

- Create cultural compatibility by finding common ground in the communication
- Accept accountability for cyclic, continuing communication that energizes feedback
- Use feedback to guard against superficial experiences that test reality

- Use feedback to address grammatical and spelling erosion of the language

Basically, the digital conversation seeks to balance its "human" limitations with the ability to promote faster connections between individuals and within groups. Focusing on the evolution of online communication is important to this discussion.

Examining Human Interaction

What happened to human interaction as social media and digital communication run rampant across the world? Technology has grown almost out of control over the past few decades. Gone is the time when people would gather in a room and talk together as the primary mode of communication. Today, the phone and the tablet dominate the way we communicate. Look around you while you are driving through town, on the highway, or at a stop light. How many of the people you observe are looking at an electronic device? How many people are texting, reading e-mail, or searching for the best location for lunch? If they are driving, they are not fully watching the road and any feedback they may get will be further distracting. Now you are at a restaurant and you see a table of five or six people together. None of them are looking at each other. Their eyes are on their phones, and the waitress is trying to get their attention about their order. Technology has their almost undivided attention.

Here is what might be a typical day when technology drives everything that happens. The alarm clock on Erica's phone rings at 5:30 a.m. She has a big meeting today so she has to get up and get going. Erica connects her phone to a Bluetooth speaker, so she can listen to her favorite tunes while getting ready. As she gets dressed, she gets a weather alert from her phone: that storm is moving out to sea, so her weekend plans may work after all. Erica gets her coffee and opens a news site on her tablet. While she is watching the news, the tablet sends a traffic alert for her route to work.

Erica is all ready for work now and heads for the car. The meeting is at a location she is not familiar with, so she pulls up "Maps" her smartphone to get directions. The phone connects music to the vehicle for her drive to the meeting. The phone rings, suspending the music, because Erica has a call. One of her colleagues is going to the same meeting and wants a ride. There is time to pick up the friend, so Erica reprograms "Maps" for that location on the way. Erica picks up Susan and asks if she is still planning the vacation trip. But Susan is not listening because she is playing a game on her phone with a Bluetooth earpiece on.

They arrive at the meeting and Erica turns the music off and puts her phone on vibrate so it will not interrupt the meeting. She meets Larry, who is running the meeting. Larry lets Erica know that this is a Skype call and there will only be four total people in the room. The call is connected, and the meeting starts. Erica and the other three people in the room monitor their phones during the meeting. A person on the Skype call asks a question for Susan, but she misses the question while working on her phone. The meeting goes well, and everyone is ready to head out to lunch. Larry asks where they can go, and they all check Google for "food near me." At lunch, there is some conversation but very little eye contact.

After lunch, Erica gives Susan a ride. Now Erica has to return to her house and get the phone charger because her phone charge has gotten very low. We can keep going, but the idea should be clear by now. Today, technology in the form of phones and tablets is getting the attention that humans used to get. This story is typical of many situations we see today. Today, many people would rather text than talk, and they prefer mediated communication to face-to-face interaction.

Whatever it is called, CMC, mediated communication, mediated discourse, or mediated interaction, we are talking about the use of information technology to interact with others. CMC does not employ the senses or facial expressions or symbolic cues. The emergence of CMC has even caused changes in our romantic relationships. Experts say that those who maintain long-distance friendships rate face-to-face communication fourth in importance behind telephone, email, and instant messaging (Wright and Webb 2011). Apparently, the real change from face-to-face to CMC was caused by access. Technology has been around for many years, but the size and price of cell phones and tablets have made it progressively more accessible to many more people.

Reading the GSMA Intelligence report "The Mobile Economy" yields information about the unbridled growth of technology and, in turn, online communication. "The mobile industry continues to scale rapidly, with a total of 3.6 billion unique mobile subscribers at the end of 2014. Half of the world's population now has a mobile subscription—up from just one in five 10 years ago. An additional one billion subscribers are predicted by 2020, taking the global penetration rate to approximately 60%. There were 7.1 billion global SIM connections at the end of 2014, and a further 243 million machine-to-machine (M2M) connections." (GSMA 2016).

A 2011 study highlighted mobile communication and its influence on the extent to which people engage face-to-face with new people in public settings (Campbell and Kwak 2011). The study examined different types of cell phone uses and found that mobile phone use enhanced information exchange with strangers. In another study, researchers demonstrated that the presence of mobile communication devices in social settings caused problems in human relationships (Przybylski and Weinstein 2013). Specifically, in conducting two experiments, the researchers uncovered evidence that the devices negatively influenced closeness, connection, and conversation quality. The research concluded that the negative influence increases when people discuss personally meaningful topics.

The relationship between mobile device availability and the quality of in-person social interactions was the subject of another study, one in which researchers discovered that conversations without the use of mobile communication technologies were rated as significantly superior to those in the presence of a mobile device (Misra, Cheng et al. 2016). People who had to communicate without the benefit of mobile devices said they had higher levels of empathetic concern, while those who used a mobile device reported lower levels (Misra, Cheng et al. 2016).

Libraries, for instance, have taken new approaches such as employing social media marketing. For instance, two Indiana researchers suggested suitable social networking strategies for academic libraries (Booker & Bandyopadhyay, 2017, pp., pg 142). The strategic approach suggested starting with creating awareness of services and activities. The next steps are to build brand strength and conduct engagements that build relationships with users. After that, libraries should provide timely support to users, encourage feedback and, when necessary, act on that feedback.

Combining the Best of Face-to-Face and Online Interactions

My hope is that FTF communicators will find ways to keep the human element involved in their online interactions, and online communicators will find a comfort level with leaving the computer to employ verbal and non-verbal to human communication. A comparison based on my research appears below (M. A. Brown, 2017, pp. 103-110):

Benefits of Face-to-Face Interactions

Face-to-face communication is important because it happens every time people meet. As hard as we may try to avoid this, it is inevitable in our society. So, even if you're totally online with social media, or you only do virtual reality, there WILL come a time when you must communicate in person. When that happens, there are emotional, physical, and environmental conditions and activities at play. You may choose to ignore them, but they are integral to the interchange and they will affect it, often even if you don't realize it. It is important to examine the use of theory and then focus on the importance of face-to-face interaction and five focus areas that are critical to making it work: comfort, control, confidence, timing, and focus.

Face-to-face communicators must:

- Feature great social context providing a collaborative relationship and creating meaning.
- Create social identity and engage in negotiations.
- Use the full communication cycle of sender, message, receiver and feedback, as well as the five senses, to minimize differences and attend to verbal and non-verbal cues.
- Work with communication partners in discourse management to repair any breaks or difficulties in the communication.

Benefits of Digital Communications

When we talk about online communication, we are referring to digital communication and social media. The explosion of online interactions is startling and ongoing. Why? Our communications are open to people we can't see, don't know, and have no reason to trust. Yet we communicate things that historically would only be shared with family and friends. We commit dependence, satisfaction, and commitment to strangers, and we tend to respond when they give us feedback. We do this because we make a decision to give most people the benefit of the doubt and we extend that benefit to people we don't know from direct experience. The next task is to demonstrate the importance of social capital and how it enhances online interactions, and will discuss the importance of expectation, value, and return on investment (EVR).

Hybrid: A New View of Communicating

Online communicators must:

- Focus on the value proposition of shared meaning that makes the communication relevant for all involved.
- Work to build trust through social capital, which illustrates how active and trusting connections with others widen our awareness and reveal commonalities that may help the interaction.
- Gain adoption and trust within an organization or community of interest.
- Ensure information is exchanged freely in a two-way interaction that allows participants to get and give recognition that matters.

This represents a relatively new view of communicating with others. Many hybrid approaches are being considered today, but the following explanations focus on aspects of two: blended learning and the hybrid shift.

Blended learning considerations are extremely relevant when discussing the hybrid approach. One study supported a graduate class by creating a blended synchronous learning environment (BSLE) which was delivered simultaneously to both online and face-to-face (F2F) students. The study followed an educational design approach to gradually implement the hybrid environment. According to study results, the BSLE environment can support online and face-to-face (F2F) students equally. This is possible by providing similar learning experiences; however, there were challenges. The study offers design principle researchers can use to address communication challenges (Wang, Quek, & Hu, 2017).

The study also argues that it is not enough to provide online resources and let students learn on their own, because considerations of lack of social presence, low participation, being impersonal, and low motivation that hinders participation must be addressed in the learning environment design (Wang et al., 2017, p. 100).

Incorporating synchronicity is recommended as a way to address the limitations of asynchronous learning resident in the proposed hybrid approach. In this context, synchronicity refers to connecting two or more people in related educational events at the same time, facilitating a synchronous learning environment. "Blended synchronous learning has the potential to integrate the respective strengths of blended learning (e.g., flexibility and convenience) and synchronous learning (e.g., immediate feedback and increased motivation), and reduce the limitations that blended asynchronous learning often has (Wang et al., 2017, p. 101)."

The practical, educational, and economic benefits of blended synchronous learning are common to the recommended hybrid approach. Blended approaches can normally be undertaken without financial support, allowing flexibility and convenience to students who have trouble getting to a class site, such as when a storm shuts down the school. Benefits of teaching, social, and cognitive presence are addressed through two-way video conferencing or even by allowing texting in class, for instance. Financially, reducing the need for classroom space can reduce costs for the school and alleviate transportation and possibly day-care costs for the student. This approach also allows a great learning opportunity for deployed military members, for instance.

The study also demonstrated desires and needs for synchronicity in education. It also pointed out that while there are benefits, there are design principles that must be incorporated into a hybrid or synchronous approach to provide the best learning environment possible. There are challenges, but none seem to be insurmountable (Wang et al., 2017, pp. 112-113).

1. Certain activities (e.g., demonstrations, group discussions) must be redesigned for online students to easily participate in;
2. The partnership strategy can be incorporated in the learning environment to facilitate the communication between online students and the instructor;
3. The instructor must pay a balanced attention to the online and classroom students;
4. Clear video communication must be ensured, and the online students preferably use voice only to communicate with the instructor as it is hard for the instructor to read text in the instructional process;
5. The learning environment must be easy to use and the students must be trained in advance to reduce possible technical difficulties;

6. The classroom should have at least two cameras and two mics with one camera focusing on the instructor and the other capturing the whole classroom. Similarly, one mic is for the instructor, and the other is for the classroom students to talk into.

Another new learning approach is called the "hybrid shift," where faculty are responsible for designing content and structure with a focus on being thoughtful, engaging, and intentional in promoting learning, effective collaboration, critical thinking, and communication skills. The hybrid shift says content and structure have been transformed as "The ubiquity of mobile smart devices and immediate online access has redefined this arrangement by 'blurring the lines' (Gikas & Grant, 2013) between technology use for formal and informal, social, and educational "activities". This has enabled a transformation of traditional face-to-face classroom spaces into online enhanced environments, bypassing faculty design and intention (Biddix, Chung, & Park, 2015; Dringus & Seagull, 2013).

The hybrid shift requires a new awareness in educators to understand activities they can create – or that are taking place without their knowledge or involvement – that can enhance online learning at any time and at any place. The "shift" focuses on using four key processes to differentiate hybrid courses for effectiveness. They are: " (1) replacement (students watch videos or complete assignments before class to reduce in-class time); (2) supplemental (students access additional materials online with no in-class time reduction); (3) emporium (students have no in-class time, but an available in-person resource center); and (4) buffet (students choose a combination of online or in-class activities)." Research on student learning differentiated by course format has shown the most promising outcomes from courses that blended online activities and face-to-face teaching, rather than focusing on fully online or in-class instruction (Biddix et al., 2015, p. 2)."

Biddix et al (2015) employed a review of a decade of studies that focused on learning environments in work at the Teaching and Learning Laboratory at MIT (Zhao & Breslow, 2013). The summary provided from that review showed an overwhelming participant preference for hybrid environments over face-to-face encounters, stating that in many cases students benefitted from the same quality of learning in both approaches, but were stronger academically in blended classrooms.

The hybrid shift represents an approach that is learner driven to allow students to complement existing instruction in a way that best fits their needs. More important, a new learning culture focused on hybrid learning allows the educator, the administrator, and the students to evolve an environment conducive to adjusting to the best of all possible approaches. A willingness on the part of faculty and institutions to add student-owned mobile devices to the learning environment that students are longing to employ is crucial to the hybrid effort.

Hybrid Approach

Whatever method is chosen, acceptance of a hybrid approach relies on understanding. In my book, *Solutions for High-Touch Communications in a High-Tech World* (M. A. Brown, 2017, pp. 195-200), the benefits, actually the necessity, for face-to-face communication is spelled out. Face-to-face has nine key characteristics:

1. Basic: Relies on the full communication process from sender to receiver to feedback, operates as a cycle, promotes shared understanding, and uses social cues.

2. Relevant: Using sound communication theory, listening to and acting on verbal and visual cues, for instance, to achieve continuous interaction between cognitive, behavioral, and environmental influences.
3. High-Touch: Create and nurture interaction relationships to achieve great collaboration that includes feedback.
4. Powerful: As we listen, understand and take action to keep the communication alive, we can be effective in terms of asking questions, changing the approach where necessary, and achieving mutual agreement.
5. Complementary: Use face-to-face and digital communication interchangeably to get the best of both worlds, still emphasizing face-to-face to get full collaboration and capability.
6. Interactive: Communication participants can find success by collaborating to work out private information and boundaries as well as control of the shared information.
7. Timely: All parties should commit to participate as long as necessary to ensure they complete a triad of communication – listen, hear, understand.
8. Compelling: The best way to know what a person wants or thinks is to ask, and when that can't be accomplished online, a face-to-face interaction offers availability of verbal and visual cues and the ability to make changes based on real-time inputs.
9. Committed: People commit to the interaction by paying attention to communication with emphasis on sending valuable messages, telling the truth, ensuring shared value for all, and remaining engaged to seek and respond to feedback.

CONCLUSION

Whether an individual prefers online communication or face-to-face interactions, there is no disputing the sheer value of digital communication. A hybrid approach provides a way to have the best of both worlds. A hybrid approach offers a new way to communicate and to adjust to variable and diverse interactions. The approach is sensitive to the environment in that there are ways to assess internal and external forces and adjust the message accordingly. The formal approach should work to counter uncertainty.

My hybrid approach uses the best of available research to identify approaches to communication that will work regardless of whether we're face-to-face or in a digital interaction. I suggest focusing on two of the six steps in my hybrid approach (M. A. Brown, 2017, pp. 207, 209).

First, interact in communication zones that are social "fields" supporting creativity and ensuring productive communication between the partners (Nardi & Whittaker, 2002). These zones are managed by creating social ties or bonds, by managing visual and verbal cues, and by cooperating in ongoing feedback. Communication zones have two key processes: (1) social bonds that allow people to connect emotionally; and (2) creating "attentional contracts" that require the parties to pay attention to messages sent and received for as long as it takes to achieve shared value (Nardi & Whittaker, 2002).

The second consideration in the hybrid approach is combining face-to-face and online interactions. Digital communications may suffer from distrust, misunderstanding, and poor decision making, but connective with good interpersonal relationships can address these issues (M. A. Brown, 2017, p. 209). Senders can achieve unexpected benefits by being creative and adapting to personal and organizational needs and objectives. When an individual in a digital interaction seeks empathy or understanding from the sender, there can be difficulties or delays in achieving that goal. The immediacy of face-to-face interaction allows senders and receivers to work together to sense needs faster and to act on them faster. This allows each partner to match their position with ideas and attitudes that resonate with personal need.

The amount of growth in communication and the multitude of ways it is accomplished signals that online communication is here to stay. By searching historical methods and practices, we can address the pros and cons, enabling us to find the best way to use it for our purposes. When measuring levels of performance, the hybrid approaches suggested here can lead to the most robust interactions. This area of communication continues to expand and deserves continued development by scholars, practitioners, and organizations.

REFERENCES

Atkinson, T., & Frechette, H. (2009). *Creating a Positive Organizational Climate in a Negative Economic OneImproving Organizational Climate to Transform Performance.* Doi: http://www. trainingindustry. com/media/2505214/creatingpositiveorgclimate_us_aug09. pdf

Bachmann, R., Knights, D., & Sydow, J. (2001). Special Issue: Trust and control in organizational relations. *Organization Studies*, *22*(2). doi:10.1177/0170840601222007

Bahrami, H. (1992). The emerging flexible organization: Perspectives from Silicon Valley. *California Management Review*, *34*(4), 33–52. doi:10.2307/41166702

Baldwin, J. A. (2018). *James Baldwin Biography and Quotes.* Retrieved from https://www.biography. com/news/james-baldwin-quotes

Biddix, J. P., Chung, C. J., & Park, H. W. (2015). The hybrid shift: Evidencing a student-driven restructuring of the college classroom. *Computers & Education*, *80*, 162–175. doi:10.1016/j.compedu.2014.08.016

Booker, L., & Bandyopadhyay, S. (2017). How academic libraries can leverage social networking to popularize their services: An empirical study. *Journal of the Indiana Academy of the Social Sciences*, *16*(2), 12.

Brown, C. (2013). *Are We Becoming More Socially Awkward? An Analysis of the Relationship Between Technological Communication Use and Social Skills in College Students.* Academic Press.

Brown, M., Sr. (2011). *Social networking and individual performance: Examining predictors of participation* (Ph.D.). Old Dominion University. Retrieved from http://proquest.umi.com/pqdweb?did=2355291551&Fmt=7&clientId=3505&RQT=309&VName=PQD

Brown, M. A. (2017). *Solutions for high-touch communications in a high-tech world.*

Brown, M. A. Sr. (2017). *Solutions for High-Touch Communications in a High-Tech World.* Hershey, PA: IGI Global. doi:10.4018/978-1-5225-1897-6

Bubaš, G. (2001). *Computer mediated communication theories and phenomena: Factors that influence collaboration over the Internet.* Paper presented at the 3rd CARNet users conference, Zagreb, Croatia.

Dringus, L. P., & Seagull, A. B. (2013). A five-year study of sustaining blended learning initiatives to enhance academic engagement in computer and information sciences campus courses. *Blended Learning: Research Perspectives, 2*, 122-140.

Evans, J. S. (1991). Strategic flexibility for high technology manoeuvres: A conceptual framework. *Journal of Management Studies*, *28*(1), 69–89. doi:10.1111/j.1467-6486.1991.tb00271.x

Frank Messina, E. D. (2011). *Two-And-A-Half Minutes To "Effective"*. Xlibris.

Gambetta, D. (1988). *Trust: making and breaking cooperative relations*. New York, NY: B. Blackwell.

Gentzler, A. L., Oberhauser, A. M., Westerman, D., & Nadorff, D. K. (2011). College students' use of electronic communication with parents: Links to loneliness, attachment, and relationship quality. *Cyberpsychology, Behavior, and Social Networking*, *14*(1-2), 71–74. doi:10.1089/cyber.2009.0409 PMID:20973676

Gheorghiţă, V., & Pădureţu, A. (2014). Social Networks And Interpersonal Communication. *System*, *2*(3), 5.

Gikas, J., & Grant, M. M. (2013). Mobile computing devices in higher education: Student perspectives on learning with cellphones, smartphones & social media. *The Internet and Higher Education*, *19*, 18–26. doi:10.1016/j.iheduc.2013.06.002

Guzley, R. M. (1992). Organizational climate and communication climate: Predictors of commitment to the organization. *Management Communication Quarterly*, *5*(4), 379–402. doi:10.1177/0893318992005004001

Jaggars, S. S. (2014). Choosing between online and face-to-face courses: Community college student voices. *American Journal of Distance Education*, *28*(1), 27–38. doi:10.1080/08923647.2014.867697

Kim, J., LaRose, R., & Peng, W. (2009). Loneliness as the cause and the effect of problematic Internet use: The relationship between Internet use and psychological well-being. *Cyberpsychology & Behavior*, *12*(4), 451–455. doi:10.1089/cpb.2008.0327 PMID:19514821

Kramer, R. M. (1996). *Trust in organizations: Frontiers of theory and research*. Thousand Oaks: Sage Publications.

Lane, C., & Bachmann, R. (1998). *Trust within and between organizations: Conceptual issues and empirical applications*. New York: Oxford University Press.

McEvily, B., Perrone, V., & Zaheer, A. (2003). Introduction to the special issue on trust in an organizational context. *Organization Science*, *14*(1), 1–4. doi:10.1287/orsc.14.1.1.12812

McKay, J. (2005, Feb.). Employers complain about communication skills. *Pittsburgh Post-Gazette*.

Moody, E. J. (2001). Internet use and its relationship to loneliness. *Cyberpsychology & Behavior*, *4*(3), 393–401. doi:10.1089/109493101300210303 PMID:11710265

Muchinsky, P. M. (1977). A Comparison of within- and Across-Subjects Analyses of the Expectancy-Valence Model for Predicting Effort. *Academy of Management Journal*, *20*(1), 154–158.

Nadkarni, S., & Narayanan, V. (2004). Strategy frames, strategic flexibility and firm performance: the moderating role of industry clockspeed. *Academy of Management Proceedings*. 10.5465/ambpp.2004.13863821

Nardi, B. A., & Whittaker, S. (2002). The place of face-to-face communication in distributed work. *Distributed Work*, 83-110.

Przybylski, A. K., & Weinstein, N. (2013). Can you connect with me now? How the presence of mobile communication technology influences face-to-face conversation quality. *Journal of Social and Personal Relationships*, *30*(3), 237–246. doi:10.1177/0265407512453827

Roberts, N., & Stockport, G. J. (2009). Defining strategic flexibility. *Global Journal of Flexible Systems Managment*, *10*(1), 27–32. doi:10.1007/BF03396553

Rousseau, D. M., Sitkin, S. B., Burt, R. S., & Camerer, C. (1998). Not so different after all: A cross-discipline view of trust. *Academy of Management. Academy of Management Review*, *23*(3), 393–404. doi:10.5465/amr.1998.926617

Sanchez, R. (1995). Strategic flexibility in product competition. *Strategic Management Journal*, *16*(S1), 135–159. doi:10.1002mj.4250160921

Sitkin, S., Sutcliffe, K., & Barrios-Choplin, J. (1992). A Dual-Capacity Model of Communication Media Choice in Organizations. *Human Communication Research*, *18*(4), 563–598. doi:10.1111/j.1468-2958.1992.tb00572.x

Slack, N. (1983). Flexibility as a manufacturing objective. *International Journal of Operations & Production Management*, *3*(3), 4–13. doi:10.1108/eb054696

Wang, Q., Quek, C. L., & Hu, X. (2017). Designing and improving a blended synchronous learning environment: An educational design research. *The International Review of Research in Open and Distributed Learning*, *18*(3). doi:10.19173/irrodl.v18i3.3034

Zhao, Y., & Breslow, L. (2013). Literature review on hybrid/blended learning. *Teaching and Learning Laboratory.* Available at: https://tll. mit. edu/sites/default/files/library/Blended_Learning_Lit_Reveiw. pdf

Chapter 12
"It's a Lot Like Ending Up on the Losing Side of a Relationship": Sports Fans, Grief, and Parasocial Divorce

Nathan John Rodriguez
Weber State University, USA

ABSTRACT

Academic interest in parasocial relationships has increased dramatically in the past decade. A subset of this broader topic concerns the relationships between sports fans and a host of athletes, coaches, and teams. An overwhelming majority of those studies examine fan reactions following a singular event rather than over an extended period of time. This chapter unpacks the ways in which college basketball fans reacted to the departure of a beloved coach over a period of nine years. Relevant reader comments were analyzed thematically and coded by an expressed grief state using the Kubler-Ross model. The term "parasocial divorce" is introduced to describe the depth of parasocial relationships for highly motivated fans. The findings reveal how fans can come to loathe a particular persona, and the conditions under which that parasocial relationship may be rehabilitated.

INTRODUCTION

Much research involving sports fans stems from theories of parasocial interaction (PSI) and parasocial relationships (PSR). The premise is that individuals develop feelings toward media personae in a similar fashion as they might develop feelings toward others with whom they interact socially. Just as relationship expectations develop over time (Planalp & Rivers, 1996), fans tend to develop expectations and assumptions regarding behavior (Cohen, 2010). Highly motivated sports fans may identify strongly with a particular team, and such devotion may be "intense, blinding fans' better judgment" regarding how to process particular events (Sanderson, 2013, p. 505). With relatively few exceptions, however, most studies have assumed a stable relationship between fans and the object of their affection (Hyatt, 2007). This is likely because most studies take a snapshot of fan behavior during a few weeks or months rather

DOI: 10.4018/978-1-7998-3355-0.ch012

than years. As a result, the depth of fan involvement and the ways in which fan loyalty may be tested remain understudied.

This study examines the complicated relationship many fans of the men's college basketball team at the University of Kansas have with former head coach Roy Williams. In an attempt to better situate the inquiry, this section offers relevant background details regarding the case study and how it fills a need in the ongoing discussion of sports fans and parasocial relationships.

At his introductory press conference on July 8, 1988, Williams articulated his hope that a similar press conference would not be required "for about 30 years" due to his loyalty, as someone who was married to "the same wife for 15 years," and owner of "the same set of golf clubs for 17, so I'm the sort of guy that if I get a place I like, and find something I like, I stick with it" (Lawrence-Journal World, 2009). Over a period of 15 years, Kansas fans had come to revere Williams for his success: In the 1990s, no school amassed more wins or a higher winning percentage (Mellinger, 2013). Although he had not won a national championship, he had reached the Final Four on multiple occasions in addition to winning a number of conference championships. Fans grew to enjoy his "Mayberry" personality: If he was upset, he might say "doggone" or "gosh darn" and he was open with his emotions following a post-season loss in a way that promoted empathy and suggested honesty (Gottesman, 1997). Williams was revered at Kansas, where fans "chanted his name before every game and he threw T-shirts into the stands. Nobody talked seriously about a statue yet, of course, but Williams could have had nearly anything" (Mellinger, 2013).

In July 2000, the University of North Carolina offered Williams, a former Assistant Coach under Dean Smith, the position of head coach. The following week was "one of the most anxious, agonizing seven-day periods in Kansas' storied basketball history" due in part to reports from multiple media outlets that Williams had accepted the position (Hartsock, 2000). Thousands of fans assembled in the university's football stadium to watch Williams's announcement via satellite, as he opened with two words: "I'm staying." Explaining his decision, Williams foreshadowed the relational metaphors that would come in the years ahead: "The North Carolina people have been fantastic with me. It's hard to have the same feeling for someone if there's been some type of rejection. I'm hoping this is not a divorce" (Hartsock, 2000).

Three years later, North Carolina renewed its request and this time, Williams accepted. A sizable segment of the Kansas fanbase felt betrayed, and a popular shirt around Lawrence referenced "Benedict Williams" (O'Neil, 2008). In 2005, Williams won the national championship at North Carolina, while Kansas was ousted in a major upset by Bucknell University in the first round of the tournament, which stagnated the grieving process for many fans. Finally, in 2008, Kansas was matched against North Carolina in the Final Four. Most analysts predicted North Carolina would advance, but after jumping out to a 40-12 lead--prompting announcer Billy Packer to comment, "this game is over" in the first half--Kansas would go on to win 84-66 (Hiestand, 2008). Two days later, during a *CBS* telecast of the national championship, Williams was shown standing and smiling in a section of Kansas fans, a Jayhawk sticker placed prominently on his black turtleneck. Kansas won the national championship, and within a matter of 48 hours, a significant contingent of Kansas fans spoke of finding closure (Armstrong, Bukaty & Gutierrez, 2013).

Kansas again faced North Carolina in the post-season in 2012 and 2013, winning both games. While some national pundits predicted a frosty reception for Williams in Kansas City, *ESPN* reported he received "rousing ovations" from Kansas fans in the two days preceding the game, and *The New York Times* noted that on game day, Kansas fans "slowly, politely...stood and cheered for Williams" (King, 2013; Borzi, 2013). In 2014, Williams would return to Lawrence to commemorate the 60th anniversary of its basketball venue, Allen Fieldhouse, alongside the other living former head coaches. He was greeted with a standing ovation and sustained applause.

Clearly, Kansas fans would not have had such a reaction in the immediate aftermath of Williams's departure. And while time may help heal perceived wounds, other external factors were involved in the process. There have been calls to examine different types of parasocial breakups (Cohen, 2003), and for longitudinal research to examine the "study of forgiveness and PSI" (Sanderson & Emmons, 2014, p. 40). This case study illumines a significant gap in the literature by demonstrating how dramatically fan attitudes can change over time.

This study is perhaps the first to examine sports fan behavior over a period of several years. Here, reader comments during a nine-year period were examined chronologically. Reader comments were categorized using the Kubler-Ross (1969) model of processing grief to better account for how fan attitudes toward Williams changed over time. The notion of parasocial divorce is introduced to highlight the depth of relational attachment fans may develop with certain media personae. To better understand some of the reasons that fans may have reacted the way in which they did, it is useful to turn to related work on parasocial relationships, sports fans, and the processing of grief.

REVIEW OF LITERATURE

Parasocial Interaction and Parasocial Relationships

As radio and television grew in popularity in the 20th century, people formed attachments to the personalities projected over the airwaves. Parasocial interaction (PSI) describes the way in which someone might feel a sense of affinity or antipathy toward a mediated figure. Horton and Wohl (1956) defined it as the "seeming face-to-face relationship between spectator and performer" (p. 215). The authors also observed the media persona "offers, above all, a continuing relationship" because the appearance of such a figure becomes a "regular and dependable event, to be counted on, planned for, and integrated into the routines of daily life" (p. 216). And this notion of dependability over time means that the parasocial relationship develops a history of "shared past experiences" that lends "additional meaning to the present" (p. 216). They concluded parasocial interaction resembled that of traditional face-to-face relationships. A recent examination of metadata from more than 250 studies found that two-thirds of all publications regarding parasocial phenomena has been produced in the last 10 years (Liebers & Schramm, 2019).

The concept of PSI is "profoundly psychological," yet received virtually no attention from psychologists in the 20th century and has generally remained in the realm of communication scholars (Giles, 2002, p. 280). In recent decades, parasocial relationships have become viewed as complementary to existing relationships (Caughey, 1984), and there is growing agreement that PSI mirrors social relationships in that people tend to use similar methods of evaluating media personae as they might in social relationships (Rubin & McHugh, 1987; Giles, 2002; Cohen, 2004). By some measures, favorite media figures are considered more favorably than good neighbors (Gleich, 1996, cited in Giles, 2002). Not all media personae are equal in the eyes of the beholder, however, and it is worth distinguishing between PSI and parasocial relationships (PSR).

As Hu (2015) pointed out, PSI and PSR have been used interchangeably in several influential publications, which has resulted in "conceptual and operational confusions" in subsequent work. PSI is "situationally bound" (Hu, 2015, p. 2) occurring "during media exposure" (Hartmann, Stuke, & Daschmann,

2008, p. 25). PSR, in contrast, is "cross-situational," (Hartmann et al., 2008, p. 26), occurring over the "long-term," and is therefore more "enduring" (Hu, 2015, p. 2). While it is worth making a distinction between the two concepts, it may be more helpful to consider the two in concert along a continuum of fan (and anti-fan) involvement. Finally, a number of authors have been perhaps too quick to argue certain findings are applicable to PSI and/or PSR generally, rather than limiting it to a particular pocket of fan studies. The potential for interactive experiences may range from "full face-to-face interaction" between fan and media persona on one end of a spectrum to a more unidirectional immersion into a cartoon series or "fictional protagonist" on the other end (Giles, 2002, pp. 286-287).

Two broad tendencies within PSI and PSR scholarship are relevant to this study. First, fans are frequently characterized--consciously or not--as static entities. In some instances, this is evident in attempts to classify fans into particular categories (Hunt, Bristol, & Bashaw, 1999). Even as Gray (2003) opened new paths to investigate fandom in terms of fans, anti-fans, and non-fans, it seemed such categories were presented as rigid; or, if fans were to become anti-fans, it was merely a temporary aberration.

For example, one of the more common methodological approaches is to query undergraduates on how they might react if their favorite television personality were taken off the air (Cohen, 2003; Hu, 2015). Because the phrase "favorite television personality" is polysemic, not only are students asked to engage in a hypothetical, they are likely considering a range of media personae. As Sun and Wu (2012) have pointed out, PSR studies have included media personae, soap opera characters, celebrities in commercials, TV shopping hosts, comedians, talk show hosts, and athletes. The growth of online forums and social media platforms eliminates the need to rely upon undergraduates responding to hypothetical prompts, and instead study bona fide fans.

Another tendency has been to examine parasocial relationships by fans where the media personae is well-liked, with far less attention paid to what happens when parasocial relationships turn sour (Hartmann et al., 2008, p. 26; Hu, 2015, p. 3). Cohen (2003) argued the end of parasocial relationships, "like the breakup of social relationships" is a "painful experience" in which an individual may exhibit symptoms "similar to those that follow the loss of a friend" (p. 200). Interpersonal relations theory has shown the perceived closeness of the relationship is predictive of the emotional distress after the relationship is ended (Bowlby, 1980; Simpson, 1987; Mearns, 1991). This particular subsection of parasocial relationships is understudied, but so too is "the passion and zeal of (sports) fans" (Klugman, 2009, p. 35). It is worth examining the extant work in the field of sports fan research.

Finally, it is worth examining image repair theory (Benoit, 1997), a subset of crisis communication studies, which argues that in the aftermath of an offensive event, an individual or entity is viewed as responsible for that event, and has a set of possible strategies to restore their image. One such strategy is bolstering (Abelson, 1959), which was described as "the obverse of denial," and an attempt to "reinforce the existence of a fact, sentiment, object or relationship" that is viewed favorably by an audience (Ware & Linkugel, 1973, p. 277). This is particularly relevant as recent research affirms that behavior by a perceived offender following an offensive act can play a pivotal role in the maintenance of a parasocial relationship, and lead to eventual forgiveness (Osterman & Hecmanczuk, 2019). Forgiveness has been characterized as having a constellation of factors, including the parasocial relationship prior to the offense, the transgression itself, and the mindset of the forgiver (Lee et al., 2018). Benoit (2018) recently turned his attention to image repair theory in the context of sports, and argued the concept of an "offensive" action should be construed broadly, including perceived sins of omission and commission through words and deeds.

Sports Fan Research

A review of research on sports fan behavior shows four key areas relevant to this study. First, as suggested by the previous section, fans create bonds with sports figures. Second, the suspense generated by sporting events results in emotional responses to incidents. Third, media attention may enhance the peaks and valleys of sports fandom relative to other media personae. Fourth, sports fans who feel they have been betrayed in some fashion may process that incident or event in ways that might be expected in other social relationships.

First, the notion of "sports fans" covers a vast terrain. As Hunt et al., (1999) explained, someone may have affinity toward the sport itself, a particular team, and/or a certain player, and a "halo effect" often means that primary affection extends or spills over into the other schema of involvement (p. 440). Broadly speaking, however, sports fans form bonds with sports figures (Fowles, 1992; Basil & Brown, 2004; Wann, 2006). Fans come to identify with teams that win, in what has been known as basking in reflected glory (BIRG), are more likely to refer to the team's performance after a win, wear the team's apparel after a win, and tend to refer to the team's victory using the first-person singular pronoun, "we" (Cialdini et al., 1976). And keeping in line with the notion of parasocial relationships as cross-situational rather than situationally bound, nongame behavior by sports personae may prompt considerable angst from a fanbase: Sanderson and Emmons (2014) referenced "The Decision" by LeBron James to leave Cleveland for Miami which caused some fans to publicly burn their jerseys of James. Subsequent research showed that Cleveland fans who had a stronger parasocial relationship with James experienced greater distress when he left the team (Bostwick & Lookadoo, 2017). Fans have feelings toward sports figures that continue after the game has been played, and "[encompass] personnel decisions" (Sanderson, 2013, p. 502). Indeed, the allegiance of fans to a particular team is often lifelong (Mastromartino, Chou & Zhang, 2017).

The element of drama in sports seems to be salient for fans and underappreciated by scholars. Hartmann et al., (2008) were perhaps the first to link the development of parasocial relationships to suspense. The suspense leading up to the eventual outcome explains in part why fans may have emotional reactions to not only sporting events but personnel decisions or incidents before and after games. Fans tend to enjoy games that feature an opposing team or athlete perceived as a rival (Bryant, Brown, Comisky, & Zillman, 1982). A study of Kentucky basketball fans revealed that many had "reported physiological responses" during games, with "emotional and physical symptoms" that are "very similar to those one would expect of an athlete who had actually participated in the contest" (Kelley & Tian, 2004, pp. 38-39). It seems the unpredictable nature of sports is, for many, a key part of the allure. For highly motivated fans, the coach can therefore come to be viewed as a conductor of the performance who may deliver joy and stave off disappointment.

Sports have assumed a "central position within culture and society" in large part because of the role of television (Real, 2011, p. 19). Television time devoted to sports grew more than 600 percent from 1960 to 1988 (Wenner, 1989). The relationship between mass media and sports has been symbiotic (Eitzen & Sage, 1989) with mass media "turning organized sports...into a full-blown social institution" and sports, in turn, acting as a "vehicle for bringing dramatic attention to new mass media forms" (Sun, Youn, & Wells, 2004, p. 4). An estimate by consulting firm AT Kearney showed sports is a $480-$620 billion industry worldwide--to put that figure in perspective, it doubles the size of the U.S. auto industry and is "seven times the size of the movie industry" (Shank & Lyberger, 2015, p. 12). Televised broadcasts today, with their enhanced production value and development of narratives are therefore "highly effective in breeding intense PSR" with sports figures (Hartmann et al., 2008, p. 26). Computer-mediated communication allows fans to consume more content. Reading sports media is shown to increase the likelihood PSR will develop (Sun & Wu, 2012, p. 143). Simply put, there are more opportunities for sports fans to indulge and/or engage in personally relevant content and communication than ever before.

In similar fashion to PSI and PSR literature, sports fan studies tend to examine parasocial behavior when it is hagiographic, or at least resembles good will. A smaller subset of work in the area has looked at the deterioration or dissolution of parasocial relationships with sports figures. Sanderson (2013) showed Brian Kelly's departure from Cincinnati to Notre Dame was a "crushing blow for fans" for fans at the former school, in part because of the uncertainty inherent in the transition (p. 488). A key takeaway from that study was "how quickly a revered group member can become despised" (p. 502). To explain what might account for such hostility, it is useful to turn to Fernquist (2000) who applied Gabennesch's (1988) theory of broken promises to sports fans. The basic notion is that "serious disappointment may result" when certain expectations have not been met (p. 332). While that study equated broken promises with wins and losses, the idea of a sports persona perceived as making and subsequently breaking a promise could hold substantial explanatory power for hostile reactions from a particular fanbase.

Grief and Divorce Studies

Interpersonal communication theories have been used to explain parasocial behavior, but few scholars have applied the concepts of grief or divorce to sports. This section therefore begins with a glance at grief studies and divorce studies before concluding with a nod to related efforts.

Kubler-Ross interviewed 400 terminally ill patients over four years and concluded "most, but not all" patients would "pass through five stages (denial, anger, bargaining, depression, and acceptance) between their awareness of serious illness and their death" (1972, p. 179). That observation continues to be "taught in medical schools, espoused by physicians, and applied in diverse contexts" (Maciejewski et al., 2007, p. 716). One clarification in recent years is to conceive of the grieving process in states rather than stages to acknowledge that grief does not necessarily progress in "distinct, sequential stages" but instead tends to "evolve and diminish in intensity over time" (Prigerson & Maciejewski, 2008, p. 437).

Divorce scholarship has shown that the process is different for "the leaver" and "the left," with raw, visceral emotions often surfacing for the latter: "The leaver's loss is more tempered, akin to grief following the death of a loved one after a long chronic illness. In contrast, the left's grief can be out of control, like the crazy, intense emotions of someone whose loved one is in the emergency room after a train wreck (Emery, 2011, p. 62).

Applications of grief and divorce to sports fans are limited. Mitrano (1999) examined metaphors used by fans of a hockey franchise that had decided to relocate and found that fans "collectively experience the same stages of grieving and loss associated with human death and divorce" (p. 151). Two other studies reviewed sports-related grief, though in a slightly different fashion: Evans and Hardy (1995) looked at athletes who felt grief after suffering an injury, and Kelley and Tian (2004) found that Kentucky basketball fans had a desire to "'dish out' grief and aggravation" by outperforming perceived rivals, and to avoid being the recipients of similar barbs (p. 51).

More broadly, it has been posited that love and hate are quite real for sports fans. Wann et. al, (1999) asked 88 college students to indicate their willingness to murder someone anonymously, and paired that with a question on their willingness to injure a star player and coach of a rival team. The results showed highly identified fans were not more aggressive generally, but only toward their sports rivals. Finally,

Klugman (2009) argued existing theories on team identification presume a "self-centered rationality at work" that fails to match or predict actual sports fan behaviors: "Thus current work on team identification indicates that something strange is occurring but it cannot yet explain why fans might behave, or consider behaving, in such excessive ways...Might love...explain something of the immoderate passions and identifications of...fans? (p. 23).

A number of scholars have called for work that examines fans over a period of time. Because so many studies examine fans at a particular moment, "Little is known about how mediated relationships can evolve...and how different stages...compare to...social relationships" (Cohen, 2010, p. 97; Click, 2007). Furthermore, there is a need for additional work on parasocial relationships in which fans actively dislike a media persona (Sun & Wu, 2012). Finally, Sanderson (2013) queried if there were "certain... sports figures...whose voluntary departures are more problematic" for fans (p. 504). This case study continues along that trajectory of inquiry by examining comments of Kansas basketball fans regarding former coach Roy Williams over a period of several years. To focus the analysis, the following research questions are posed:

- **RQ1:** In what manner did Kansas basketball fans discuss Roy Williams in an online setting?
- **RQ2:** Does the Kubler-Ross grief model explain the dynamics of fan attitudes toward Roy Williams?

METHODS

This project continues a recent trend in scholarship that uses qualitative methods to explain how sports fans respond to particular events (Kassing & Sanderson, 2009; Sanderson, 2013; Sanderson & Emmons, 2014). Here, the comment section of KUsports.com, the online iteration of the sports section of the daily newspaper in Lawrence, Kansas, was selected for analysis. The website was selected due to relatively high traffic, free registration and searchable archives. In an email interview, a former KUsports.com online editor indicated the site has approximately 500,000 unique visitors each month, and more than 250,000 registered users, though many of the latter are inactive (Newell, 2013).

One shortcoming is the site did not enable comments until 2005 (Hittle, 2009), so there is no online record of immediate fan reaction to Williams's departure. Therefore, only articles that referenced Williams in the headline after 2005 were selected, which resulted in the analysis of 995 reader comments on 17 articles from 2006-2014. The average of 59 reader comments per article shows that while Williams was not a frequent subject for KUsports.com writers, even brief wire reports generated substantial commentary from readers.

To answer RQ1, a thematic analysis was applied using a constant comparative method (Glaser & Strauss, 1967). The articles were reviewed chronologically to identify the presence of themes at various times. To determine what might be considered distinctive, Foss (2009) suggested the intensity and frequency of various features of the discourse should help guide the interpretation (p. 389). In other words, the repetition of certain themes in reader comments over a period of time suggests those feelings are significant to that particular group and should be documented. In a sense, this thematic analysis was a necessary precursor to help determine whether the Kubler-Ross model might have explanatory power, which follows suggestions that "emergent data-driven" methods can complement "more common theory-based" methods (Dick, 2010, pp. 410-411).

To answer RQ2, the Kubler-Ross (1969) model of processing grief was applied to each post. This follows Sanderson and Cheong (2010), who were among the first to apply the Kubler-Ross model beyond a face-to-face context, as they studied fan reactions to Michael Jackson's death, and coded responses based on the expressed grief state. Denial is present in a statement that something "cannot be true" and is characterized by disbelief that something has occurred (Kubler-Ross, 1969, p. 38). Anger is typically linked with statements indicating resentment, while Bargaining is typically an attempt at a short-term "deal" to stave off the inevitable (Kubler-Ross, 1969). Depression is constituted by statements that express personal shame or feelings of guilt, while Acceptance is seen in comments that one is neither "depressed [n]or angry" (Kubler-Ross, 1969, p. 112).

Baym (2009) observed generalizability in qualitative work is "neither relevant nor possible" given the fluid nature of reality, and the goal should therefore be to generate analyses with thick descriptions that could then be compared to other contexts (p. 175, 186). It is natural any such inquiry will be "suggestive, incomplete and inconclusive" (Charmaz, 2005). The hope is to achieve limited triangulation where certain patterns are identified and might be of utility to researchers in a similar area of interest (Lincoln, Lynham, & Guba, 2000).

Finally, Markham (2004) suggested online posts should not be corrected for spelling, punctuation, or grammar: "We literally reconfigure these people when we edit the sentences, because for many of them, these messages are a deliberate presentation of self" (p. 153). The insertion of "[sic]" therefore seemed unnecessary since some participants actively intend to deploy conversational heuristics in their posts (Rodriguez, 2016). In this spirit, posts were not edited for content; those that were truncated for concision are appended as "clipped."

RESULTS

Kansas basketball fans characterized Roy Williams in a variety of ways following his voluntary departure from the university. Had this study examined a snapshot of articles from 2006-2007, the tenor of such characterizations by fans would seem volatile and openly hostile. Had this study examined the past few years, a sampling of comments would seem to indicate nostalgic affection. Indeed, the most prominent characteristic of the fan discussions studied is that vitriol toward Williams dissipated dramatically after Kansas defeated North Carolina in the 2008 Final Four--and just 48 hours later, Williams was shown on the *CBS* national broadcast wearing a Jayhawk sticker. That night, Kansas would go on to win its first national championship since 1988. For Kansas fans who would forgive and accept Williams, that 48-hour period in April 2008 was by far the most frequently referenced constellation of events to change attitudes regarding his departure.

Three basic themes and attitudinal tendencies emerged during the analysis. The first is non-basketball related and concerns health issues Williams has had. The second involves the notion of broken promises, as fans characterized their dissatisfaction as stemming not from the idea of Williams leaving, but the manner in which his departure occurred. The final theme extends the notion of parasocial breakups to include parasocial divorce. The way in which each of these themes evolved chronologically suggested the potential utility of the Kubler-Ross model, which demonstrated a substantial dissolution of anger in 2008 that would eventually give way to acceptance of his departure.

Beyond Basketball: Health and Well-Wishing

In the years Williams coached at Kansas, fans had grown accustomed to seeing Williams experience vertigo during games, and he would be rendered incapacitated for a few moments before collecting himself and returning to a seat on the bench. Three years after leaving Kansas, a 140-word wire report documented that Williams had a pinched sciatic nerve and could require back surgery ("Tar Heels," 2006). The first three posts are succinct examples of the type of vitriol exhibited by the fans:

Maybe a KU BBall fan practices voodoo? Maybe he got it from patting himself on the back. (Mr_Missive, 2006)

Nobody cares about Benedict Roy around here anymore. Please confine your reporting to KU interests and not traitorous narcissists who could care less about the University of Kansas. (PsychHawk, 2006)

We new Roy had no spine the day he took the job Dean Smith wanted him to take. I forsee surgery, and maybe Kramer can drop a Junior Mint in before they sew him up to create some more issues!!!!!!!! I HATE ROY. I HOPE YOUR BACK ROTS TO A PILE OF DUST YOU TRAITOR! (cstevenday, 2006)

A few participants offered more moderate analyses. One individual was particularly prescient regarding the eventual conditions for reconciliation:

Agreed. I never hated Roy for leaving but like most, I was disappointed. I hope some day he returns to AFH to a standing ovation. I know if I were in attendance I would be on my feet. Sadly, I don't see that happening any time soon. I don't think Roy will ever be completely forgiven until CS wins it all. I hope that will happen soon. (seattlehawk_78, 2006)

Others attempted to encourage fans to remember their feelings for Williams before he departed:

...get over your hatred of Roy...I daresay no one on this board had anything but love for Roy 6 years ago... (Michael Foulston, 2006, clipped)

While much of the content in these posts is indefensible, it should be noted that sciatica, as characterized by the report, was not an urgent or particularly serious medical situation for Williams, who joked about the potential impact on his golf game.

In another minor incident in January 2008, Williams tripped over a cord in his office, and hit his head on the door ("Williams falls," 2008). The wire report indicated Williams "was on his feet unaided immediately after" and required five stitches. Some of the comments lacked empathy, ranging from "Good" (KanKu, 2008), to "WHO THE @#$% GIVES A @#$% ABOUT ROY WILLIAMS?" (vanillajoe, 2008) to "I hope the door is O.K." (yates33333, 2008). Another contingent of fans were curious as to why the article was posted. A vocal Williams supporter chastised others: "That kinda 'class' belongs at KState. It has no place here" (JoeRoss, 2008). And just one other expressed concern about his long-term health: "I really worry about Roy ever since he almost fainted when he stood up on the sidelines those few times before he left KU. I think it has happened once or twice at UNC too. Then the vertigo thing this summer. Now this. I really hope he is not hiding or ignoring a bigger problem" (Mohawk69, 2008).

This is in stark contrast to a September 2012 article that documented Williams having to undergo surgery to remove a tumor on his right kidney, with doctors unsure of whether it was malignant (Bedore, 2012). This time, the response was one of overwhelming support for Williams, notwithstanding a few misguided attempts at levity. Of the 73 reader comments, just one could reasonably be interpreted as vitriolic, as the KUsports.com staff removed it for violating the terms of the site. The first response to the redacted comment received eight "thumbs-up" from users--in stark contrast to the vast majority of comments, which do not generate any "thumbs-up":

Seriously dude (or whatever you are) ???

If you have a problem with Williams, so be it--but this is not the time to spew your poison. Nothing like kicking a man when he's down.

I'm glad that you're not the chosen spokesperson for KU fans. Roy was a GREAT coach here (and at UNC) and he was a GREAT spokesperson and representative for the University of Kansas while he was here. Heck -- he was a great spokesperson and rep for us AFTER he left too.

Get over it. (nuleafjhawk, 2012)

It is entirely possible that some of the earlier responses in the first two instances were more hostile because the participants did not perceive the situation as dire. Then again, antipathy toward Williams at that time may have superseded empathy in a number of scenarios. This attitude change from anger to acceptance will be discussed later, but first the notion of broken promises may help explain some of the vitriol.

Broken Promises

Williams was viewed for years by a sizable segment of Kansas fans as having broken his promise of staying at Kansas, not because he left but the way in which he left. One person commented that no one could blame Williams for returning to his home in North Carolina, but it became reasonable to hold this perceived broken promise against him. Some felt his comments about "never leaving" were indicative of a contrived "personality" (fansincewilt, 2008). Another attempted to persuade "people who like Roy and wish him well" by reminding them of two factors: "...he lied when he left, no matter how you spin it, he lied. Second, you can't wish him well because the two schools recruit a lot of the same people, and if they get the ones we want, it hurts us in the long run, so quit wishing the heels & UNC-Roy well..." (Jim Pendleton, 2008, clipped).

For fans who presented their opinions as a reflection of the broader fanbase, it was not a problem with Williams leaving, only how he did so: "Its not that he left. It is HOW he left. That is what we had the issue with..." (pitthawker, 2008, clipped):

...What they do find fault in is the fact that he said he was staying and then he left. It is A LOT like ending up on the losing side of a relationship. When someone that you care for deeply chooses another over you, it hurts. It brings to mind some credibility issues regarding the one who left. I don't fault him for leaving. I fault him for how he left. (JayCeph, 2007, clipped)

Finally, for another individual, having Williams turn down the North Carolina job a first time and openly expressing his disgust at it being mentioned a second time seemed to reaffirm the coach's loyalty to Kansas in a way that is characterized as renewing marriage vows:

Any of you who don't have a problem with our former coach...consider this situation. You're married for fifteen years, have lots of good times and children...and because of that, you'll always love them. But shortly after *renewing your vows*, your spouse runs off with her high school sweetheart, and then hits the freaking lottery!! You can honestly tell me that you still wouldn't be just a little angry? (sevenyearhawk, 2008)

That particular observation helps bridge the notion of broken promises and parasocial divorce. But what the last few sentences hint at with "high school sweetheart" and winning the lottery is the notion of experiencing success. Another aspect of the parasocial relationship fans had with Williams was the idea or expectation that he would find postseason success and bring a national championship to the university. In that way, his leaving before delivering a championship--and subsequently winning the 2005 national championship at North Carolina--could be interpreted as violating an unspoken agreement that Williams shared with the Jayhawk faithful. The first broken promise of staying at Kansas is far more overt and identifiable, but it is possible that for some fans, uncertainty regarding the future led to a heightened sense of betrayal.

Processing Parasocial Divorce: The Leaver and the Left

A sizable segment of participants compared Williams's departure to that of a significant other ending a relationship. A sampling of such commentary prior to April 2008 finds fans comparing it to "being dumped by your boyfriend who really seemed to care!" (aquakej, 2006, clipped). Many referred to the promises made by Williams, and indicated they now thought of him "like an ex" (Kirk, 2008, clipped) or a "cheating ex" (Kevin Long, 2008, clipped). A couple of participants expounded upon the analogy:

A parallel has been made many times of the 'jilted lover' and if we were to use that one today, then no matter how much you cared for that person, they still chose to leave you and then go on to a very public and 'successful' relationship with their new fling. That will never feel good. 15 years at KU is one thing and only worth so much... (JayCeph, 2008, clipped)

Some fans explained their bitterness toward Williams as a result of the strong parasocial ties that had developed during his tenure. His combination of his personality and success in the program tended to strengthen that bond, as one participant elaborated in a lengthy post:

The reason people, myself included, were so bitter when Roy left, was because he was such a part of our KU lives. He was involved, loving to his players and fans, and wore his emotions on his sleeve. The nasty things people say are going to happen because of the love we felt for Roy. This game is not about Roy and us, it is about the players on the floor, but the media had no choice but to bring up what every KU fan was thinking, a chance to play Roy and his Tarheels. I like to think of Roy as more of a wife you had children with, telling you she loved you and would stay with you regardless of who came knocking, then, 3 years later, she says, "honey, you are great and I gave you 15 great years, but I have to go home to my old husband who I love very deeply right now." If that happened to you, or one of your buddies, no matter how good the wife was, you would say some pretty nasty things about her. I agree with what all of you are saying. Yes we need to let it go, but it says a lot about Roy, that some of us just can't. Yes, we have remarried to a much better person, and we are happy again, but the pain is going to resurface, no matter how much we try and fight it, especially now, so lets get all the words out, so come game time, we can just focus on rooting our Jayhawks and our much better person, Coach Self, to a National Championship game. Love you Roy, and I understand why you left, but that does not mean it didn't hurt, and some of us will enjoy beating your tarheels back to your "homeland". GO HAWKS (truehawkfan, 2008)

Because of the frequency of comments from fans that compared Williams leaving to being on the receiving end of a divorce--or at least a break-up from a long-term relationship, it seemed worthwhile to engage in a secondary, purposive analysis that analyzed each post in terms of an expressed grief state. When reviewed chronologically and over a period of several years, it would seem the expressions of parasocial loss as articulated by Kansas fans tend to align with the way individuals might process grief in a social relationship.

Grief Analysis: From Anger to Acceptance

It was anticipated many posts would contain an expression of more than one grief state, yet this was not the case as 88 percent of posts were coded using a single state. It is beyond the scope of this endeavor to craft a distinction between depression and anger, as outward expressions of anger may conceal a depressed state. Nearly all of the posts concerning Williams fell on either side of a prominent anger/acceptance divide.

The previous section may have suggested the types of posts that would be coded as exhibiting anger or acceptance. An example of each is as follows:

Anger: "Roy Williams is no longer our coach...No one likes to hear about X-girlfriends and I do not want to hear about Benedict Williams. (BABOY, 2006, clipped)

Acceptance: "He proved his love for KU when he wore the KU sticker during the 2008 NC game, much to the chagrin of UNC fans." (brooksmd, 2012)

Of the 995 posts examined, less than 1 percent expressed an emotion that could be interpreted as exhibiting depression, and nearly all these focused on what was characterized as an incessant local media obsession with Williams. No posts exhibited "bargaining," which was not particularly surprising given that no participants mentioned or suggested any possibility of Williams returning to coach at Kansas. Perhaps if KUsports.com had user comments enabled at the time of Williams's departure, more posts would have exhibited characteristics of either denial or depression.

To demonstrate just how rapidly fan sentiment changed following the 2008 game, a review of the five articles published the week before and after are telling: Prior to the game, 60 percent expressed anger and 40 percent expressed acceptance; After the game, 16 percent expressed anger and 84 percent expressed acceptance. Sanderson (2013) showed how quickly a beloved group member could become despised, and these findings revealed how quickly a parasocial relationship could be repaired.

It is nearly cliché in sports to claim: "winning covers up mistakes," meaning that players, coaches, and fans tend to move forward after a victory rather than dwelling upon missed opportunities. This study shows a victory over a former coach can facilitate the grieving process for fans. In this case, winning--and the passage of time--led to Kansas fans forgiving and accepting Roy Williams.

Broken promises theory helped explain the anger that fans initially felt toward Williams. He was perceived as first going back on his claim that he was staying at Kansas and would not return to North Carolina. The second broken promise was that Williams would, in his oft-repeated phrase, "keep knockin' on that [post-season] door" and deliver a national championship to Kansas fans. Kansas would go on to defeat Williams, and the analysis demonstrated the shift from anger to acceptance was rapid. It may not be possible to piece together a broken promise, but Williams attending the next Kansas game wearing a Jayhawk sticker was seen as demonstrating his continued affection for the school and its fans, which would ameliorate the first broken promise. The perceived broken promise of post-season success was resolved when head coach Bill Self won the national championship for the Kansas faithful. Not only did this seemingly absolve Williams of that particular promise, it also seemed to remove any uncertainty Kansas fans had about their current head coach.

Had Kansas had lost to North Carolina in 2008, the results might have been markedly different. Fans could have clung to feelings of anger toward Williams, which may have been heightened by his besting of their current coach, Bill Self. Indeed, a frequent observation by fans since that particular game is they had relinquished ill will toward Williams in part because of the belief that Self is a better coach by virtue of his head-to-head success against Williams and his post-season success. It is quite likely this perception hastened the parasocial reconciliation process.

Recent posts indicate the vast majority (90 percent) of fans have come to terms with Williams's departure. The remaining 10 percent of fans may never forgive Williams, but when vitriolic comments surface, they are now quickly met by responses from others who challenge the content and suggestions to "get over it."

DISCUSSION AND FUTURE RESEARCH

This study introduced the notion of parasocial divorce, which broadens our understanding of the parasocial breakup aspect of PSI and PSR. Parasocial breakup had been applied to a hypothetical scenario of a favorite television personality being taken off the air. For vocal members of a fanbase, however, when a well-liked sports persona voluntarily leaves for another (perhaps rival) team, the metaphor of divorce is perhaps more apt. A breakup concept is more apropos in sports if the personnel move was not perceived as voluntary. But when a fanbase expresses a deep sense of personal betrayal regarding a sports persona leaving the team, both the strength of verbiage and the presence of relational metaphors suggest the concept of divorce may be more fitting.

These findings therefore affirmed an early observation from Horton and Wohl (1956) that the relationship between a fan and a media persona is "experienced as of the same order as, and related to, the network of actual social relations" and is "even more the case when the persona becomes a common object to the members of the primary groups in which the spectator carries on his everyday life" (p. 228).

A great deal has been written about pockets of fandom and the normal states of behavior and practice. What remains undertheorized is what happens when a fanbase experiences a crisis. This essay suggests grief studies can lend explanatory power to conceptualize changes in fandom over time. It also answers calls to examine "the verbal abuse of opposing...coaches through anonymous postings on internet forums and discussion boards" (Dalakas & Melancon, 2012, p. 56).

The antipathy fans felt toward Williams was partly explained by Gabennesch's (1988) broken promises theory. It has been shown that social violations and trust violations are "more offensive" than minor and major moral violations by both friends and media figures (Cohen, 2010, p. 105). This case study reaffirmed just "how quickly a revered group member can become despised" (Sanderson, 2013, p. 502). It also revealed how quickly a persona perceived as betraying the group could once again become accepted.

Sanderson and Emmons (2014) argued a number of variables influence whether forgiveness will occur, including the perceived seriousness of the transgression and the perceived firmness of a particular promise. Cohen (2003, 2004) has shown the mourning process is similar in interpersonal and mediated relationships. In this particular case, the passage of several years, a head-to-head victory, a symbolic gesture by Williams, and post-season success by his successor converged to expedite the parasocial grieving process for Kansas basketball fans. The long-term nature of the analysis demonstrated the capacity for forgiveness and eventual acceptance of a sports figure. Marwick and Boyd conceptualize celebrity as "an organic and ever-changing performative practice" (2011, p. 140). This analysis suggests such a view might rightly be applied to fandom as well.

There are limits to these findings. First, the interpretive nature of inquiry means another researcher could emerge with different findings. Second, these results concerned highly motivated sports fans with well-established parasocial relationships that perceived certain promises had been made and broken. Third, the potential for and speed of parasocial reconciliation was enabled by a constellation of factors in 2008. If, for example, Williams had defeated Kansas, it is quite likely that anger or bitterness would have continued or been renewed, rather than giving way to eventual forgiveness and acceptance. Now that the initial groundwork for PSI and PSR has been laid out in recent years, there is a greater need to situate studies within a proper, more limited context.

Additional long-term examinations of fan behavior are needed. Individuals who have fallen from grace (e.g., Joe Paterno, Lance Armstrong), could provide interesting case studies to examine how fans process bad news. Grief studies could certainly be applied in such an analysis. For example, the relationship between LeBron James and Cleveland Cavaliers fans is filled with peaks and valleys, and examining the way in which fans characterized him at different stages of that relationship could help reveal the nuances of anger and acceptance.

Future work could also examine the multi-generational nature of devoted sports fans. It seems reasonable to presume that sports programs (or franchises) that enjoy a tradition of success are more likely to have fans that "pass on" their affinity for a sports entity to a child, grandchild, or extended family member. This phenomenon may enhance parasocial dynamics with sports personae, as behavior seen as violating the group not only represents an affront to the individual sports fan, but to their family and social circles.

This analysis is among the first to examine sports fans attitudes over an extended period of time. Any other approach in this particular case study would have been a mere snapshot of a fluid process that would fail to reveal a fuller picture--one of strong relational dynamics with a sports persona. That parasocial relationship, in which Kansas fans perceived Williams as breaking his promise(s), is better viewed through the lens of grief studies and can be characterized as parasocial divorce.

These findings enhance our understanding of the types of apology—in this case, bolstering—that might be deemed acceptable by a scorned fanbase. Reconciliation and acceptance may never occur for some Kansas fans, but it has for most at this time. That process was facilitated by external factors that remediated Williams's perceived transgressions with the fanbase. This study adopted a holistic approach to fandom by recognizing the fluidity of relational dynamics in parasocial relationships with sports figures. The passage of time may be part of the forgiveness equation in parasocial divorce, combined with behavior that is deemed acceptable, and an attempt at image repair that is deemed authentic. Reader comments suggested the depths of fan devotion exceed what might be considered rational behavior and demonstrated the concepts of love, separation, grief and divorce are, in certain cases apropos in characterizing fan attitudes toward sports figures.

REFERENCES

Abelson, R. P. (1959). Modes of resolution of belief dilemmas. *The Journal of Conflict Resolution, 3*(4), 343–352. doi:10.1177/002200275900300403

Annette, N. (2011). To see and be seen: Celebrity practice on Twitter. *Convergence, 17*(2), 139–158. doi:10.1177/1354856510394539

Armstrong, D., Bukaty, N., & Gutierrez, J. (2013, Mar. 9). *The Border Patrol* [Radio Broadcast]. Kansas City: WHB.

Basil, M. D., & Brown, W. J. (2004). Magic Johnson and Mark McGwire: The power of identification with sports celebrities. In L. R. Kahle & C. Riley (Eds.), *Sports Marketing and the Psychology of Marketing Communication* (pp. 159–171). Hillsdale, NJ: Lawrence Erlbaum Associates.

Baym, N. K. (2009). What constitutes quality in qualitative internet research? In A. Markham & N. K. Baym (Eds.), *Internet inquiry: Conversations about method* (pp. 173–189). Los Angeles: Sage Publications; doi:10.4135/9781483329086.n16

Bedore, G. (1988, July 8). North Carolina assistant coach is KU's choice. *Lawrence Journal-World.* Retrieved from http://www2.kusports.com/news/1988/jul/08/north_carolina_assistant/

Bedore, G. (2003, April 11). Will Roy stay or will he go? *Lawrence Journal-World.* Retrieved from http://www2.kusports.com/news/2003/apr/11/will_roy_stay/

Bedore, G. (2012, March 23). Williams still fan of Jayhawks. *Lawrence Journal-World.* Retrieved from http://www2.kusports.com/news/2012/mar/23/roy-williams-still-fan-jayhawks/

Benoit, W. L. (1997). Image repair discourse and crisis communication. *Public Relations Review, 23*(2), 177–186. doi:10.1016/S0363-8111(97)90023-0

Benoit, W. L. (2018). Image Repair Theory in Sport. In A. C. Billings, W. T. Coombs, & K. A. Brown (Eds.), *Reputational Challenges in Sport: Theory and Application, 13.* New York: Routledge. doi:10.4324/9781315165608-3

Billings, A. C. (2012). *Sports media: Transformation, integration, consumption.* New York: Routledge. doi:10.4324/9780203832790

Borzi, P. (2013, March 25). Kansas surges past North Carolina and Williams. *The New York Times.* Retrieved from https://www.nytimes.com/2013/03/25/sports/ncaabasketball/kansas-surges-past-north-carolina-and-coach-roy-williams.html?_r=0

Bostwick, E. N., & Lookadoo, K. L. (2017). The Return of the King: How Cleveland Reunited With LeBron James After a Parasocial Breakup. *Communication & Sport, 5*(6), 689–711. doi:10.1177/2167479516659460

Bowlby, J. (1980). Attachment and Loss: Vol. III. *Loss.* New York: Basic Books.

Brown, N. A., & Billings, A. C. (2012). Sports fans as crisis communicators on social media websites. *Public Relations Review, 39*(1), 74–81. doi:10.1016/j.pubrev.2012.09.012

Bryant, J., Comisky, P., & Zillmann, D. (1981). The appeal of rough-and-tumble play in televised professional football. *Communication Quarterly*, *29*(4), 256–262. doi:10.1080/01463378109369413

Caughey, J. L. (1984). *Imaginary social worlds: A cultural approach*. Lincoln: University of Nebraska Press.

Charmaz, K. (2005). Grounded theory in the 21st century: Applications for advancing social justice studies. In N. K. Denzin & Y. S. Lincoln (Eds.), *The Sage handbook of qualitative research* (3rd ed., pp. 507–536). Thousand Oaks, CA: Sage Publications.

Cialdini, R. B., Borden, R. J., Thorne, A., Walker, M. R., Freeman, S., & Sloan, L. R. (1976). Basking in reflected glory: Three (football) field studies. *Journal of Personality and Social Psychology*, *34*(3), 366–375. doi:10.1037/0022-3514.34.3.366

Click, M. A. (2007). Untidy: Fan response to the soiling of Martha Stewart's spotless image. In Fandom: Identities and Communities in a Mediated World (pp. 301–315). New York: NYU.

Cohen, E. L. (2010). Expectancy violations in relationships with friends and media figures. *Communication Research Reports*, *27*(2), 91–111. doi:10.1080/08824091003737836

Cohen, J. (2003). Parasocial breakups: Measuring individual differences in responses to the dissolution of parasocial relationships. *Mass Communication & Society*, *6*(2), 191–202. doi:10.1207/S15327825MCS0602_5

Cohen, J. (2004). Parasocial break-up from favorite television characters: The role of attachment styles and relationship intensity. *Journal of Social and Personal Relationships*, *21*(2), 187–202. doi:10.1177/0265407504041374

Dalakas, V., & Melancon, J. P. (2012). Fan identification, schadenfreude toward hated rivals, and the mediating effects of importance of winning index. *Journal of Services Marketing*, *26*(1), 51–59. doi:10.1108/08876041211199724

Dick, B. (2010). What can grounded theorists and action researchers learn from each other? In The Sage Handbook of Grounded Theory Paperback Edition (pp. 398–416). Thousand Oaks, CA: Sage Publications.

Eitzen, S. D., & Sage, G. H. (1989). *Sociology of Northern American sport* (4th ed.). Dubuque, IA: Brown.

Emery, R. E. (2011). *Renegotiating family relationships: Divorce, child custody, and mediation* (2nd ed.). New York: Guilford Press.

Evans, L., & Hardy, L. (1995). Sport injury and grief responses: A review. *Journal of Sport & Exercise Psychology*, *17*(3), 227–245. doi:10.1123/jsep.17.3.227

Fernquist, R. M. (2000). An aggregate analysis of professional sports, suicide, and homicide rates: 30 U.S. metropolitan areas, 1971-1990. *Aggression and Violent Behavior*, *5*(4), 329–341. doi:10.1016/S1359-1789(98)00012-3

Foss, S. K. (2009). *Rhetorical criticism: Exploration and practice* (4th ed.). Long Grove, IL: Waveland Press.

Fowles, J. (1992). *Star struck: Celebrity performers and the American public*. Washington, DC: Smithsonian Institution Press.

Gabennesch, H. (1988). When promises fail: A theory of temporal fluctuations in suicide. *Social Forces*, *67*(1), 129–145. doi:10.2307/2579103

Giles, D. C. (2002). Parasocial interaction: A review of the literature and a model for future research. *Media Psychology*, *4*(3), 279–305. doi:10.1207/S1532785XMEP0403_04

Glaser, B. G., & Strauss, A. L. (1967). *The discovery of grounded theory*. Hawthorne, NY: Aldine.

Gleich, U. (1996). Sind Fernsehpersonen die "Freunde" des Zuschauers? Ein Vergleich zwischen parasozialen und realen sozialen Beziehungen [Are media figures "friends" of the viewers? A comparsion between parasocial and real social relationships]. In P. Vorderer (Ed.), *Fernsehen als "Beziehingskiste": Parasoziale beziehungen und interaktionen mit TV-personen* (pp. 113–144). Opladen: Westdeutscher Verlag. doi:10.1007/978-3-322-83274-0_12

Gottesman, A. (1997, March 22). Arizona 85, Kansas 82: 'cats run wild, scald Jayhawks--superior speed ignites tournament shocker. *Chicago Tribune*. Retrieved from http://articles.chicagotribune.com/1997-03-22/sports/9703220149_1_jayhawks-fourth-seeded-arizona-southeast-regional-semifinal

Gray, J. (2003). New audiences, new textualities: Anti-fans and non-fans. *International Journal of Cultural Studies*, *6*(1), 64–81. doi:10.1177/1367877903006001004

Hartmann, T., Stuke, D., & Daschmann, G. (2008). Positive parasocial relationships with drivers affect suspense in racing sport spectators. *Journal of Media Psychology*, *20*(1), 24–34. doi:10.1027/1864-1105.20.1.24

Hartsock, A. (2000, July 7). "I'm staying": Roy opts to remain at Kansas. *Lawrence Journal-World*. Retrieved from http://www2.kusports.com/news/2000/jul/07/im_staying/

Hiestand, M. (2008, April 7). CBS' Packer says honesty is still his policy. *USA Today*. Retrieved from https://www.usatoday.com/sports/columnist/hiestand-tv/2008-04-06-packer_N.htm

Hittle, S. (2009, July 23). In the world of online comments, there are plenty of opinions, but few names. *Lawrence Journal-World*. Retrieved from http://www2.ljworld.com/news/2009/jul/23/plenty-opinions-few-names/

Horton, D., & Richard Wohl, R. (1956). Mass communication and parasocial interaction: Observations on intimacy at a distance. *Psychiatry*, *19*, 215–229. doi:10.1080/00332747.1956.11023049 PMID:13359569

Hu, M. (2015). The influence of a scandal on parasocial relationship, parasocial interaction, and parasocial breakup. *Psychology of Popular Media Culture*. doi:10.1037/ppm0000068

Hunt, K. A., Bristol, T., & Edward Bashaw, R. (1999). A conceptual approach to classifying sports fans. *Journal of Services Marketing*, *13*(6), 439–452. doi:10.1108/08876049910298720

Jensen, T. (2008, April 5). Final Four 2008: Kansas vs North Carolina betting preview. *Point-Spreads*. Retrieved from http://www.point-spreads.com/college-basketball/040506-final-four-2008-kansas-vs-north-carolina-betting-preview.html

Kassing, J. W., & Sanderson, J. (2009). You're the kind of guy that we all want for a drinking buddy: Expressions of parasocial interaction on floydlandis.com. *Western Journal of Communication, 73*(2), 182–203. doi:10.1080/10570310902856063

Kelley, S. W., & Tian, K. (2004). Fanatical consumption: An investigation of the behavior of sports fans through textual data. In Sports marketing and the psychology of marketing communication (pp. 27–66). Mahwah, NJ: Lawrence Erlbaum Associates.

King, J. (2013, May 3). Roy vs. Kansas feels different this time. *ESPN.com.* Retrieved from http://espn.go.com/blog/north-carolina-basketball/post/_/id/13278/roy-vs-kansas-feels-different-this-time

Klugman, M. (2009). Loves, suffering and identification: The passions of Australian Football League fans. *The International Journal of the History of Sport, 26*(1), 21–44. doi:10.1080/09523360802500071

Kubler-Ross, E. (1969). *On death and dying*. New York: MacMillan.

Kubler-Ross, E. (1972). On death and dying. *Journal of the American Medical Association, 221*(2), 174–179. doi:10.1001/jama.1972.03200150040010 PMID:5067627

Lawrence Journal-World. (2006, June 24). *Tar Heels coach battling bad back*. Retrieved from http://www2.kusports.com/news/2006/jun/24/tar_heels_coach_battling_bad_back/

Lawrence Journal-World. (2008, January 12). *Williams Falls, Needs Stitches*. Retrieved from http://www2.kusports.com/news/2008/jan/12/williams_falls_needs_stitches/?mens_basketball

Lawrence Journal-World. (2009, July 8). *KUSports.com Jayhawk flashback: July 8, 1988, Roy Williams introductory press conference* [Video file]. Retrieved from https://www.youtube.com/watch?v=bnpT7AhUuwk

Lee, S. H., Simkins, T. J., Luster, S., & Chowdhury, S. A. (2018). Forgiving sports celebrities with ethical transgressions: The role of parasocial relationships, ethical intent and regulatory focus mindset. *Journal of Global Sport Management, 3*(2), 124–145. doi:10.1080/24704067.2018.1441737

Liebers, N., & Schramm, H. (2019). Parasocial interactions and relationship with media characters—An inventory of 60 years of research. *Communication Research Trends, 38*(2), 4–31.

Lincoln, Y. S., Lynham, S. A., & Guba, E. G. (2000). Paradigmatic controversies contradictions and emerging confluences. In The handbook of qualitative research (4th ed., pp. 163–188). Beverly Hills, CA: Sage Publications.

Maciejewski, P. K., Zhang, B., Block, S. D., & Prigerson, H. G. (2007). An empirical examination of the stage theory of grief. *Journal of the American Medical Association, 297*(7), 716–723. doi:10.1001/jama.297.7.716 PMID:17312291

Markham, A. N. (2004). Representation in online ethnographies: A matter of context sensitivity. In Online Social Research: Methods, Issues, and Ethics (pp. 141–155). New York: Peter Lang.

Mastromartino, B., Chou, W. W., & Zhang, J. J. (2017). The passion that unites us all - The culture and consumption of sport fans. In C. L. Wang (Ed.), *Exploring the rise of fandom in contemporary consumer culture* (pp. 52–70). Hershey, PA: IGI Global.

Mearns, J. (1991). Coping with a breakup: Negative mood regulation expectancies and depression following the end of a romantic relationship. *Journal of Personality and Social Psychology*, *60*(2), 327–334. doi:10.1037/0022-3514.60.2.327 PMID:2016673

Mellinger, S. (2013, May 3). Roy Williams may be here, but this is Bill Self's time at KU. *The Kansas City Star*. Retrieved from https://www.kansascity.com/2013/03/21/4135991/carolina-may-be-here-but-this.html

Mitrano, J. R. (1999). 'Sudden death' of hockey in Hartford: Sports fans and franchise relocation. *Sociology of Sport Journal*, *16*(2), 134–154. doi:10.1123sj.16.2.134

O'Neil, D. (2008, March 16). Kansas-Carolina, Ivy revenge and mid-major matchups are 'serendipity.' *ESPN.com*. Retrieved from: http://sports.espn.go.com/ncb/ncaatourney08/columns/story?columnist=oneil_dana&id=3297190

Osterman, L. L., & Hecmanczuk, T. A. (2019). Parasocial forgiveness: The roles of parasocial closeness and offense perceptions. *Journal of Social and Personal Relationships*, *§§§*, 0265407519879511.

Planalp, S., & Rivers, M. (1996). Changes in knowledge of personal relationships. In Knowledge structures in close relationships: A social psychological approach (pp. 299–324). Mahwah, NJ: Lawrence Erlbaum Associates.

Prigerson, H. G., & Maciejewski, P. K. (2008). Grief and acceptance as opposite sides of the same coin: Setting a research agenda to study peaceful acceptance of loss. *The British Journal of Psychiatry*, *193*(6), 435–437. doi:10.1192/bjp.bp.108.053157 PMID:19043142

Real, M. (2011). Theorizing the sports-television dream marriage: Why sports fit television so well. In Sports media: Transformation, integration, consumption (pp. 19-39). Academic Press.

Rodriguez, N. J. (2016). Vaccine-Hesitant Justifications: 'Too Many, Too Soon,' Narrative Persuasion and the Conflation of Expertise. *Global Qualitative Nursing Research*, *3*, 3. doi:10.1177/2333393616663304 PMID:28508015

Rubin, R. B., & McHugh, M. P. (1987). Development of parasocial interaction relationships. *Journal of Broadcasting & Electronic Media*, *31*(3), 279–292. doi:10.1080/08838158709386664

Sanderson, J. (2013). From loving the hero to despising the villain: Sports fans, Facebook, and social identity threats. *Mass Communication & Society*, *16*(4), 487–509. doi:10.1080/15205436.2012.730650

Sanderson, J., & Cheong, P. H. (2010). Tweeting prayers and communicating grief over Michael Jackson online. *Bulletin of Science, Technology & Society*, *30*(5), 328–340. doi:10.1177/0270467610380010

Sanderson, J., & Emmons, B. (2014). Extending and withholding forgiveness to Josh Hamilton: Exploring forgiveness within parasocial interaction. *Communication & Sport*, *2*(1), 24–47. doi:10.1177/2167479513482306

senorpalmer. (2009, December 17). *Roy Williams could give a sh!t about North Carolina* [Video file]. Retrieved from https://www.youtube.com/watch?v=pymcjOpiII8

Shank, M. D., & Lyberger, M. R. (2015). *Sports Marketing: A Strategic Perspective*. New York: Routledge.

Simpson, J. A. (1987). The dissolution of romantic relationships: Factors involved in relationship stability and emotional distress. *Journal of Personality and Social Psychology, 53*(4), 683–692. doi:10.1037/0022-3514.53.4.683

Sun, T., & Wu, G. (2012). Influence of personality traits on parasocial relationship with sports celebrities: A hierarchical approach. *Journal of Consumer Behaviour, 11*(2), 136–146. doi:10.1002/cb.1378

Sun, T., Youn, S., & Wells, W. D. (2004). Exploration of consumption and communication communities in sports marketing. In Sports Marketing and the Psychology of Marketing Communication (pp. 3–26). Mahwah, NJ: Lawrence Erlbaum Associates.

Wann, D. L. (2006). The causes and consequences of sport team identification. In Handbook of sports and media (pp. 331–352). Mahwah, NJ: Lawrence Erlbaum Associates.

Wann, D. L., Peterson, R. R., Cothran, C., & Dykes, M. (1999). Sport fan aggression and anonymity: The importance of team identification. *Social Behavior and Personality, 27*(6), 597–602. doi:10.2224bp.1999.27.6.597

Ware, B. L., & Linkugel, W. A. (1973). They spoke in defense of themselves: On the generic criticism of apologia. *The Quarterly Journal of Speech, 59*(3), 273–283. doi:10.1080/00335637309383176

Wenner, L. A. (1989). Media, sports, & society: The research agenda. In L. A. Wenner (Ed.), *Media, sports, & society* (pp. 13–48). Newbury Park, CA: Sage Publications.

Chapter 13
A Model for the Creation of Academic Activities Based on Visits

Antonio Sarasa-Cabezuelo
https://orcid.org/0000-0003-3698-7954
Universidad Complutense de Madrid, Spain

José Luis Fernández-Vindel
Universidad Nacional de Educación a Distancia, Spain

ABSTRACT

A typical activity of some university studies is to make visits to places outside the university where students must observe certain elements and take notes of what has been observed. Normally these visits are carried out the instructions of a guide that has been made by the teacher where it is indicated in detail how to carry out the visit: what should be seen and observed, what type of information should be retrieved, and the type of report that should be done of the visit. Likewise, these activities have some evaluation mechanism associated. The creation of these activities consumes a lot of time for the teacher. This chapter describes a proposal to automate the creation of such activities using an application that would act as an added value service that would process the information available in open data repositories and linked data in order to offer an editor/publisher of activities of visits.

INTRODUCTION

In some university studies it is common to carry out activities that consist of visiting a place and making observations or taking notes of certain elements seen during the visit. Normally, the activity is carried out outside the scope of the class, and to carry it out there is a guide that indicates in detail how to carry it out: what to visit, what details have to be observed, what type of information is necessary to extract… The guide is written by the teacher. Likewise, two variants can be found in the way of carrying out the activity. In some cases, the proposed visit is common for all students, and in other cases, each student

DOI: 10.4018/978-1-7998-3355-0.ch013

or group of students has a particularized visit that is different from all the others. As part of the activity, it is possible also to perform two independent evaluation tasks: a) Write a report about the visit, b) Fill out an evaluation test where it is asked something about the visit.

The structure of the activity described constitutes a pattern that can be found in many studies such as Art History, Botany, Library Science, Architecture, Fine Arts ... For example in Art History this activity pattern is materialized in visits to museums where students They should check the paintings of a specific painter. In the guide, it is indicated which paintings should be visited, in which pictorial aspects to be observed (colors used, geometric shapes, type of characters that appear ...), what documentation readings prior to the visit should be done, what questions should be answered once the activity has been carried out, and other aspects.

These types of activities are critical in the formation of a student since they give him a very practical vision of the subject he is studying. However, the design of these activities pose a problem for teachers in terms of temporary cost since they require the consultation of different sources of information to locate the specific elements that should be visited and the characteristics of the elements on which they should be observed, the creation of the visit guide or post-visit evaluation (report or evaluation tests about the visit). This cost increases if it each student or group of students must perform a different activity. Likewise, if several visits are made throughout the course, the cost in the worst case may be impossible to bear.

In this sense, in order to reduce the cost of the process of creating these activities, its creation could be automated. This is feasible given that, as indicated above, the structure of this type of activity responds to a common pattern, where the only thing that changes is the domain of knowledge. For example, when a professor of Art History wants to make an activity of this type, the steps that must follow are: a) Select a painter or artist that the visit will deal with, b) Find the museums that exhibit painters' paintings on which the visit to be planned is concerned, c) Find out if the type of paintings of the selected painters in a specific museum can be used to cover the objectives of the proposed visit, d) Select the paintings that will constitute the visit, e) Write the guide of the visit, f) Write the evaluation associated with the visit, g) Correct the evaluation of the visit. Although the process can be automated, it would be necessary to find particular resources for each learning domain. These resources are the basis for configuring the student's visit. In the previous example are resources: museums, painters and paintings.

On the other hand, there are currently countless repositories of digital information on very diverse topics that can be freely exploited. Specifically, it is worth noting the open data repositories (Huijboom et al, 2011) and the linked data (Bizer, 2009) repositories. These types of repositories respond to two different but complementary approaches to publishing information. In the case of open data, access to information is based on REST type web services (Masse, 2011) that are associated with web resources that are in the repository. In this sense, the information is normally recovered in the form of files in open formats such as JSON, CSV or XML. The invocation of the service is similar to accessing a web page (an URL is used, which can be complemented with some search condition, executed on a web browser), which instead of returning an html page, returns a file in some format open. On the other hand, in the case of linked data, access to information is done using a query language called SPARQL (Rakhmawati et al, 2013) that allows to navigate and retrieve information that is encoded according to a domain ontology. The information thus encoded has a graph structure, so that a query in SPARQL essentially allows to navigate the information graph and select certain branches or nodes of the graph (which constitute the result of the query). Another feature of this last type of repositories is that the information retrieved is usually provided in the form of RDF files (Shadbolt et al, 2010), although in some cases it can also be retrieved in other open formats (JSON, CSV or XML).

In this article it is proposed to take advantage of the information from digital repositories (Larson,2010) described in order to automate the creation of activities that consist of conducting a visit according to the activity pattern shown above. The role played by the repository would be to offer the primary resources with which to build the activity of the visit.

There are a wide variety of content creation tools that could be used to generate activities of visits such as text editors or digital content generation tools. A widely used example is the exeLearning tool that allows to create standardized learning objects where the content of the object is the aggregation of different resources such as images, videos ... and the use of information aggregation templates. The main limitation of these tools with respect to the stated objective is that they do not allow modeling and implementing the activity pattern described. It would be possible to create the activity, however, the use and recovery of the resources used in the visit guide should be obtained and handled in a manual way, making the process more inefficient. In addition, a content creation tool also does not provide the evaluation and student management part.

The article is structured as follows. In first section, an introduction is done about open data repositories and linked data repositories. Also it is introduced the concept of value-added services In the next section, the structure of visit activity structure is formalized and a life cycle proposal for the creation of activities according to the pattern defined in terms of tasks and actors is described. In the next section, it is described a case of use of the development of activities in the context of the area of Art History. And finally in the last sections a set of conclusions and lines of future work are proposed.

BACKGROUND

Open Data Initiative

Open data (Janssen et al, 2012) is an initiative that arises in the field of large public and private institutions that aims to make available to anyone the data generated in their daily activity. In this way, anyone can exploit the data in various ways such as performing analyzes, prediction models, obtaining trends.... To access the data, these are stored in digital repositories that offer various services to potential users. These services include search and recovery mechanisms. The search can be based on keywords associated with the data sets, or by browsing through a taxonomy that has been used to characterize them. There are also data catalogs where all available data sets are collected. With respect to the recovery service, there are two ways in general. One way is to download the data set directly by clicking on a link in the repository. And the other way is the recovery via a REST web services API (Bodle, 2011). A web service of this type is a URI that is referencing a web resource, so invoking it returns the resource as a result. The API offers a catalog of this type of services (Hausenblas et al, 2010). The main advantage of an API is that it is possible to access resources in a transparent way and from the point of view of the institution allows to make its resources public without having to show the internal functioning of the system. APIs usually constitute a query language, since URIs can be accompanied by key-value pairs that represent conditions or ways of configuring the web service. These pairs are separated in the URI by the symbol "&". When the web service is invoked, it returns as a result the associated web resource in the form of an open format file such as JSON, XML, CSV ... There are a wide variety of public and private institutions that are participating in this initiative. For example, in Spain at the level of the government of Spain (Sanchez, 2015) is the open data portal (https://datos.gob.es/) which compiles a catalog of Spanish public institutions (Ferrer-Sapena et al, 2011)that provide open data sets from their activity. At the level of private institutions, large companies in the technology sector (Russell, 2013) stand out such as Twitter (https://developer.twitter.com/en/docs/api-reference-index), Facebook (https://developers.facebook.com/docs) / graph-api / using-graph-api), LinkedIn (https://developer.linkedin.com/docs/rest-api) or Google's multiple open data APIs (Mueller, 2006). Note that in recent years some of these companies have begun to charge a fee for using these services and accessing stored data. A very interesting aspect of these service APIs (Hausenblas, 2009) is that they facilitate the creation of a type of computer application that is characterized because the persistence of the data is in a third party rather than in the application itself. In this way, all maintenance tasks of the data sets are delegated to the institution that offers them. However, this is also a disadvantage because if the institution stops offering these data or modifies the way of access, the applications created based on them will cease to function or will have to be modified. This situation has already happened for example with applications that made use of the data returned by Google APIs, when transforming into payment services, they have also changed the way of access, having to modify the code of the programs that made use of the same.

Linked Data Initiative

Linked data (Heath et al, 2011) is an initiative that arises in the field of semantic web (Verborgh et al, 2012). The objective is to interconnect data sets through semantic relationships that facilitate the search and retrieval of information. To describe these relationships, tuples of the Subject-Property-Value type described using a standard called RDF (Kahan et al, 2002) are used. These tuples can be chained giving rise to a graph of relationships that can be navigated to retrieve related information. The great advantage offered by linked data (Hartig et al, 2010) is that they constitute a large database distributed throughout the web because the information that is interconnected does not have to be in the same physical place. Therefore the exploration of the data linked in similar to the navigation that is done through the web links found on a web page. To consult this graph, a standard query language called SPARQL (Perez et al, 2006) has been defined that acts on documents written in RDF. Syntactically the language is very similar to the SQL language. As a result of the execution of a query in SPARQL, a set of RDF triples that satisfy the search is returned. Normally the institutions that offer access to their linked data provide what is called a SPARQL access point (Quilitz et al, 2008) that offers various services to the user who wants to consult the data such as a SPARQL editor, frequent queries, documentation or or pieces of code in different languages of programming to invoke the queries embedded in a program. Likewise, these access portals allow retrieving the data of the queries in various open formats such as JSON, XML or CSV. Also, the use of SPARQL queries allows the creation of the same type of computer applications above described where data persistence is maintained by a third party. In recent years, a large number of institutions have published their data in the form of linked data. Some examples would be the National Library of Spain (http://datos.bne.es/inicio.html), the European project Europeana (https://pro.europeana.eu/page/sparql) or Wikidata (https://query.wikidata.org/).

Value-Added Services

Value-added services (Lee et al, 2012) refer to a type of computer applications that are characterized in that they offer functionalities built on the data provided by third parties. Normally these data are not useful in the state they are in and require some type of processing to be useful. In this sense, essentially a value-added service takes the data of a third party and processes it in some way that is useful to a type of user. In recent years, these types of applications have multiplied encouraged by the phenomenon called Big Data (Kitchin, 2014) that has made available to users huge amounts of information that can be analyzed and processed. Technologically, these applications use the REST-type web services APIs (Srivastava et, 2003) that offer the repositories of the data that is exploited. In this way the applications combine various invocations of web services to retrieve data that are subsequently processed to implement the functionality offered. A historical precedent to web services is called Mashup (Benslimane et al, 2008). The main advantage offered by this type of applications is that the maintenance of the persistence of the information is performed by a third party so that the value offered by the application is the way in which the data is processed and the results are displayed. On the other hand, another advantage of these applications is that the exploitation of information is very simple, given that it follows a producer-consumer model (Srivastava et al, 2003). It also has the advantage of the wide variety of applications that can be made from the same set of data, since the only thing to design are forms of processing and visualization that are useful for a certain type of user. Each of them will lead to a different value-added service. There are a large number of examples of applications (Labrinidis et al, 2012) of this type such as hotel search websites, airplanes, shows ... such as Trivago (https://www.trivago.es), Booking (https://www.booking.com), Kayak (https://www.kayak.es) ... All these applications take the data of third parties and offer a value-added service that consists in facilitating the search that best suits the needs of each user. However, the data used does not belong to the application, it is data offered by the establishments, which are aggregated in a concrete and useful way for the user.

A MODEL FOR THE DEVELOPMENT OF ACTIVITIES

This section will describe a proposal of life cycle model for the creation of visit-based activities. As described above, there are many university studies that carry out learning activities based on visiting a place where students must observe a set of elements and characteristics of them. These activities follow a common pattern of how they are created by the teacher. Thus the teacher normally performs the following tasks:

1. Find the place where the visit is going to take place.
2. Find in the place the elements that should be visited and determine what characteristics should be observed by the students.
3. Create the guide that details how the visit should be carried out: elements that should be observed and information that should be extracted, and how the report should be generated by students as a result of the visit.
4. Generate the evaluation test associated with the visit.
5. Perform the correction of the report generated by the students and the assessment test.

This pattern can be abstracted in a life cycle model that allows structuring the production of activities and implementing a sustainable production process. The key point to create this model is the availability of resources that can be used by the teacher to create activities and assessment tests. These resources can be found in open and linked data repositories. In this way, the teacher is released having to search for resources, he will only have to select them. To define the model (see Figure 1), the following elements have to be considered: a) Actors involved and b) Tasks performed by the actors.

With respect to the actors, 3 types of actors are going to be distinguished: teacher, student and computer engineer. The teacher is the actor who will perform most of the tasks of the model and who will be responsible for creating and publishing the activities, and correcting the assessment tests. The student will be responsible for consuming the activities created by the teacher and taking the assessment test. Finally, the computer engineer will be responsible for performing the necessary encodings to invoke the web services or SPARQL queries of the open data and linked data repositories. The content of these repositories will be used by the teacher to create the activities.

And with respect to the activities developed by the actors, 8 types are defined:

1. Select the type of activity. The tasks performed in an activity can be varied: read a book, view a video, and listen to an audition... And the activity could be structured in different ways for which it could have been previously created templates. Each template would have a different structure so that the teacher would only have to indicate what content should be inserted in each part of the template. For this, the teacher would have a template repository where he could choose a template or create new templates.
2. Creation of the activity. Once it has been selected the template that corresponds to the type of activity, fill in the template with content. These contents are extracted from the repositories of open data and linked data to which it is possible to access. If the content of the teacher is not found among the contents of these repositories, it is possible also add it directly from the local disk. Once the activity is created, the teacher will publish it so that it is available to students. In the creation process, the activity could also be kept in a draft state, if the teacher has started to create the activity but has not finished it.
3. Creation of the evaluation test. After creating the activity, the teacher must create a test and evaluation associated with the activity. For this, it will also be available evaluation test templates in which it will be only necessary to indicate the contents that should be included in the template. However, these templates will have as particularity that they will have different types of questions from which

Figure 1. A model for the development of activities

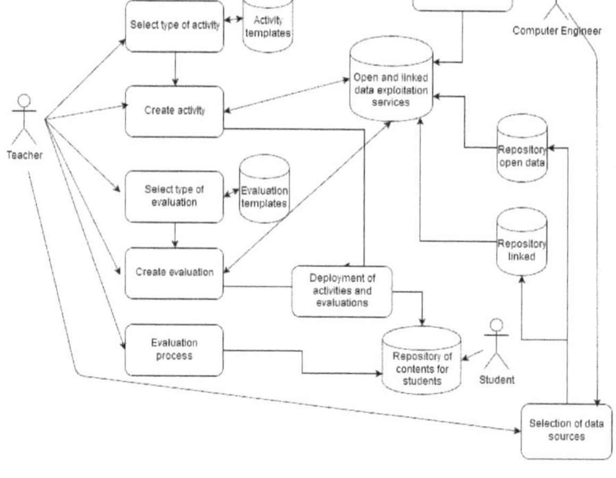

the teacher can choose: free text questions, multiple choice test questions, relationship questions ... As in the case of activities, the teacher There will be a repository of evaluation templates, in which it will be able to choose the template wanted or if it does not exist, it is possible to create new templates.

4. Creation of the evaluation test. Once it has been selected the corresponding template, the teacher will indicate the contents to be included and select the types of questions. To do this, it will make use of the open data and linked data repositories that are accessed. Once the evaluation test is finished, it will publish and display the test so that students can access it. As in the previous case, in the creation process the test may also be in a draft state at some point if it has not been completed by the teacher.
5. Deployment of activities and evaluation tests. As previously mentioned, when the teacher finishes the creation of the activities and the evaluation tests, he publishes and displays them. This action results in them being available to students. An evaluation test cannot be carried out until the activity has been carried out by the student. Likewise, the next planned activity cannot be accessed if the previous activities and evaluation tests have not been carried out.
6. Evaluation process. The evaluation tests carried out by the students are stored in a repository. The teacher will access this repository to evaluate the tests performed by the students. In this sense, a grade will be assigned, which will be associated to the student's account. In this way, when the student enters his account he will be able to see the activities carried out, and the qualifications of the evaluation tests he has carried out. Likewise, it is also possible to see the evaluation tests that are pending evaluation by the teacher.
7. Selection of data sources. A key task in the proposed process model is the selection of the data sources to be used. To do this, the teacher will select which sources they want to use, and will register them so that the computer engineer can make the connection with those sources.
8. Implementation of the exploitation services of the data sources. Once the teacher has selected the data sources he will use, the computer engineer must make the connection with them. To do this, it will have to schedule invocations of web services in the case of open data repositories or SPARQL queries if it is linked data repositories. It is also possible that the information could be found in other data sources as a relational database for which it should be implemented SQL queries. In either case, these implementations will result in obtaining the information from the repositories and their storage in local persistence systems. Likewise, as part of the engineer's task, he must create the necessary programs so that the teacher can select the recovered contents in a friendly way, through intuitive searches and a flexible selection mechanism.

DEVELOPMENT OF ACTIVITIES FOR THE AREA OF ART HISTORY

Open Data Portal of the City Council of Madrid

The city council of Madrid has an open data portal (https://datos.madrid.es/portal/site/egob/) where it is possible to access a catalog of 435 datasets about the activity of the city. The catalog can be downloaded in several formats such as RDF or CSV. Each dataset has associated information about the data included: frequency, number of total downloads, formats in which it is available, date on which it was included in the catalog and other information. Datasets are ordered using conceptual categories. In order to search on the portal, there are two different ways. One possibility is to navigate on the catalog using the categories. And other possibility is to use a filtering system. In this case, it is able to be setting a set of filters using the elements shown in the link "Filter by...". Using these filters, the search is limited by formats, categories, date of incorporation, name.... The data recovered can be download directly using a file in an open format or it is able to use a web service embedded in a program in some programming language.

The Open Data Portal of the Municipal Transport Company.

The Municipal Transport Company of Madrid has an open data portal [29] that can be found at https://opendata.emtmadrid.es/Home. The goal of the portal is to provide information on the activity of the services offered by the EMT (bus, public bike, mobility, and parking). Access to data can be done in two different ways. The first way is through an API-REST web services. In order to use it, it is necessary to register. The record returns as an e-mail message with two values: the client identifier and a pass-word. Next, you have to choose the service you want to use. There are main services: BUS, GEO, MEDIA, INFOPARKING and BICIMAD. Each service has an associated set of web services that, when invoked, return a result that is encoded in an XML document. For example, the BUS service has the "GetRoute-Lines" service which, given the number of a line or set of bus lines, obtains the itinerary of a line (or several lines separated by the character pipe (|)), with the vertices to build the routes of the route and the UTM coordinates of the road axes and the stop codes. An important feature of these services is that the data used is retrieved in real time from sensors that are located in buses, bicycles or parking. The other alternative way of accessing the data is through a catalog of data. These data have been previously processed so it does not support any query, they can simply be downloaded in a file in XML format. There are two types: about general information about the services of the EMT such as incidents in the lines, schedules, or schedules, or specific information of each of the bus lines of the EMT. In any of the cases, the data used is not recovered in real time. Likewise, the portal offers a repository of mobile applications that have been built on the data offered by the open data portal.

The Linked Data Portal of Wikidata

Wikidata (Erxleben et al, 2014) is a repository of linked data .It is supported by Wikipedia (Lehmann et al, 2015), Wikimedia Commons and other initiatives related to Wikimedia (Auer et al, 2007). The goal is to create an enormous graph of knowledge linked. Each element of information has an identifier, a label, a description and a set of aliases. The identifier is unique and it is formed by a letter Q followed by a number. The description shows characteristics about the element and consists of a property and a value. A property is identified by a letter P followed by a number. The more interesting feature about the properties is that it is able to link to external databases. In particular, there are links that connect with content in wikis such as Wikipedia, Wikibooks or Wikiquote. The information can be displayed in any language, and always it will display the most up-to-date data. There are several ways to access to data: integrated tools, external tools or programming interfaces. One of the more used tool is the SPARQL endpoint (Vrandečić, 2012). With this tool, it is possible to write queries in SPARQL and recover the results in several formats such as CSV or JSON. Also, it offers another services: the coding of the invocation of queries in several programming languages, the possibility of associating a url with the results of a query to share it, a set of examples of SPARQL queries, a visual assistant to build the queries, several visual browsers on the linked data, and a wide variety of help documentation.

An Application to Create Activities

The objective of this experiment was to show how it can be exploited of data from open data stores and linked data stores to generate activities based on visits in the context of Art History. These are two sources of information that are consulted in a different way and that contain information and complementary data. In this sense, it was defined the following characteristics:

- The application must retrieve all the available information about the painting museums of Madrid, as well as the painters and the paintings that the museums have. This data will be extracted from an open data warehouse belonging to the Madrid City Council and a linked data warehouse belonging to Wikidata. The information retrieved should be processed to be combined in a consistent manner.
- The application must also retrieve information on how to get to a specific museum using the municipal buses of the city from a specific point in the city. This information will be extracted from an open data warehouse belonging to the Municipal Transport Company.
- The data will be recovered and processed by an application created in Python, which will be responsible for connecting to each of the aforementioned data stores, making the queries, processing the data and generating a web application that allows the user to navigate over the recovered data.
- In addition, the web application will allow to create activities based in visits to the museums and evaluation tests.

In order to do this application, a Python program has been created whose main characteristics are the following:

1. The resources that are used by the teacher to generate the activities are recovered from two different data repositories. For a part of the repository of linked data Wikidata. This repository contains information about the museums and paintings of the main museums in Spain. In particular, it is possible to access photographs of museums, paintings and texts that describe the history of the museum or the biography of the painter. To retrieve this information, a set of SPARQL queries have been designed that retrieve the following information from each museum: image, location, opening date, website, founder and street. However, when making the queries it has been detected that some information fields are not available for all the information elements. For example, paintings appear where their location is not indicated or when they were painted. This is because Wikidata is a collaborative project and sometimes there are authors who do not enter all the data because they are not available.
2. In order to complement the information retrieved from Wikidata, the open data portal of the Madrid city council has also been used. Several data sets have been consulted on this portal that include information about museums, paintings and painters, with which the information retrieved from Wikidata has been complemented. In this way it has been possible to have more complete information. Access to this information has been made through REST type web services that are embedded within the Python code. These services are launched at the same time as SPARQL queries on Wikidata, and with the results obtained they are stored together in the tables of a relational database.

3. Finally, use has been made of the open data portal of the EMT (Municipal Transport Company). This portal has been used to obtain the route to reach a museum using buses. This information has been used by the teacher to fill in the activities and provide students with instructions on how to get to the museums. To retrieve this information, web services of type REST have been used. Unlike the other portals, these services are dynamically executed when the teacher creates the activity since it is not known a priori from which points the route will be created to the museum that appears in the activity.

Both in the case of SPARQL queries and in the case of web services, the results are obtained in JSON files. These files are processed to retrieve the information embedded in them, and it is stored combined in a relational database. To consolidate the information, each JSON document is processed and the museum-to-museum information is processed, and that information that is not found in the Wikidata JSON document is completed with the information retrieved from the Madrid open data portal. Note that there are some cases in which the information does not appear in any of the sources, in which case, the information field is empty. In this way when the teacher has to select some element of the museum, the origin of the information will be transparent. Then, in order to work from the Python program, the information is stored in a dictionary that contains the consolidated information that has been retrieved from the database. In this dictionary the following information is stored from each museum: name, image, latitude and longitude, street, postal code, description, timetables, website, inauguration, number of collections and founder. The dictionary takes as a key the name of the museum and as a value another dictionary that contains the information mentioned above.

The main page of the web application shows 4 options: Museums ("Museos"), Painters ("Pintores"), Route ("Ruta") and Activities ("Actividades"). When you click on the Museums option ("Museums") a new web page is generated that shows a list of all the museums in Madrid (see Figure 2) that you have retrieved from the SPARQL queries made and the web services executed. Each museum shows an image and certain information about the museum: website, latitude and longitude of its location, the street where it is located, the postal code, a brief description, the opening hours and the curious fact about when it was inaugurated. More information is really available that could be displayed but with the aim of not overloading the presentation too much, this representative information has been chosen.

Figure 2. Option that shows the information recovered about the museums of Madrid

When it is clicked on the painter's option, a new page is generated where a drop-down appears with all the museums. Then it is possible to choose a specific museum (see Figure 3.a). Next, a text box appears where it can be inserted a painter's name. When it is clicked on Send ("Enviar"), then what is done is to check if the introduced painter has paintings in the museum that has been selected. If there is not any paint, it is indicated with a message that no paint has been found. And in case of finding the name of the painter, what is done is to recover all the paintings that the museum has made by this painter. As a result, a list of all paintings by the selected painter is shown (see Figure 3.b). Each image shows an image and some data on it such as: title, width, height, collection to which it belongs and a web link to the museum page where the image has been recovered. As in the case of museums, more information is available but only the most characteristic has been shown.

Figure 3. a) Search form; b) Recovery of the paintings of "Francisco de Goya" of the "Museo del Prado"

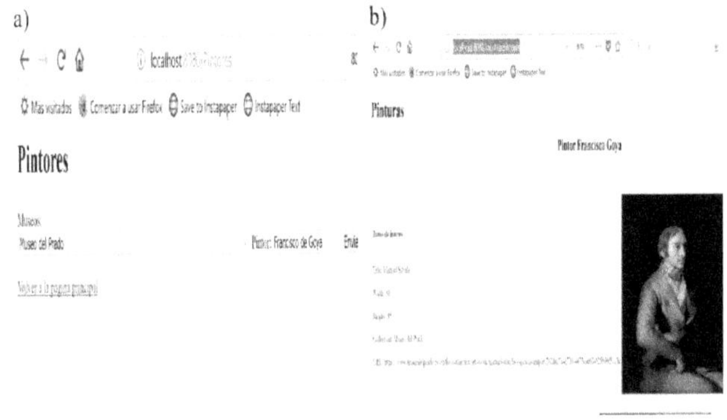

When the route option is pressed, a page is generated where there is a drop-down where it is possible to select one of the museums in Madrid that have been recovered in the SPARQL query. Next, there is a text box where it is possible to insert the name of a street and the number of the street. Then click on the Send ("Enviar") button (see Figure 4). As a result, the program launches a web service that connects to the EMT open data portal. The processing performed by this service consists in finding the set of buses that should be taken to go from the indicated street to the selected museum. The web service returns as a result a json document in which the list of buses, bus stops and travel times corresponding to the journey between the museum and the requested street appears. It also indicates the approximate

Figure 4. Search form to "Museo del Prado"

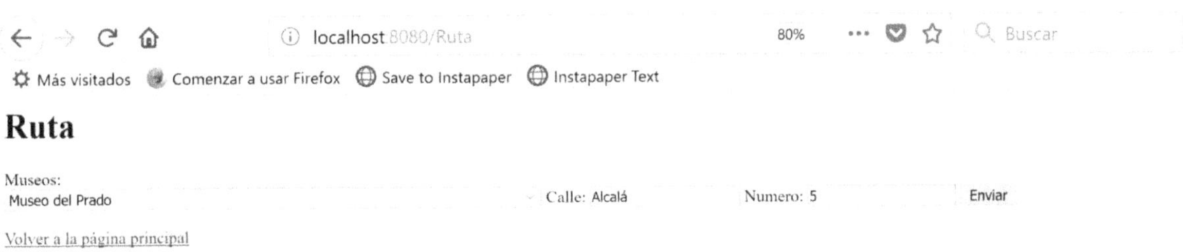

time that the calculated route will last at the moment the web service is launched. The formatted JSON document information will be displayed on the web page.

Finally there is the option of activities. When this option is pressed, a new web page is generated in which a randomized evaluation test appears (see Figure 5). For this, what the application does is to take a picture of a museum or painting according to the choice that would have been made in the previous pages. It also takes a randomized assessment test template. There are several evaluation templates with different types of questions: relationship questions, false / true questions, test type questions with several options ... Depending on the template that it has choosen, it generates the appropriate type of question, using the information that it has been retrieved from the Museum or painting. Then the student can answer the question. All assessment test templates are questions that have a single true answer so they can be automatically resolved by the program without teacher intervention. Once the student answers the question, they are automatically shown the true answer. Then, it is able to continue to solve more questions or if it is wanted to leave the question page and return to the beginning. The result of the questions is stored in the account associated with the student.

Figure 5. Page of the educational activity

CONCLUSION

This article has presented a proposal for the creation of activities based on visits to institutions. These types of activities are very common in certain university studies, and involve a great cost and effort for the teacher since a large number of sources of information must be reviewed. The article describes that all these activities, regardless of the area of knowledge, follow a common pattern in the process of creating them. Based on this fact, the article proposes a life cycle model that tries to model this pattern. In this way, the main actors that intervene in the productive process of the activities are identified, and the tasks that they must perform must be defined, and how both the tasks and the different actors involved must be coordinated. Another result of the article is the proposal that is used as resources to create the activities, which can be retrieved from open and linked data repositories. The advantage of using this type of repositories is the ease of retrieving information, its transparent character, and the fact of being free. Also using this type of repositories it is possible to create a type of application that implements the concept of value-added service, and whose main characteristic is to exploit information that is maintained by a third party and in the main contribution is the way in which that information It is processed and exploited. On the other hand, the article presents an example of the application of the proposal to create activities according to the model presented. In this case, it is a web application that takes the information from 3 open and linked data repositories: the open data portal of the Madrid city council, the open data portal of the EMT and the linked data portal Wikidata. The purpose of the application is to recover all the information available from the museums of Madrid and the paintings that contain those museums. Based on the information retrieved, a set of user-oriented services have been implemented. Essentially, the application allows you to navigate through museums and through the paintings that each museum has, taking as its entry point the name of a painter who is used to recover the paintings. Likewise, the application is capable of generating randomized evaluation tests by taking the information it has recovered and the museums and paintings that the user has visited. These questions have only one correct answer so they can be resolved by the program without teacher intervention. The grades obtained are stored in the user's account.

The main limitations presented by the proposal presented in the article is the total implementation of the model described. Another important limitation is the type of templates used to generate the evaluation tests, since the number of types of questions that can be asked are numerous, and those that allow randomness are not so many. There would also be another limitation on the templates used to present information about museums and paintings.

FUTURE RESEARCH DIRECTIONS

The main lines of future work are the following:

1. Create an intelligent system that can raise new evaluation questions based on user responses.
2. Create a mobile application from which access and interaction with the application is allowed.
3. Extend the tool that has been currently implemented so that you can retrieve information from other museums in Spain.
4. Create an advanced editor that allows you to manage different data sources and combine them.
5. Add intelligence to the system so that it is able to create activities semi-automatically.
6. Refine the proposed process model with the objective of adapting it to any area of knowledge.

REFERENCES

Auer, S., Bizer, C., Kobilarov, G., Lehmann, J., Cyganiak, R., & Ives, Z. (2007). Dbpedia: A nucleus for a web of open data. In *Lecture Notes in Computer Science: vol. 4825. Proceedings of the 6th International Semantic Web Conference and the 2nd Asian Semantic Web Conference*. Springer. 10.1007/978-3-540-76298-0_52

Benslimane, D., Dustdar, S., & Sheth, A. (2008). Services mashups: The new generation of web applications. *IEEE Internet Computing*, *12*(5), 13–15. doi:10.1109/MIC.2008.110

Bizer, C. (2009). The emerging web of linked data. *IEEE Intelligent Systems*, *24*(5), 87–92. doi:10.1109/MIS.2009.102

Bodle, R. (2011). Regimes of sharing: Open APIs, interoperability, and Facebook. *Information Communication and Society*, *14*(3), 320–337. doi:10.1080/1369118X.2010.542825

Erxleben, F., Günther, M., Krötzsch, M., Mendez, J., & Vrandečić, D. (2014). Introducing Wikidata to the linked data web. In *Lecture Notes in Computer Science: vol. 8796. Proceedings of the 13th International Semantic Web Conference*. Springer. 10.1007/978-3-319-11964-9_4

Ferrer-Sapena, A., Peset, F., & Aleixandre-Benavent, R. (2011). Acceso a los datos públicos y su reutilización: Open data y open government. *El Profesional de la Información*, *20*(3), 260–269. doi:10.3145/epi.2011.may.03

Hartig, O., & Langegger, A. (2010). A database perspective on consuming linked data on the web. *Datenbank-Spektrum*, *10*(2), 57–66. doi:10.100713222-010-0021-7

Hausenblas, M. (2009). Exploiting linked data to build web applications. *IEEE Internet Computing*, *13*(4), 68–73. doi:10.1109/MIC.2009.79

Hausenblas, M., & Karnstedt, M. (2010). Understanding linked open data as a web-scale database. *Proceedings of the second conference on Advances in Databases Knowledge and Data Applications*, 56-61. Accessed 20 November 2018: https://aran.library.nuigalway.ie/handle/10379/1127

Heath, T., & Bizer, C. (2011). Linked data: Evolving the web into a global data space. *Synthesis Lectures on the Semantic Web: Theory and Technology*, *1*(1), 1-136. doi:10.2200/S00334ED1V01Y201102WBE001

Huijboom, N., & Van den Broek, T. (2011). Open data: an international comparison of strategies. *European Journal of ePractice*, *12*(1), 4-16.

Janssen, M., Charalabidis, Y., & Zuiderwijk, A. (2012). Benefits, adoption barriers and myths of open data and open government. *Information Systems Management*, *29*(4), 258–268. doi:10.1080/10580530.2012.716740

Kahan, J., Koivunen, M. R., Prud'Hommeaux, E., & Swick, R. R. (2002). Annotea: An open RDF infrastructure for shared Web annotations. *Computer Networks*, *39*(5), 589–608. doi:10.1016/S1389-1286(02)00220-7

Kitchin, R. (2014). *The data revolution: Big data, open data, data infrastructures and their consequences*. Sage.

Labrinidis, A., & Jagadish, H. V. (2012). Challenges and opportunities with big data. *VLDB Endowment*, *5*(12), 2032–2033. doi:10.14778/2367502.2367572

Larson, R. R. (2010). Introduction to information retrieval. *Journal of the American Society for Information Science and Technology*, *61*(4), 852–853. doi:10.1002/asi.21234

Lee, Y. J., & Kim, J. S. (2012). Automatic web api composition for semantic data mashups. In *Proceedings of Fourth International Conference on Computational Intelligence and Communication Networks*. IEEE Computer Society. 10.1109/CICN.2012.56

Lehmann, J., Isele, R., Jakob, M., Jentzsch, A., Kontokostas, D., Mendes, P. N., ... Bizer, C. (2015). DBpedia–a large-scale, multilingual knowledge base extracted from Wikipedia. *Semantic Web*, *6*(2), 167–195. doi:10.3233/SW-140134

Masse, M. (2011). *REST API Design Rulebook: Designing Consistent RESTful Web Service Interfaces*. Sebastopol: O'Reilly Media, Inc.

Mueller, J. P. (2006). *Mining Google? Web Services: Building Applications with the Google? API*. New York: John Wiley & Sons.

Pérez, J., Arenas, M., & Gutierrez, C. (2006). Semantics and Complexity of SPARQL. *Lecture Notes in Computer Science, 4273*, 30-43. 10.1007/11926078_3

Quilitz, B., & Leser, U. (2008). Querying distributed RDF data sources with SPARQL. In *Lecture Notes in Computer Science: vol 5021. Proceedings of European Semantic Web Conference*. Springer. 10.1007/978-3-540-68234-9_39

Rakhmawati, N. A., Umbrich, J., Karnstedt, M., Hasnain, A., & Hausenblas, M. (2013). *Querying over Federated SPARQL Endpoints. A State of the Art Survey*. Technical Publication. Accessed 20 November 2018: https://arxiv.org/pdf/1306.1723

Russell, M. A. (2013). *Mining the Social Web: Data Mining Facebook, Twitter, LinkedIn, Google+, GitHub, and More*. Sebastopol: O'Reilly Media, Inc.

Sanchez, V. M. (2015). Análisis de proyectos "Open Data" en España: propuestas de mejora. *Cuadernos de Gestión de Información, 5*(1), 53-72. Accessed 20 November 2018: https://revistas.um.es/gesinfo/article/view/232231

Shadbolt, N., & Gibbins, N. (2010). Resource Description Framework (RDF). In *Encyclopedia of library and information sciences* (pp. 4539-4547). Accessed 20 November 2018: https://eprints.soton.ac.uk/268264/

Srivastava, B., & Koehler, J. (2003). Web service composition-current solutions and open problems. *Proceedings of ICAPS 2003 workshop on Planning for Web Services*, 28-35.

Verborgh, R., Steiner, T., Van Deursen, D., Coppens, S., Vallés, J. G., & Van de Walle, R. (2012). Functional descriptions as the bridge between hypermedia APIs and the Semantic Web. In *Proceedings of the third international workshop on restful design*. ACM. 10.1145/2307819.2307828

Vrandečić, D. (2012). Wikidata: A new platform for collaborative data collection. In *Proceedings of the 21st International Conference on World Wide Web*. ACM.

ADDITIONAL READING

Booshehri, M., & Luksch, P. (2016, November). Towards linked open data enabled ontology learning from text. In *Proceedings of the 18th International Conference on Information Integration and Web-based Applications and Services* (pp. 252-256). ACM. 10.1145/3011141.3011184

Elzein, N. M., Majid, M. A., Hashem, I. A. T., Yaqoob, I., Alaba, F. A., & Imran, M. (2018). Managing big RDF data in clouds: Challenges, opportunities, and solutions. *Sustainable Cities and Society*, *39*, 375–386. doi:10.1016/j.scs.2018.02.019

Hossain, M. A., Dwivedi, Y. K., & Rana, N. P. (2016). State-of-the-art in open data research: Insights from existing literature and a research agenda. *Journal of Organizational Computing and Electronic Commerce*, *26*(1-2), 14–40. doi:10.1080/10919392.2015.1124007

Michel, F., Faron-Zucker, C., Corby, O., & Gandon, F. (2019, May). Enabling Automatic Discovery and Querying of Web APIs at Web Scale using Linked Data Standards. In *Companion Proceedings of The 2019 World Wide Web Conference* (pp. 883-892). ACM. 10.1145/3308560.3317073

Modi, K. J., Garg, S., & Chaudhary, S. (2019). An Integrated Framework for RESTful Web Services Using Linked Open Data. *International Journal of Grid and High Performance Computing*, *11*(2), 24–49. doi:10.4018/IJGHPC.2019040102

Pauwels, P., McGlinn, K., Törmä, S., & Beetz, J. (2018). Linked Data. In *Building Information Modeling* (pp. 181–197). Cham: Springer. doi:10.1007/978-3-319-92862-3_10

Radulovic, F., Mihindukulasooriya, N., García-Castro, R., & Gómez-Pérez, A. (2018). A comprehensive quality model for linked data. *Semantic Web*, *9*(1), 3–24. doi:10.3233/SW-170267

Spahiu, B., Xie, C., Rula, A., Maurino, A., & Cai, H. (2016, May). Profiling similarity links in linked open data. In 2016 IEEE 32nd International Conference on Data Engineering Workshops (ICDEW) (pp. 103-108). IEEE. 10.1109/ICDEW.2016.7495626

Sultana, M., Paul, P. P., & Gavrilova, M. L. (2014). Online user interaction traits in web-based social biometrics. In *Computer Vision and Image Processing in Intelligent Systems and Multimedia Technologies* (pp. 177–190). IGI Global. doi:10.4018/978-1-4666-6030-4.ch009

KEY TERMS AND DEFINITIONS

Digital Repository: It is a computer application that allows you to store information and offers different services to the user. Essentially it allows searching and retrieving stored information.

Linked Data: It is an initiative that aims to relate data and information to create a large semantic network that can be consulted.

Open Data: It is an initiative that aims to provide the data generated in the institutions so that anyone can use them to exploit them.

RDF: It is a language that allows to represent knowledge using triplets of the subject-predicate-object type.

SPARQL: It is a query language on documents described in RDF.

Web Service: It is a way to implement services on the web, which are associated with web resources.

Wikidata: It is an initiative supported by Wikimedia that maintains a repository of linked data.

Chapter 14
Interactive Communication Systems and Technologies for Effective E-Learning

Aisha A. Al-Ajeel
Kuwait University, Kuwait

Lujain A. Al-Anjari
Kuwait University, Kuwait

Muhammad Sarfraz
https://orcid.org/0000-0003-3196-9132
Kuwait University, Kuwait

ABSTRACT

Interactive communication systems and technologies are big motivation for e-learning nowadays. This chapter is dedicated for studying the effect of e-learning worldwide. The state of Kuwait is considered as a special case for this study. Two research methods have been used in this study. One method uses a questionnaire for an input of 104 students to study their level of acceptance and their attitudes towards e-learning. Other method has been set to collect interviews with 21 instructors who are involved in e-learning at undergraduate and graduate levels. The participants among students and instructors have been randomly chosen from different Kuwait institutions. A detailed analysis of the input has triggered the best practices to change the people's behavior towards e-learning and find a possible solution to bridge the gap between system makers and the users.

INTRODUCTION

E-learning is one of the most preferred systems of learning nowadays (Rosenberg 2001). Web-based learning or in other words online learning has the same meaning as e-learning since they all refer to the same point, which is using technology platforms to study or improve in the area of interest. e-Learning

DOI: 10.4018/978-1-7998-3355-0.ch014

endorses active and independent learning and it links various resources in several formats. The growth of the 2020 e-learning market is incredible. It has been estimated that the market will grow at a 5% compound annual growth rate between now and the year 2024, at which point it will likely exceed $200 billion (Hamed 2012). It was also predicted that the market will climb to $325 billion by the year 2025 (Kennoel 2013).

The world is being exposed to innovative technologies, and curiosity around them is blooming. In addition, people who have passion and are eager to learn and develop are constantly searching for new styles of learning. One example of these new styles of learning is blended learning (Anthony 2013), (Paul 2007), (Mehmet 2011), which is an approach to education that combines online educational materials and opportunities for interaction online with traditional place-based classroom methods (Sung 2009).

Over the past few years, e-learning systems have made an innovative transformation for learning that attracted the attention of researchers. "Providing rapid access to specific knowledge and information, it offers online instruction that can be delivered anytime and anywhere through a wide range of electronic learning solutions such as Web-based courseware, online discussion groups, live virtual classes, video and audio streaming, Web chat, online simulations, and virtual mentoring" (Ebru 2009).

But, in the current world of academia where traditional learning occurs, there are certain studying barriers like lack of information sources, accessibility to books, time consuming search for information, low quality and un-updated information, which are fiscally draining for schools and busy classrooms. Therefore, the e-learning field is becoming one of the most attractive ways to people who would like to learn and stay away from these kinds of boundaries. What we are trying to do is seek the alignment of building an e-Learning system where learning keeps up with rapid technologies changes (Jean-Francois 2018).

Unfortunately, in the State of Kuwait, some of the learning programs are not meeting demand which caused a stability in the learning curve. Currently, Kuwait is behind other countries who are utilizing technology and applying e-learning such as the United States of America, India, China and South Korea (Hallberg 2017). The gap between Kuwait and those countries here is that they are meeting the needs of the user mostly and people are able to adopt the new innovation and transition faster.

It is a common practice today that the teaching community directs students online to do their various educational assignments and tasks. This exposes the tracking of online behavior. This also, because of digital technologies, exposes opportunities to the private world. Hence, an element of digital surveillance, through e-learning, will help to have better control in the education (Faith 2016).

Consequently, this chapter will study the effect of e-learning on our lives where rapid transformations in learning environment and rapid shifts in technology are the biggest challenges we face today. We aim to create awareness for people to dig more into web-based learning and change mindsets and systems to think about designing a better future for our upcoming generations.

E-Learning

What is e-Learning? Different people view it from different perspectives (Anthony 2016). Figure 1 explains some of the views. Different authors, in different times, have elaborated the meaning of e-learning. E-learning is learning conducted via electronic media, typically on the Internet (Ebru 2009). The use of Internet technologies to deliver a broad array of solutions that enhance knowledge and performance (Isabelle 2017). Understanding e-learning is simple. E-learning is learning utilizing electronic technologies to access educational curriculum outside of a traditional classroom. In most cases, it refers to a course,

Figure 1. Views on e-Learning

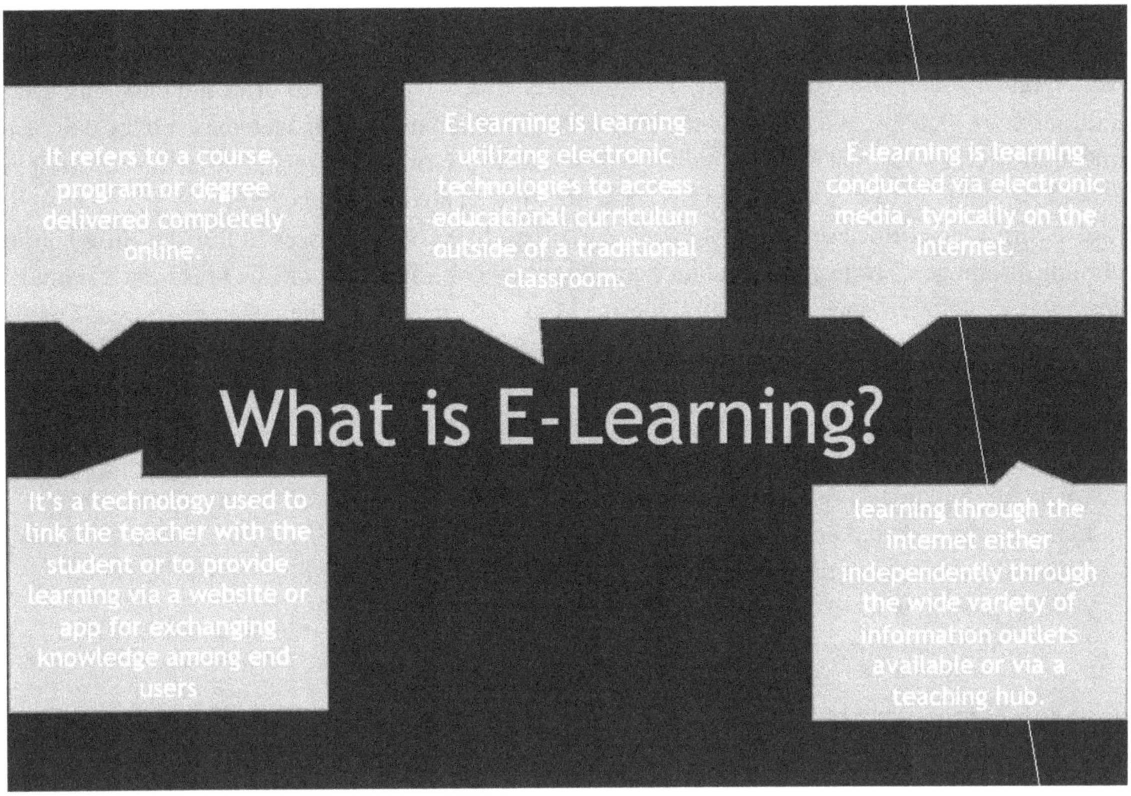

program or degree delivered completely online. There are many terms used to describe learning that is delivered online, via the internet, ranging from distance education, to computerized electronic learning, online learning, internet learning and many others (History of ELearning). Learning is going to a next level, it is being reformed. It will emphasize 'how' not 'what'. Online learning can be delivered through a variety of tools, the most effective possibly being the online learning platform. Businesses, associations, and other organizations use online learning platforms to deliver training to employees, customers, and members (Ebru 2009).

Learning trends will impact the online learning platform market, affecting how businesses use their platforms. E-Learning is an interesting way of studying and learning something new. There are many forms of e-learning used currently to transfer experience and knowledge. Numerous of e-learning methods are seen and used in different schools, colleges and other institutes. Some of those e-learning methods are web-based learning, computer-based training, webinars, virtual classroom, mobile learning, video-based learning and custom e-learning. There are different types of e-learning offered by institutions which include online courses, online degree programs, blended learning courses, blended learning degree programs and joint online learning with other institutes (Anthony 2013), (Paul 2007), (Mehmet 2011).

e-Learning is the future. Market research firm Global Industry Analysts projected "E Learning" would reach $107 Billion in 2015 and it did. Now, Research and Markets forecasts show triple the revenue of 2015 – e-learning will grow to $325 Billion by 2025. In 2013, Lynda.com, the online learning giant and arguably the 800-pound gorilla in the e-learning space, had just accepted $103 million in venture funds.

Two years later, in 2015, they were acquired by LinkedIn for $1.5 Billion. It is now called LinkedIn Learning. This acquisition and the trends within hiring, workplace skills, and training are a big reason why we are seeing these revenue projections. People need to keep learning to succeed in these times.

By embracing technology in its many forms, universities will be able to offer life-changing access to millions more globally. But that's not the only prize. Through deeper engagements and local industry partnerships world-wide, top colleges will be able to create a virtuous cycle that advances research and collaborative thinking to tackle some of the most pressing challenges we face today.

SoloLearn is an online e-learning platform that offers free coding classes in thirteen different programming disciplines. The learning content is jointly created by SoloLearn and, increasingly, community contributors. SoloLearn apps are available on Android, iOS, and web. The courses are free. Figure 2 shows a snapshot about some important and exciting information on the SoloLearn. One can see, there are millions of users and downloads. There are enormous amount of courses offered by SoloLearn. All of the courses have the following useful and impressive features:

- Available Anytime & Anywhere!
- Learn on the web and on the go.
- Available on all major devices and platforms.
- Always pick up where you left off.
- More simple and enjoyable than ever!

Research Methodology

Two research methods have been used in this study. The first method used is an e-survey which was designed and distributed to students to study their level of acceptance and their attitudes towards e-learning. We have distributed an e-survey for students using Google Docs as a tool. The e-survey was distributed through links using Twitter and WhatsApp. It was distributed among 107 students in different Kuwait institutions including Kuwait University, Gulf University for Science & Technology, The American University of the Middle East and American University of Kuwait. All of the students responded completely except 3. So, the input of 104 students is presented and analyzed in Section 4.

The second method of research implemented in this chapter is to collect information from interviews with instructors who used e-Learning systems, articles regarding e-Learning, books, published researches. An interview questionnaire was carefully designed and interviews were arranged by face to face meetings with instructors. As many as 21 instructors were approached in different Kuwait institutions including Kuwait University, Gulf University for Science & Technology, The American University of the Middle East and American University of Kuwait. Everybody participated and responded to their best. So, the input of 21 instructors is presented and analyzed in Section 5.

Analysis for Student Data

A Questionnaire Survey was designed and posted through social media for the university student community. Number of surveys which were successfully collected is 104 students. Data was analyzed using graphs and diagrams. Females took the higher percentage with more than three quarters (77.9%) whereas the rest was from the opposite gender with a percentage of (22.1%). The ages of the students were broken into three categories. The highest percentage (47.1%) goes to the students with the ages between 17

Figure 2. Some interesting information on SoloLearn

to 22 years old. The ages between 23 to 28 took the second highest score with a (42.3%). The ages of 29 years old and above were the least number of the people who have done the survey with a (10.6%).

Based on the survey submitted to students in Kuwait from different fields and majors, some of them introduced e-learning as learning through the internet either independently through the wide variety of information outlets available or via a teaching hub. While some students familiarized it as a technology used to link the teacher with the student or to provide learning via a website or application for exchanging knowledge among end-users as learning from far distance which can be accessed from anywhere and anytime. Others say that taking courses or programs online instead of classrooms, being able to view lectures, tutorial, and extra material on the website using online education platforms is e-learning.

It was noticed that as many as 64.4% are aware of the meaning of e-learning. Some of them described it as learning online and using electronic technologies, learning through internet either independently through the wide variety of information outlets available or via teaching hub, and some described it as learning from distance and etc. However, 35.6% of the students were not found aware of e-learning which are the targeted segment to teach them the know-how of electronic learning and spread awareness across them.

Most of the students agreed that people who have the passion and are eager to learn and develop are constantly searching for new styles of learning. A percentage of 95% of agreement was reflected which means that if a student has the willingness to learn, s/he will be able to do whatever it takes to get the knowledge needed or wanted through any new style of learning. That will make entering and announcing new styles of learning to these kinds of people easier. This will cause an inclination in student knowledge development and new generation way of thinking.

"Learning is going to a next level?" This question is not concise on showing exactly that learning is going to which level and where does it stand right now? It makes the survey answering person wondering what that sentence means. It makes the survey answering person wondering what that sentence means. This question was a way to see whether people have a positive view on the future of learning or they are finding something that makes learning stands at the same level. Although, majority of the survey (88.5%) agreed that learning will take to next level. But, it was discovered that 11.5% of the community did not agree. So, from the perspective of the nature of question, it is a big percentage that we have to look at. This percentage might affect the future of learning. In order to develop and get learning to a higher level or get these people the awareness of new types of learning and new ways of learning that just appeared, we should consider finding the reason why they think that learning is not going to a next level and be able to solve the issue or doubt of disagreement choice.

Today, it is easier to gather information online and in a fast way, 97.1% of the students agreed. This statement approved that most of the upcoming generation has an awareness on how to use the web on getting information easily and in a faster way than it used to be before. This also approved the statement of "Learning is going to a next level" since most of the students are aware of the online search way for information. Nonetheless, it will help on introducing e-learning in an easier way in future.

For searching on the amount of time for the usage of electronic devices, a very high percentage (96.2%) of the population was found that people think/sure that they or the community around them are using electronic devices not less than two hours a day. That approves the statement where it says "People who have the passion are eager to learn and develop are constantly searching for new styles of learning". Few years ago, it was rarely to see someone who would hold electronic device in hand. Today, most of the people have their own appliances in hand and it is strange if they don't carry one. This means that most people are eager and passionate to learn if there is an added value to their own needs or wants (Frida 2017), (David 2014), (Harriet 2015).

In this study, the student community with a percentage of 89.4% finds that e-learning is an interesting way to study and learn something new. With this high number of interest on e-learning, it means that students have the passion and eager to learn something new. Therefore, there should be someone who can try to understand the students and teachers needs and do a system that fits both users. That will make the state of Kuwait studying field go to a higher level and allow users to reach with more resources of knowledge. This also indicates the facts that technology is everywhere and information is available worldwide by people using it.

When asked about "Do you use any of the following trends for learning?" the following choices were provided to check in case of "Yes" answer.

- Mobile apps such as (YouTube and Google)
- Gamification and Game-Based Learning
- Social Learning such as Twitter, LinkedIn, and Instagram
- Virtual reality and augmented reality

Out of 104 respondents, a very high majority of 99 responded "Yes" from surveys conducted on students from different universities in Kuwait. This is shown in Table 1. An extremely high number of students who have used e-learning with a full awareness that they are getting something new (added value) from these kinds of trends. Nonetheless, mobile apps such as (YouTube and Google web browser) are being used the most by students with a 92.9%. Virtual reality and Augmented reality were least used by 15.2%.

Table 1. Facts for trends in learning

#	Trends for learning	Number of responses	%
1	Mobile apps such as (YouTube and Google)	92	92.9
2	Gamification and Game-Based Learning	23	23.2
3	Social Learning such as Twitter, LinkedIn, and Instagram	56	56.6
4	Virtual reality and augmented reality	15	15.2

The study found out that most of the students (92.3%) use their own electronic devices such as mobile phones and laptops for searching information they need mostly. Only very few of them (7.7%) go to a bookstore / library to search for information. In here, we can conclude that the new generation is preferring to get information online rather than going to a bookstore to get some information they need or want. It was also noted that the students are submitting most of their assignments paperless, see Table 2. By using email, blackboard or other online portal which is a great sign for students and teacher's awareness of e-learning. Also, this helps on approving that students are interested in e-learning. So, they have used online learning since they are already using an electronic device to submit their assignments.

Table 2. Facts for ways people are submitting their assignments

#	Ways of submitting assignments	Number of responses	%
1	By paper	15	14.4
2	Paperless: by email, blackboard or other online portal	21	20.2
3	Both ways	68	65.4

Table 3. Impact of e-learning on educational journey

Weight (1 is lowest, 5 is highest)	No. of responses	Total Weight	%
1	3	3	2.9
2	7	14	6.7
3	25	75	24
4	29	116	27.9
5	40	200	38.5

E-learning has a very high impact on students learning journey as it is shown in Table 3. Since e-learning has an excellent impact on learning nowadays by most students votes, there should be more ways that allow students to study online. By embracing e-learning, Kuwait will take a step further in learning field and a new knowledgeable generation will come forward.

In Figure 3, while studying the obstacles of learning growth in Kuwait which are adjusting to change, lack of prioritization, lack of goal, learning environment, lack of technical skills and boredom. The highest effect of obstacles among them all was boredom. Whereas, lack of technical skills took the least affect as an obstacle towards learning. Thus, there should be a better way to improve systems of learning and try to make them more fun to use.

Analysis for Instructor Interview Data

The second method adopted in this research is by collecting information from interviews with instructors who used e-Learning systems, articles regarding e-Learning, books, and published researches. An interview was designed and the interviews were collected face to face through appointments as well as through media. Number of interviews which were successfully collected is 21. The interviewees who participated were picked from different Kuwait institutions including Kuwait University, Gulf University for Science & Technology, The American University of the Middle East and American University of Kuwait. Everybody participated and responded to their best.

As far as demography is concerned, instructors from various fields of specialization and Universities are selected. Their Work Experience is varying over different years including: 0 – 3, 3 – 7, 7 – 15, and more than 15 years. None of the instructor in the age group 0 – 3 participated. However, the largest number of participants belonged to the age group of 15 years and above. Both of the genders (Male and Female) for the age groups of 21 – 30, 31 – 40, 41 – 50, and more than 50 years participated in the interviews. Majority of the instructors (16) were male and 5 of them were females. The highest number of participants (11) belonged to the age group 31 – 40 whereas the second highest number (6) was the age group of greater the 50. Their highest educational degrees were considered as Master and Doctorate. Majority of the instructors (16) were Doctorate and 5 of them were with Master degrees.

Majority of the interviewees agreed that people who have passion and are eager to learn and develop, are constantly searching for new styles of learning. Majority of the interviewees (20) agreed and expressed that learning widens the scope and helps going further. It trends towards information search using search engines and gather material for development. Technology is always evolving and, with the

Figure 3. Demonstration of the data facts for intensity of obstacles of learning in Kuwait

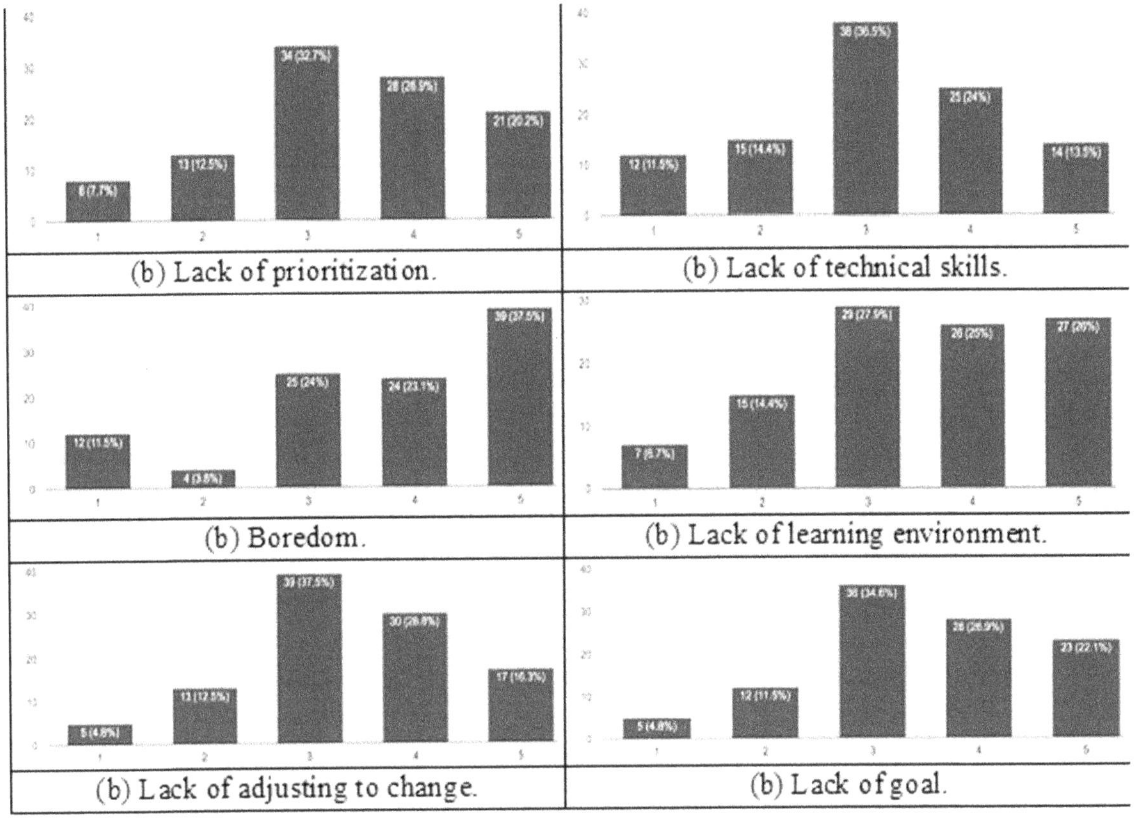

new technologies, one can very obviously progress. People are increasingly relying on the internet to carry out several tasks and are improving in the skills.

Majority of the interviewees provided their opinions that, in the current world of academia where traditional learning occurs, there are certain studying barriers like: lack of information sources, accessibility to books, time consuming search for information, low quality and un-updated information. It was said that, with the availability and improvement of technology, information is available everywhere, it is about how to access information. Many educational platforms are available to students and they can download the necessary items they require. Access to information today is much easier than before and universities provide many resources on-line.

All the participants agreed that they are involved in using the new learning trends. These include Microlearning, Video-Based Learning, Social Learning, Content Curation, Artificial Intelligence, Learner Assistance, Gamification, Game-Based Learning, Virtual Reality and Augmented Reality. They agreed that online instructions can be delivered anytime and anywhere through a wide range of electronic learning solutions. They pointed out many of them including Web-based courseware, Canvas, Online discussion groups, Live virtual classes, Video and audio streaming, Web chat, online simulations, and Virtual mentoring. People are highly of the view to support blended learning, which is an approach in education that combines online educational materials and opportunities for online interaction with traditional place-based classroom methods.

Table 4. Intensity of obstacles of learning in Kuwait

#	Obstacles	Total Weight	Average
1.	Adjusting to change	80	3.809524
2.	Lack of goal	85	4.047619
3.	Lack of prioritization	89	4.238095
4.	Learning environment	76	3.619048
5.	Lack of technical skills	75	3.571429
6.	Boredom	64	3.2

To support the e-learning in Kuwait, there are various opinions which include to overcome: Lack of research, Lack of training/knowledge/ facilities and resources, invest in training and resources, introducing the technology early enough to students, change in the culture, lack of on-line courses that are approved by Kuwait university, lack of on-line degrees, training of faculty and others.

Various obstacles of learning, like adjusting to change, lack of goal, lack of prioritization, learning environment, lack of technical skills, and boredom, were considered to be measured. The idea was to find out, to which extent each obstacle of learning is affecting the overall learning growth in Kuwait. A Likert scale (1 for very low effect and 5 for very high effect) was adopted to make measures for the intensity of obstacle of learning. One can see in Table 4 that Lack of prioritization is the biggest obstacle of learning whereas the "Boredom" is the lowest.

Majority of the instructors answered very positively to the question "If you support developing learning in Kuwait, what do you think is the impact that e-learning would have on education?" They provided very high score, see Table 5, for developing learning in Kuwait. They agreed very strongly that e-learning will have impact on education in Kuwait.

An input was seek to find out the Strengths, Weaknesses, Opportunities and Threats (SWOT) to e-learning. The participants, very generously, provided their thoughts which have enriched this research to determine stronger SWOT analysis. The input provided is shown in Table 6.

Suggestions and Recommendation

When asked about "If you have any additional comments or suggestions, feel free to express them", very interesting ideas were received. Overall, it is reflected that e-learning is instrumental, dynamic, available, affordable, and effective for any society. It is easier to gather information online and in a fast way.

Table 5. Intensity of obstacles of learning in Kuwait

Weight	Number of responses	Total Weight
1	0	0
2	1	2
3	6	18
4	9	36
5	5	25

Table 6. SWOT analysis

Strengths	Weaknesses
· Information can be available any time that is convenient to students · Accessible · Expand learning options · Speed · Interactive · Overcome and eliminate distance obstacles · Variability of fields discipline · Different way to understand the topic by going to different resources in different format.	· Disconnections, no face to face interaction · Systems not easy to use · Less instructor feedbacks · Technology changes · Student willingness and Faculty's ability · Culture. · Student retention · Difficult to adapt · Follow ups and trainings are needed · Authentication
Opportunities	**Threats**
· Enhance learning · Create easier educational systems · More class material coverage · Increasing media types as new tech emerges · Workshops, Trainings and Conferences · Short courses - newly developed · Hub for researching for graduate studies · New technologies are always revolving · New updates on content delivery. · Develop new skills. · Diversity in students and teachers.	· Cheating · Student boredom and lack of discipline (They only flip through slides) · Politics · Assessment of student will not be accurate · Privacy and security issues · Exhausting training · Copy rights issues of vendor selection · Hacking. · Difficult to assess performance

Specifically, it is more useful for those who want to enhance their knowledge or skills in specific areas, with desired timings, limited budget and some awareness of IT skills. It impacts on people's learning. It is preferably useful for a mature class of community, for example, graduate students, employees in businesses. It is also useful, but up to a limited extent, for primary, elementary and high school students. Even university undergraduate can also benefit in a specific way.

There is a need to have an interactive online platform between instructors and undergraduate students that allows the instructors to adapt to student's needs. It was observed and analyzed, it is effective to have e-learning, web based or mobile based systems. However, it would be more effective to step forward for Virtual Reality, Augmented Reality and Artificial Intelligence. Tools are there, we just need to look for them. There is need to invest a good budget in the area of e-learning.

CONCLUSION

E-learning is not just a change of technology. It is part of a redefinition of how we as a species transmit knowledge, skills, and values to younger generations of workers and students. Changing people's mindsets and behaviors towards e-Learning will need time, there are further steps need to be followed to seek the alignment of building e-learning system. To bridge the gap, adopt new technologies, have special budget for such projects, and assign this task to committees within the college. It is also needed to build a system that is dynamic and not rigid allowing it to easily accommodate changes. We need to be constantly updated with new technologies and shorten adoption time. Provide the need by checking all the training needs of our workforce in private and public sector. Set the right incentives, and have monitoring mechanisms for learners. To improve information retention, engagement, and teaching, blended learning is the key. we need to introduce blending learning in the college. It will be a great start by building a pilot project on a special segment and test its efficiency with a right group of subject matter experts. In this way, it needs to be approached with caution, the "learning" culture and attitudes towards learning need to be more responsive to new methods in order for them to be effective.

REFERENCES

Boninger, F., & Molnar, A. (2016). *Learning to Be Watched: Surveillance Culture at School. The Eighteenth Annual Report on Schoolhouse Commercializing Trends, 2014-2015*. Boulder: National Education Policy Center, School of Education, University of Colorado.

Clover, I. (2017). Advantages and Disadvantages of eLearning. *eLearning Industry*. Retrieved January 29, 2020, from https://elearningindustry.com/advantages-and-disadvantages-of-elearning

D'Abreo, K., Pimento, J., & D'Souza, D. (2013). *E-Learning Website Solution*. Project Report. Computer Engineering Department, Xavier Institute of Engineering, University of Mumbai. http://jonathanpimento.com/Demos/project/Docs/report.pdf

David, N. (2014). One-Third of U.S. Students Use School-Issued Devices. *The Journal*. https://thejournal.com/articles/2014/04/08/a-third-of-secondary-students-use-school-issued-mobile-devices.aspx

Fazlollahtabar, H., & Abbasi, A. (2012). A Computer Integrated Framework for E-Learning Control Systems Based on Data Flow Diagrams. *Journal of Control Engineering and Technology*, *2*(2), 97–105.

Frida, Cardozo, Gebhart, Gullo, & Kalia. (2017, Apr. 13). *Spying on Students. School-issued devices and student privacy*. Executive summary.

Ginns, P., & Ellis, R. (2007). Quality in Blended Learning: Exploring the Relationships between on-Line and Face-to-Face Teaching and Learning. *The Internet and Higher Education*, *10*(1), 53–64. doi:10.1016/j.iheduc.2006.10.003

Hallberg, A. (2017). The Top 4 Countries That Have Developed eLearning. *eLearning Industry*. https://elearningindustry.com/countries-that-have-developed-elearning-top-4

Harriet, T. (2015). Google's Chromebooks make up half of U.S. devices sold. *CNBC*. http://www.cnbc.com/2015/12/03/googles-chromebooks-make-up-half-of-us-classroom-devices.html

History of ELearning & Interactive Books. (n.d.). *Interactive eBooks Articles*. The Institute of Progressive Education and Learning. Retrieved January 29, 2020, from http://institute-of-progressive-education-and-learning.org/elearning-i/history-of-elearning/

Jean-Francois, E. (2018). *Transnational Perspectives on Innovation in Teaching and Learning Technologies*. Koninklijke Brill, NV. doi:10.1163/9789004366077

Kelly, A. E., & Seppala, M. (2016). Changing Policies Concerning Student Privacy and Ethics in Online Education. *International Journal of Education Technology*, *6*(8), 652–655. doi:10.7763/IJIET.2016.V6.768

Ocak, M. A. (2011). Why are faculty members not teaching blended courses? Insights from faculty members. *Computers & Education*, *56*(3), 689–699. doi:10.1016/j.compedu.2010.10.011

Özpolat, E., & Akar, G. B. (2009). Automatic Detection of Learning Styles for an e-Learning System. *Computers & Education*, *53*(2), 355–367. doi:10.1016/j.compedu.2009.02.018

Park, S. Y. (2009). An Analysis of the Technology Acceptance Model in Understanding University Students' Behavioral Intention to Use e-Learning. *Journal of Educational Technology & Society*, *12*(3), 150–162. www.jstor.org/stable/jeductechsoci.12.3.150

Picciano, A. G., Dziuban, C. D., & Graham, C. R. (2013). Blended learning: Research perspectives, Volume 2. Routledge.

Rosenberg, M. J. (2001). *E-learning: Strategies for delivering knowledge in the digital age*. New York: McGraw-Hill.

Compilation of References

Abdel-Fattah, M. (2005). Arabic Sign Language: A Perspective. *Journal of Deaf Studies and Deaf Education*, *10*(2), 212–221. doi:10.1093/deafed/eni007 PMID:15778217

Abdo, M., Hamdy, A., Salem, S., & Saad, E. M. (2015). Arabic alphabet and numbers sign language recognition. *International Journal of Advanced Computer Science and Applications*, *6*(11), 209–214.

Abelson, R. P. (1959). Modes of resolution of belief dilemmas. *The Journal of Conflict Resolution*, *3*(4), 343–352. doi:10.1177/002200275900300403

Abrams, R. (2015). *USA Today Recognizes Tierra Resources in Leading the Charge to Prevent the Next Katrina*. Retrieved September 29, 2019, from https://tierraresourcesllc.com/usa-today-recognizes-tierra-resources-leading-charge-prevent-next-katrina

Access-Disability Support Unit. (2016). *Glossary of Conditions/ Impairments/ Disability*. Malta: University of Malta.

Adam, B., Juan, C., & Dawn, S. (2009). Secure Content Sniffing for Web Browsers, or How to Stop Papers from Reviewing Themselves. *IEEE Symposium on Security and Privacy, 30*(8), 360-371. doi: 10.1109/SP.2009.3

Adams Becker, S., Cummins, M., Davis, A., Freeman, A., Hall Giesinger, C., & Ananthanarayanan, V. (2017). *NMC Horizon Report: 2017 Higher Education Edition*. Austin, TX: The New Media Consortium.

Adrienne, P. F., Elizabeth, H., Serge, E., Ariel, H., Erika, C., & David, W. (2012). Android permissions: user attention, comprehension, and behavior. *Eighth Symposium on Usable Privacy and Security*, (3). doi: 10.1145/2335356.2335360

Ahmad, M. A., Leo, K. W., Mohamad, G. H. P., Ahmad, A., Hashim, S. A., Chulan, R. M., & Baijan, A. H. (2017, September). *Development of the automated bunker door by using a microcontroller-system*. Paper presented at theIOP Conference Series: Materials Science and Engineering, Selangor, Malaysia.

Ahmed, A. (2010). Online privacy concerns among social networks' users. *Cross-Cultural Communication*, *6*(4), 74–89.

Ait Omar, D., Outanoute, M., Baslam, M., Fakir, M., & Bouikhalne, B. (2017). Joint Price and QoS Competition with Bounded Rational Customers. In *Networked Systems: 5th International Conference, NETYS 2017, Marrakech, Morocco, May 17-19, 2017, Proceedings*. Springer International Publishing. 10.1007/978-3-319-59647-1_33

Ait Omar, D., El Amrani, M., Baslam, M., & Fakir, M. (2019a). A Game-Theoretic Approach for the Internet Content Distribution Chain. In A. Podelski & F. Taïani (Eds.), *Networked Systems* (pp. 270–285). Cham: Springer International Publishing. doi:10.1007/978-3-030-05529-5_18

Ait Omar, D., Garmani, H., El Amrani, M., Baslam, M., & Fakir, M. (2019b). A Customer Confusion Environment in Telecommunication Networks: Analysis and Policy Impact. *International Journal of Cooperative Information Systems*, *28*(02), 1930002. doi:10.1142/S021884301930002X

Ait Omar, D., Outanoute, M., Baslam, M., Fakir, M., & Bouikhalenc, B. (2019c). On understanding price-QoS war for competitive market and confused consumers. *Computing*, *101*(9), 1327–1348. doi:10.100700607-018-0642-5

Akash, D. (2015). *Impact of international terrorism on commodity markets*. LAP Lambert Academic Publishing.

Akyol, Z., Garrison, D. R., & Ozden, M. Y. (2009). Online and blended communities of inquiry: Exploring the developmental and perceptional differences. *International Review of Research in Open and Distance Learning*, *10*(6), 65–83. doi:10.19173/irrodl.v10i6.765

Al Jenaibi, B. N. A. (2011). Use of social media in the United Arab Emirates: An initial study. *Global Media Journal*, *1*(2), 3–27. Retrieved from http://www.gmjme.com/gmj_custom_files/volume1_issue2/articles_in_english/volume1-issue2-article-3-27.pdf

Aldaihani, M. (2011). *A comparative study of inclusive education in Kuwait and England* (Doctoral dissertation). University of Birmingham.

Al-Hilawani, Y. (2009). Perspectives and reviews on deafness in the State of Kuwait and the United Arab Emirates. *Deaf people around the world: Educational and social perspectives*, 119-132.

Al-Meqbel, A., & Al-Baghli, H. (2015). The prevalence of hearing impairment in high-risk infants in Kuwait. *Auditory and Vestibular Research*, *24*(1), 11-16.

Al-Sebeih, K., Al-Kandari, M., Al-Awadi, S. A., Hegazy, F. F., Al-Khamees, G. A., Naguib, K. K., & Al-Dabbous, R. M. (2014). Connexin 26 gene mutations in non-syndromic hearing loss among Kuwaiti patients. *Medical Principles and Practice*, *23*(1), 74–79. doi:10.1159/000348304 PMID:24080506

Alt, D. (2018). Students' wellbeing, fear of missing out, and social media engagement for leisure in higher education learning environments: Research and reviews. *Current Psychology, 37*(1), 128-138. http://dx.doi.org.ezproxy.uttyler.edu:2048/10.1007/s12144-016-9496-1

Altman, E. (2012). In which content to specialize? a game theoretic analysis. In *International Conference on Research in Networking*. Springer.

Amin, O., Said, H., Samy, A., & Mohammed, H. K. (2015, December). HMM based automatic Arabic sign language translator using Kinect. *Tenth International Conference on Computer Engineering & Systems*, 389-392. 10.1109/ICCES.2015.7393081

Amrani, M. E., Omar, D. A., Baslam, M., & Minaoui, B. (2018b). A Game Theoretic Approach to Optimize Transmission Strategies for Mobile Applications. In *2018 6th International Conference on Wireless Networks and Mobile Communications (WINCOM)*. IEEE. 10.1109/WINCOM.2018.8629618

Amrani, M. E., Garmani, H., Baslam, M., Minaoui, B., & Ayachi, R. E. (2018a). Cournot Game of ISP under Net Neutrality. *International Journal of Control and Automation, 11*(7), 16. doi:10.14257/ijca.2018.11.7.07

Android WebView. (2018). Retrieved from https://developer.android.com/reference/ android/webkit/WebView.html

Annette, N. (2011). To see and be seen: Celebrity practice on Twitter. *Convergence*, *17*(2), 139–158. doi:10.1177/1354856510394539

Ante, S. E., & Knutson, R. (2013, July). Good luck leaving your wireless phone plan. *Wall Street Journal*.

Antonakos, E., Roussos, A., & Zafeiriou, S. (2015, May). A survey on mouth modeling and analysis for sign language recognition. *11th IEEE International Conference and Workshops on Automatic Face and Gesture Recognition,* 1, 1-7. 10.1109/FG.2015.7163162

Aouiti, N., & Jemni, M. (2016, July). Classifiers in Arab Gloss Annotation System for Arabic Sign Language. *International Conference on Computers Helping People with Special Needs*, 400-406. 10.1007/978-3-319-41267-2_56

AP. (2015). *Alligator program.* Retrieved September 29, 2019, from https://www.wlf.louisiana.gov/wildlife/alligator-program

Arab sign-language family. (2013). Retrieved in 17 July 2019, from https://en.wikipedia.org/wiki/Arab_sign-language_family

Arab Youth Survey. (2019). *A Call for Reform.* A White Paper on the findings of the 11th annual ASDA'A BCW Arab Youth Survey 2019. Retrieved from: file:///C:/Users/Z10509/Downloads/2019%20Arab%20Youth%20Survey.pdf

Aragay, N., Barrios, M., Ramirez-Gendrau, I., Caballero, A.G., Garrido, G., Ramos-Grille, I., … Vallès, V. (2018). Impulsivity profiles in pathological slot machine gamblers. In Comprehensive Psychiatry (vol. 83, pp. 79-83). Elsevier Inc.

Aragon, S. R. (2003). Creating social presence in online environments. *New Directions for Adult and Continuing Education, 100*(100), 57–68. doi:10.1002/ace.119

Aranjo, B., Soori, P. K., & Talukder, P. (2012, July). *Stepper motor drives for robotic applications.* Paper presented at thePower Engineering and Optimization Conference, Melaka, Malaysia.

Arif, A., Rizvi, S. T. H., Jawaid, I., Waleed, M. A., & Shakeel, M. R. (2016). Techno-Talk: An American Sign Language (ASL) Translator. *International Conference on Control, Decision and Information Technologies*, 665-670. 10.1109/CoDIT.2016.7593642

Armstrong, D., Bukaty, N., & Gutierrez, J. (2013, Mar. 9). *The Border Patrol* [Radio Broadcast]. Kansas City: WHB.

Arnold, N., & Paulus, T. (2010). Using a social networking site for experiential learning: Appropriating, lurking, modeling and community building. *Internet and Higher Education, 13*(4), 188–196. doi:10.1016/j.iheduc.2010.04.002

Arroyo, I., Woolf, B., Royer, J. M., & Tai, M. (2009). Affective gendered learning companions. In *Proceedings of the International Conference on Artificial Intelligence in Education* (pp. 41-48). Brighton, UK: IOS Press.

Assaleh, K., Shanableh, T., Fanaswala, M., Amin, F., & Bajaj, H. (2010). Continuous Arabic Sign Language Recognition in User Dependent Mode. *Journal of Intelligent Learning Systems and Applications, 2*(01), 19–27. doi:10.4236/jilsa.2010.21003

Atkinson, T., & Frechette, H. (2009). *Creating a Positive Organizational Climate in a Negative Economic OneImproving Organizational Climate to Transform Performance.* Doi: http://www. trainingindustry. com/media/2505214/creating-positiveorgclimate_us_aug09. pdf

Atwood, J., Eicholtz, M., & Farrell, J. (2012). *American Sign Language Recognition System. In Artificial Intelligence and Machine Learning for Engineering Design.* Dept. of Mechanical Engineering, Carnegie Mellon University.

Auer, S., Bizer, C., Kobilarov, G., Lehmann, J., Cyganiak, R., & Ives, Z. (2007). Dbpedia: A nucleus for a web of open data. In *Lecture Notes in Computer Science: vol. 4825. Proceedings of the 6th International Semantic Web Conference and the 2nd Asian Semantic Web Conference*. Springer. 10.1007/978-3-540-76298-0_52

Auer, M. M., & Griffiths, M. D. (2015). Testing normative and self-appraisal feedback in an online slot-machine pop-up in a real-world setting. In S. Kopp (Ed.), *Frontiers in psychology* (pp. 6–339). Bielefeld University. doi:10.3389/fpsyg.2015.00339

Australian Web Industry Association. (2014). The accessibility of Cloud computing – Current And Future Trends. Author.

Avia, M. D., Sánchez-Bernardos, M. L., Sanz, J., Carrillo, J., & Rojo, N. (1998). Self-presentation strategies and the five-factor model. *Journal of Research in Personality*, *32*(1), 108–114. doi:10.1006/jrpe.1997.2205

Bachmann, R., Knights, D., & Sydow, J. (2001). Special Issue: Trust and control in organizational relations. *Organization Studies*, *22*(2). doi:10.1177/0170840601222007

Bada, E., & Meneses, C. (2012). Una propuesta para la extracción de preguntas y respuetas. *XVII Congreso Internacional de Informática Educativa (TISE).*

Bahrami, H. (1992). The emerging flexible organization: Perspectives from Silicon Valley. *California Management Review*, *34*(4), 33–52. doi:10.2307/41166702

Bailenson, J., Yee, N., Blascovich, J., & Guadagno, R. E. (2008). Transformed social interaction in mediated interpersonal communication. In E. A. Konijn, S. Utz, M. Tanis, & S. B. Barnes (Eds.), *Mediated interpersonal communication* (pp. 78–99). New York, NY: Routledge.

Baisley-Nodine, E., Ritzhaupt, A., & Antonenko, P. (2018). Exploring social presence within an online course using Twitter. *E-Learning and Digital Media*, *15*(5), 235–253. doi:10.1177/2042753018786004

Baldassarri, S., Cerezo, E., & Royo-Santas, F. (2009, August). Automatic translation system to Spanish Sign Language with a virtual interpreter. In *IFIP Conference on Human-Computer Interaction*, 196-199. Springer. 10.1007/978-3-642-03655-2_23

Baldwin, J. A. (2018). *James Baldwin Biography and Quotes*. Retrieved from https://www.biography.com/news/james-baldwin-quotes

Barbieri Delgado, A. F., Castro Arenas, R. Y., Gonzales Remache, T. M., Raffo Lopez Torres, M. A., & Ratti Scudellari, M. C. (2016). *Mejorar el proceso de enseñanza aprendizaje para el desarrollo de las habilidades emocionales de los niños y niñas de 5 años del aula Amarilla de la Institución Educativa cuna jardín Armatambo del distrito de Chorrillos perteneciente a la UGEL 07.* Academic Press.

Barney, J. B. (1991). Firm resources and sustained competitive advantage. *Journal of Management*, *17*(1), 99–120. doi:10.1177/014920639101700108

Basil, M. D., & Brown, W. J. (2004). Magic Johnson and Mark McGwire: The power of identification with sports celebrities. In L. R. Kahle & C. Riley (Eds.), *Sports Marketing and the Psychology of Marketing Communication* (pp. 159–171). Hillsdale, NJ: Lawrence Erlbaum Associates.

Baslam, M., El-Azouzi, R., Sabir, E., & Echabbi, L. (2011). Market share game with adversarial access providers: A neutral and a non-neutral network analysis. In *Network Games, Control and Optimization (NetGCooP), 2011 5th International Conference on.* IEEE.

Baslam, M., Echabbi, L., El-Azouzi, R., & Sabir, E. (2012). Joint Price and QoS Market Share Game with Adversarial Service Providers and Migrating Customers. In R. Jain & R. Kannan (Eds.), *Game Theory for Networks* (pp. 642–657). Springer Berlin Heidelberg.

Baslam, M., El-Azouzi, R., Sabir, E., & Echabbi, L. (2011). Market share game with adversarial Access providers : A neutral and a non-neutral network analysis. *International Conference on NETwork Games, Control and Optimization (NetGCooP 2011)*, 1-6.

Baslam, M., El-Azouzi, R., Sabir, E., Echabbi, L., & Bouyakhf, E.-H. (2013a). A Game Theoretic Analysis of Price-QoS Market Share in Presence of Adversarial Service Providers. In H. Hanappi (Ed.), *Game Theory Relaunched.* InTech. doi:10.5772/54380

Baurick, T. (2018). The Drowing coast. *New York Times*. Retrieved September 25, 2019, from https://www.nytimes.com/interactive/2018/02/24/us/louisiana-wetlands-insects.html

Baym, N. K. (2009). What constitutes quality in qualitative internet research? In A. Markham & N. K. Baym (Eds.), *Internet inquiry: Conversations about method* (pp. 173–189). Los Angeles: Sage Publications; doi:10.4135/9781483329086.n16

Bedore, G. (1988, July 8). North Carolina assistant coach is KU's choice. *Lawrence Journal-World*. Retrieved from http://www2.kusports.com/news/1988/jul/08/north_carolina_assistant/

Bedore, G. (2003, April 11). Will Roy stay or will he go? *Lawrence Journal-World*. Retrieved from http://www2.kusports.com/news/2003/apr/11/will_roy_stay/

Bedore, G. (2012, March 23). Williams still fan of Jayhawks. *Lawrence Journal-World*. Retrieved from http://www2.kusports.com/news/2012/mar/23/roy-williams-still-fan-jayhawks/

Bekiroglu, E. (2008). Microcontroller-based full control of ultrasonic motor with frequency and voltage adjusting. *Sensors and Actuators. A, Physical, 141*(1), 151–159. doi:10.1016/j.sna.2007.07.018

Benoit, W. L. (1997). Image repair discourse and crisis communication. *Public Relations Review, 23*(2), 177–186. doi:10.1016/S0363-8111(97)90023-0

Benoit, W. L. (2018). Image Repair Theory in Sport. In A. C. Billings, W. T. Coombs, & K. A. Brown (Eds.), *Reputational Challenges in Sport: Theory and Application, 13*. New York: Routledge. doi:10.4324/9781315165608-3

Benslimane, D., Dustdar, S., & Sheth, A. (2008). Services mashups: The new generation of web applications. *IEEE Internet Computing, 12*(5), 13–15. doi:10.1109/MIC.2008.110

Biddix, J. P., Chung, C. J., & Park, H. W. (2015). The hybrid shift: Evidencing a student-driven restructuring of the college classroom. *Computers & Education, 80*, 162–175. doi:10.1016/j.compedu.2014.08.016

Billings, A. C. (2012). *Sports media: Transformation, integration, consumption*. New York: Routledge. doi:10.4324/9780203832790

Birnbaum, M. G. (2008). Taking Goffman on a tour of Facebook: College students and the presentations of self in a mediated digital environment. *Dissertation Abstracts International*, Section A, 69.

Bishop, C. M. (2006). *Pattern recognition and machine learning*. New York: Springer.

Biswas, G., Roscoe, R., Jeong, H., & Sulcer, B. (2009). Promoting self-regulated learning skills in agent-based learning environments. In *Proceedings of the 17th International Conference on Computers in Education* (pp. 67-74). Hong Kong: Asia-Pacific Society for Computers in Education.

Bizer, C. (2009). The emerging web of linked data. *IEEE Intelligent Systems, 24*(5), 87–92. doi:10.1109/MIS.2009.102

Blanchard, T. (2014). Runway show features student designers. *LSU Ag Center News*. Retrieved May 21, 2018, from https://www.lsuagcenter.com/news_archive/2014/may/headline_news/Runway-show-features-student-designers.htm

Bodle, R. (2011). Regimes of sharing: Open APIs, interoperability, and Facebook. *Information Communication and Society, 14*(3), 320–337. doi:10.1080/1369118X.2010.542825

Boninger, F., & Molnar, A. (2016). *Learning to Be Watched: Surveillance Culture at School. The Eighteenth Annual Report on Schoolhouse Commercializing Trends, 2014-2015*. Boulder: National Education Policy Center, School of Education, University of Colorado.

Booker, L., & Bandyopadhyay, S. (2017). How academic libraries can leverage social networking to popularize their services: An empirical study. *Journal of the Indiana Academy of the Social Sciences*, *16*(2), 12.

Bortree, D. (2005). Presentation of self on the Web: An ethnographic study of teenage girls' weblogs. *Education Communication and Information*, *5*(1), 25–39. doi:10.1080/14636310500061102

Borzi, P. (2013, March 25). Kansas surges past North Carolina and Williams. *The New York Times*. Retrieved from https://www.nytimes.com/2013/03/25/sports/ncaabasketball/kansas-surges-past-north-carolina-and-coach-roy-williams.html?_r=0

Bostwick, E. N., & Lookadoo, K. L. (2017). The Return of the King: How Cleveland Reunited With LeBron James After a Parasocial Breakup. *Communication & Sport*, *5*(6), 689–711. doi:10.1177/2167479516659460

Boulares, M., & Jemni, M. (2012, April). Mobile sign language translation system for deaf community. *Proceedings of the international cross-disciplinary conference on web accessibility*, 37. 10.1145/2207016.2207049

Bourlakis, M. A., & Weightman, P. W. (Eds.). (2004). *Food supply chain management*. Blackwell Publication.

Bowe, F. G. (2002). Deaf and Hard of Hearing Americans' Instant Messaging and E-Mail Use: A National Survey. *American Annals of the Deaf*, *147*(4), 6–10. doi:10.1353/aad.2012.0251 PMID:12592800

Bowlby, J. (1980). Attachment and Loss: Vol. III. *Loss*. New York: Basic Books.

Boyd, D. (2007). Why youth (heart) social network sites: The role of networked publics in teenage social life. In D. Buckingham (Ed.), *Youth, identity, and digital media* (pp. 119–142). Cambridge, MA: MIT Press.

Brackmann, C., Barone, D., Casali, A., Boucinha, R., & Muñoz-Hernández, S. (2016). Computational thinking: Panorama of the Americas. *International Symposium on Computers in Education (SIIE)*, 1-6.

Brito, M. (2016). *Slot machine game with symbol lock-in*. US Patent 9,520,031.

Brown, C. (2013). *Are We Becoming More Socially Awkward? An Analysis of the Relationship Between Technological Communication Use and Social Skills in College Students*. Academic Press.

Brown, M. A. (2017). *Solutions for high-touch communications in a high-tech world*.

Brown, M., Sr. (2011). *Social networking and individual performance: Examining predictors of participation* (Ph.D.). Old Dominion University. Retrieved from http://proquest.umi.com/pqdweb?did=2355291551&Fmt=7&clientId=3505&RQT=309&VName=PQD

Brown, M. A. Sr. (2017). *Solutions for High-Touch Communications in a High-Tech World*. Hershey, PA: IGI Global. doi:10.4018/978-1-5225-1897-6

Brown, N. A., & Billings, A. C. (2012). Sports fans as crisis communicators on social media websites. *Public Relations Review*, *39*(1), 74–81. doi:10.1016/j.pubrev.2012.09.012

Bryant, J., Comisky, P., & Zillmann, D. (1981). The appeal of rough-and-tumble play in televised professional football. *Communication Quarterly*, *29*(4), 256–262. doi:10.1080/01463378109369413

Bubaš, G. (2001). *Computer mediated communication theories and phenomena: Factors that influence collaboration over the Internet*. Paper presented at the 3rd CARNet users conference, Zagreb, Croatia.

Busse, C., Schleper, M. C., Niu, M., & Wagner, S. M. (2016). Supplier development for sustainability: Contextual barriers in global supply chains. *International Journal of Physical Distribution & Logistics Management*, *46*(5), 442–468. doi:10.1108/IJPDLM-12-2015-0300

Caimei, W., Yan, X., Wenchao, H., Huihua, X., Jianmeng, H., & Cheng, S. (2017). A Verified Secure Protocol Model of OAuth Dynamic Client Registration. *3rd International Conference on Big Data Computing and Communications (BIGCOM),* 106 – 110. doi: 10.1109/BIGCOM.2017.50

Cain, J., & Fink, J. L. III. (2010). Legal and ethical issues regarding social media and pharmacy education. *American Journal of Pharmaceutical Education, 74*(10), 1–8. doi:10.5688/aj7410184 PMID:21436925

Campe, S., & Denner, J. (2015). Programming Games for Learning: A Research Synthesis. *Annual meeting of the American Educational Research Association*, Chicago, IL.

Castaldo, S., & Dagnino, B. (Eds.). (2010). *Coopetition: winning strategies for the 21st century*. Edward Elgar Publishing.

Caughey, J. L. (1984). *Imaginary social worlds: A cultural approach*. Lincoln: University of Nebraska Press.

CBEEE. (2015). *Can't Beat 'EM, Eat 'EM*. Retrieved September 25, 2019, from http://www.cantbeatemeatem.us

Charlie, C., Ben, L., Benjamin, G. Z., & Christian, S. (2011). ZOZZLE: Fast and precise in- browser JavaScript malware detection. *Proceedings of the 20th USENIX Conference on Security*. Retrieved from https://usenix.org/conference/usenix-security-11/zozzle-fast-and-precise-browser-javascript-malware-detection

Charmaz, K. (2005). Grounded theory in the 21st century: Applications for advancing social justice studies. In N. K. Denzin & Y. S. Lincoln (Eds.), *The Sage handbook of qualitative research* (3rd ed., pp. 507–536). Thousand Oaks, CA: Sage Publications.

Chase, C., Chin, D., Oppezzo, M., & Schwartz, D. (2009). Teachable agents and the protégé effect: Increasing the effort towards learning. *Journal of Science Education and Technology, 18*(4), 334–352. doi:10.100710956-009-9180-4

Chebrolu, K., Raman, B., & Rao, R. R. (2005). A network layer approach to enable TCP over multiple interfaces. *Wireless Networks, 11*(5), 637–650. doi:10.100711276-005-3518-5

Chebrolu, K., & Rao, R. R. (2006). Bandwidth aggregation for real-time applications in heterogeneous wireless networks. *IEEE Transactions on Mobile Computing, 5*(4), 388–403. doi:10.1109/TMC.2006.1599407

Chen, Y. S. (2016). E-Commerce in International Franchising. In Encyclopedia of E-Commerce Development, Implementation, and Management (pp. 578-590). IGI Global. doi:10.4018/978-1-4666-9787-4.ch042

Chen, Y., Watson, E., & Azevedo, R. F. L. (2011). Soft landings curriculum of US-China entrepreneurship. *China Currents, 10*(2).

Chen, Y., Watson, E., Liu, C., Cornachione, E., & Wu, S. (2010). *Soft landing curriculum of entrepreneurship in emerging markets.* In Academy of International Business 2010 Annual Meeting, Rio de Janeiro, Brazil.

Chen, D. (2004). Young children who are deaf-blind: Implications for professionals in deaf and hard of hearing services. *The Volta Review, 104*(4), 273–284.

Chen, Y. S. (2018). Utilizing Information Science and Technology in Franchise Organizations. In *Encyclopedia of Information Science and Technology* (4th ed., pp. 4822–4835). IGI Global. doi:10.4018/978-1-5225-2255-3.ch418

Chen, Y. S., Davutyan, N., & Ersoy, İ. (2016). A "Flying High, Landing Soft" platform for supplier diversity. In *Handbook of Research on Global Supply Chain Management* (pp. 72–84). IGI Global. doi:10.4018/978-1-4666-9639-6.ch004

Chen, Y. S., Watson, E., & Azevedo, R. F. (2014). E-strategy and soft landings for franchising in emerging markets. In *Trends in E-Business, E-Services, and E-Commerce: Impact of Technology on Goods, Services, and Business Transactions* (pp. 148–159). IGI Global. doi:10.4018/978-1-4666-4510-3.ch008

Cialdini, R. B., Borden, R. J., Thorne, A., Walker, M. R., Freeman, S., & Sloan, L. R. (1976). Basking in reflected glory: Three (football) field studies. *Journal of Personality and Social Psychology*, *34*(3), 366–375. doi:10.1037/0022-3514.34.3.366

Clement, J. (2019, July). *Reach of leading social media and networking sites used by teenagers and young adults in the United States as of February 2017.* Retrieved from https://www.statista.com/statistics/199242/social-media-and-networking-sites-used-by-us-teenagers/

Click, M. A. (2007). Untidy: Fan response to the soiling of Martha Stewart's spotless image. In Fandom: Identities and Communities in a Mediated World (pp. 301–315). New York: NYU.

Clover, I. (2017). Advantages and Disadvantages of eLearning. *eLearning Industry*. Retrieved January 29, 2020, from https://elearningindustry.com/advantages-and-disadvantages-of-elearning

Cohen, E. L. (2010). Expectancy violations in relationships with friends and media figures. *Communication Research Reports*, *27*(2), 91–111. doi:10.1080/08824091003737836

Cohen, J. (2003). Parasocial breakups: Measuring individual differences in responses to the dissolution of parasocial relationships. *Mass Communication & Society*, *6*(2), 191–202. doi:10.1207/S15327825MCS0602_5

Cohen, J. (2004). Parasocial break-up from favorite television characters: The role of attachment styles and relationship intensity. *Journal of Social and Personal Relationships*, *21*(2), 187–202. doi:10.1177/0265407504041374

Cooperhouse, L., & Surgi, M. L. (2014). *Maximizing your food incubation program.* In National Business Incubation Association 28th Conference, New Orleans, LA.

Coucheney, P., Maille, P., & Tuffin, B. (2013). Impact of Competition Between ISPs on the Net Neutrality Debate. *IEEE eTransactions on Network and Service Management*, *10*(4), 425–433. doi:10.1109/TNSM.2013.090313.120326

Craft, C. (2015). *Creating and restoring wetlands: From theory to practice*. Elsevier.

Croucher, S. M. (2011). Social networking and cultural adaptation: A theoretical model. *Journal of International and Intercultural Communication*, *4*(4), 259–264. doi:10.1080/17513057.2011.598046

D'Abreo, K., Pimento, J., & D'Souza, D. (2013). *E-Learning Website Solution.* Project Report. Computer Engineering Department, Xavier Institute of Engineering, University of Mumbai. http://jonathanpimento.com/Demos/project/Docs/report.pdf

Dagnino, G. B., & Padula, G. (2009). Coopetition strategy. *Coopetition strategy: Theory, experiments and cases*, 25-43.

Dalakas, V., & Melancon, J. P. (2012). Fan identification, schadenfreude toward hated rivals, and the mediating effects of importance of winning index. *Journal of Services Marketing*, *26*(1), 51–59. doi:10.1108/08876041211199724

Daniel, F., Ralf, K., & Guido, S. (2016). A Comprehensive Formal Security Analysis of OAuth 2.0. *CCS '16 Proceedings of the 2016 ACM SIGSAC Conference on Computer and Communications Security*, 1204-1215, doi: 10.1145/2976749.2978385

David, N. (2014). One-Third of U.S. Students Use School-Issued Devices. *The Journal*. https://thejournal.com/articles/2014/04/08/a-third-of-secondary-students-use-school-issued-mobile-devices.aspx

Davis, R. O. (2018). The impact of pedagogical agent gesturing in multimedia learning environments: A meta-analysis. *Educational Research Review*, *24*, 193–209. doi:10.1016/j.edurev.2018.05.002

Deaf Australia Inc. (2010). *Next generation National Relay Service Campaign. How does VRS work?* Retrieved in May 2018 https://sites.google.com/site/nrscampaign/resources/how-does-it-work

Deaf in Kuwait. (2019). Retrieved in 17 July 2019, from https://joshuaproject.net/people_groups/19007/KU

Deafwebsites.com. (2013 Sep 26). *The History of Sign Language*. Retrieved in May 2018. http://www.deafwebsites.com/sign-language/history-sign-language.html

DeAndrea, D. C., & Walther, J. B. (2011). Attribution for inconsistencies between online and offline self-presentations. *Communication Research*, *36*(6), 805–825. doi:10.1177/0093650210385340

Deep, K., & Thakur, M. (2007). A new mutation operator for real coded genetic algorithms. *Applied Mathematics and Computation*, *193*(1), 211–230. doi:10.1016/j.amc.2007.03.046

Delello, J. A., & Consalvo, A. L. (2019). "I Found Myself Retweeting": Using Twitter Chats to Build Professional Learning Networks. In J. Yoon & P. Semingson (Eds.), *Educational Technology and Resources for Synchronous Learning in Higher Education* (pp. 88–108). Hershey, PA: IGI Global. doi:10.4018/978-1-5225-7567-2.ch005

Delello, J. A., & McWhorter, R. R. (2014). Creating virtual communities of practice with the visual social media platform Pinterest. *International Journal of Social Media and Interactive Learning Environments*, *2*(3), 216–236. doi:10.1504/IJSMILE.2014.064205

Delello, J. A., & McWhorter, R. R. (2014). New visual social media for the higher education classroom. In G. Mallia (Ed.), *The social classroom: Integrating social network use in education* (pp. 368–393). Hershey, PA: IGI Global. doi:10.4018/978-1-4666-4904-0.ch019

Delello, J. A., McWhorter, R. R., & Camp, K. M. (2015). Using social media as a tool for learning: A multi-disciplinary study. *International Journal on E-Learning*, *14*(2), 163–180.

Dick, B. (2010). What can grounded theorists and action researchers learn from each other? In The Sage Handbook of Grounded Theory Paperback Edition (pp. 398–416). Thousand Oaks, CA: Sage Publications.

Digital. (2019). *Report: The United Arab Emirates*. Retrieved from: https://image.slidesharecdn.com/datareportal-20190131gd100digital2019unitedarabemiratesv02-190131132748/95/digital-2019-united-arab-emirates-january-2019-v02-2-638.jpg?cb=1548941304

DiVerniero, R. A., & Hosek, A. M. (2011). Students' perceptions and communicative management of instructors' online self-disclosure. *Communication Quarterly*, *59*(4), 428–449. doi:10.1080/01463373.2011.597275

Dix, A., Finlay, J., Abowd, G., & Beale, R. (2004). *Human-Computer Interaction*. Prentice Hall.

Dominick, J. R. (1999). Who do you think you are? Personal home pages and self-presentation on the World Wide Web. *Journalism & Mass Communication Quarterly*, *76*(4), 646–658. doi:10.1177/107769909907600403

Donath, J., & Boyd, D. (2004). Public displays of connection. *BT Technology Journal*, *22*(4), 71–82. doi:10.1023/B:BTTJ.0000047585.06264.cc

Dongwan, S., Huiping, Y., & Une, R. (2013). Supporting visual security cues for WebView-based Android apps. *28th Annual ACM Symposium on Applied Computing, ser. SAC '13. New York,* 1867–1876. doi: 10.1145/2480362.2480709

Dongwan, S., & Rodrigo, L. (2011). An empirical study of visual security cues to prevent the ssl stripping attack. *27th Annual Computer Security Applications Conference, ser. ACSAC '11. New York,* 287–296. doi: 10.1145/2076732.2076773

Dow. (2015). *Going Back to School: Ambassadors for Science*. Retrieved September 29, 2019, from https://corporate.dow.com/en-us/science-and-sustainability/2025-goals/earth-day/stem-outreach

Dringus, L. P., & Seagull, A. B. (2013). A five-year study of sustaining blended learning initiatives to enhance academic engagement in computer and information sciences campus courses. *Blended Learning: Research Perspectives, 2*, 122-140.

Dubai School of Government. (2013). *The Arab World Online report: Trends in Internet usage in the Arab region*. Retrieved from: http://www.dsg.ae/en/Publication/Pdf_En/424201311017185100000.pdf

Dunlap, J. C., & Lowenthal, P. R. (2009). Tweeting the night away: Using Twitter to enhance social presence. *Journal of Information Systems Education*, *20*(2).

Educause. (2006). *7 things you should know about Facebook*. Retrieved from https://net.educause.edu/ir/library/pdf/ELI7017.pdf

Edward, J., & Policy, P. (2012). Community food hubs. *The Ralph W. Voorhees Center for Civic Engagement at Rutgers*. Retrieved May 21, 2018, from https://rwv.rutgers.edu/wp-content/uploads/2013/07/FoodHubFinalReport.pdf

Eitzen, S. D., & Sage, G. H. (1989). *Sociology of Northern American sport* (4th ed.). Dubuque, IA: Brown.

Ekman, P. (1972). Universals and cultural differences in facil expression of emotion. In *Nebraska Symposium on Motivation* (vol. 19). Lincoln, NE: University of Nebraska Press.

Ekman, P. (1999). Facial expressions. In T. Dalgleish & M. Power (Eds.), *Handbook of cognition and emotion* (pp. 301–320). New York, NY: John Wiley & Sons.

Elements, C. - W3C Working Group Note. (2018). Retrieved from https://www.w3.org/TR/custom-elements

Emery, R. E. (2011). *Renegotiating family relationships: Divorce, child custody, and mediation* (2nd ed.). New York: Guilford Press.

Erika, C., & David, W. (2014). Bifocals: Analyzing webview vulnerabilities in android applications. *Information Security Applications - 14th International Workshop, WISA*, 8267. doi: 10.1007/978-3-319-05149-9_9

Erol, Y., Balik, H. H., Inal, S., & Karabulut, D. (2007). Safe and Secure PIC Based Remote Control Application for Intelligent Home. *International Journal of Computer Science and Network Security*, *7*, 179–182.

Erxleben, F., Günther, M., Krötzsch, M., Mendez, J., & Vrandečić, D. (2014). Introducing Wikidata to the linked data web. In *Lecture Notes in Computer Science: vol. 8796. Proceedings of the 13th International Semantic Web Conference*. Springer. 10.1007/978-3-319-11964-9_4

Eteokleous-Grigoriou, N., & Ktoridou, D. (2014). Social networking for educational purposes: The development of social-cultural skills through special interest groups. In G. Mallia (Ed.), *The social classroom: Integrating social network use in education* (pp. 394–416). Hershey, PA: IGI Global. doi:10.4018/978-1-4666-4904-0.ch020

Evans, J. S. (1991). Strategic flexibility for high technology manoeuvres: A conceptual framework. *Journal of Management Studies*, *28*(1), 69–89. doi:10.1111/j.1467-6486.1991.tb00271.x

Evans, L., & Hardy, L. (1995). Sport injury and grief responses: A review. *Journal of Sport & Exercise Psychology*, *17*(3), 227–245. doi:10.1123/jsep.17.3.227

Facebook App in Tizen. (2018). Retrieved from https://developer.tizen.org/ko/development/articles/facebook-apptizen

Faculty Focus. (2011). *Social media usage trends among higher education faculty*. Faculty Focus Special Report. Retrieved from http://cdn.facultyfocus.com/wp-content/uploads/2011/09/2011-social-media-report.pdf

Fadi, M., & Mohamed, S. (2016). Hardening the OAuth-WebView Implementations in Android Applications by Re-Factoring the Chromium Library. *2nd IEEE International Conference on Collaboration and Internet Computing*, 196–205. doi: 10.1109/CIC.2016.036

Fazlollahtabar, H., & Abbasi, A. (2012). A Computer Integrated Framework for E-Learning Control Systems Based on Data Flow Diagrams. *Journal of Control Engineering and Technology*, *2*(2), 97–105.

Fernquist, R. M. (2000). An aggregate analysis of professional sports, suicide, and homicide rates: 30 U.S. metropolitan areas, 1971-1990. *Aggression and Violent Behavior*, *5*(4), 329–341. doi:10.1016/S1359-1789(98)00012-3

Ferrer-Sapena, A., Peset, F., & Aleixandre-Benavent, R. (2011). Acceso a los datos públicos y su reutilización: Open data y open government. *El Profesional de la Información*, *20*(3), 260–269. doi:10.3145/epi.2011.may.03

Florida Telecommunications Relay, Inc. (1990). *Federal Legislation Americans with Disabilities Act ADA 1990 Title 4 (Telecommunications) Deaf, Hard of Hearing, Deaf/Blind.* Speech Disabled Relay Service.

Folger, T. (2013). Rising seas. *National Geographic*, *224*(3), 30–59.

Foss, S. K. (2009). *Rhetorical criticism: Exploration and practice* (4th ed.). Long Grove, IL: Waveland Press.

Fowles, J. (1992). *Star struck: Celebrity performers and the American public.* Washington, DC: Smithsonian Institution Press.

Francisco, C., & Karen, P. L. (2011). *Security Analysis of Double Redirection Protocols.* Retrieved from https://bit.ly/2o5eggP

Frank Messina, E. D. (2011). *Two-And-A-Half Minutes To "Effective"*. Xlibris.

Freudenburg, W. R., & Gramling, R. (2012). *Blowout in the Gulf: The BP oil spill disaster and the future of energy in America.* MIT.

Frida, Cardozo, Gebhart, Gullo, & Kalia. (2017, Apr. 13). *Spying on Students. School-issued devices and student privacy.* Executive summary.

Gabay, D., & Moulin, H. (1978). *On the uniqueness and stability of Nash's equilibrium in non cooperative games.* Université Paris IX-Dauphine, Centre de recherche de mathématiques de la decision.

Gabennesch, H. (1988). When promises fail: A theory of temporal fluctuations in suicide. *Social Forces*, *67*(1), 129–145. doi:10.2307/2579103

Gaffey, B. E., Cromartie, J. B., & Seaboldt, M. R. (2012). *Evaluating communication technologies for the deaf and hard of hearing.*

Gambetta, D. (1988). *Trust: making and breaking cooperative relations.* New York, NY: B. Blackwell.

Garmani, H., Ait Omar, D., El Amrani, M., Baslam, M., & Jourhmane, M. (2019a). Analysis of a dynamics duopoly game with two content providers. *Chaos, Solitons, and Fractals*, *109466*. doi:10.1016/j.chaos.2019.109466

Garmani, H., Baslam, M., & Jourhmane, M. (2018a). Analysis of Competition between CPs Fronting Advertisers. *International Journal of Advanced Science and Technology*, *117*, 53–66. doi:10.14257/ijast.2018.117.05

Garmani, H., Baslam, M., & Jourhmane, M. (2018b). Caching Games between ISP in Information Centric Network. *International Journal of Control and Automation*, *11*(4), 125–142. doi:10.14257/ijca.2018.11.4.12

Garmani, H., Omar, D. A., Amrani, M. E., Baslam, M., & Jourhmane, M. (2019b). Analysis of sponsored content in the case of competing content providers. *International Journal of Control and Automation*, *12*(10), 14. doi:10.33832/ijca.2019.12.10.04

Garmani, H., Outanoute, M., Baslam, M., & Jourhmane, M. (2019b). New Competition-Based Approach for Caching Popular Content in ICN. In A. Podelski & F. Taïani (Eds.), *Networked Systems* (pp. 286–300). Cham: Springer International Publishing. doi:10.1007/978-3-030-05529-5_19

Garrison, D. R., Anderson, T., & Archer, W. (2000). Critical inquiry in a text based environment: Computer conferencing in higher education. *The Internet and Higher Education*, *2*(2/3), 87–105.

Garrison, R. D. (2017). *21st century e-learning*. Routledge.

Gates, W. (1999). *Business @ the speed of thought*. Warner Books. doi:10.1111/1467-8616.00097

Gentzler, A. L., Oberhauser, A. M., Westerman, D., & Nadorff, D. K. (2011). College students' use of electronic communication with parents: Links to loneliness, attachment, and relationship quality. *Cyberpsychology, Behavior, and Social Networking*, *14*(1-2), 71–74. doi:10.1089/cyber.2009.0409 PMID:20973676

Gheorghiță, V., & Pădurețu, A. (2014). Social Networks And Interpersonal Communication. *System*, *2*(3), 5.

Gibbs, J. L., Ellison, N. B., & Heino, R. D. (2006). Self-presentation in online personas: The role of anticipated future interaction, self-disclosure, and perceived success in Internet dating. *Communication Research*, *33*(2), 152–177. doi:10.1177/0093650205285368

Gikas, J., & Grant, M. M. (2013). Mobile computing devices in higher education: Student perspectives on learning with cellphones, smartphones & social media. *The Internet and Higher Education*, *19*, 18–26. doi:10.1016/j.iheduc.2013.06.002

Gil'Adi, D. (1998). *La Inteligencia Emocional en Práctica*. Carácas: Interamericana de Venezuela.

Giles, D. C. (2002). Parasocial interaction: A review of the literature and a model for future research. *Media Psychology*, *4*(3), 279–305. doi:10.1207/S1532785XMEP0403_04

Ginns, P., & Ellis, R. (2007). Quality in Blended Learning: Exploring the Relationships between on-Line and Face-to-Face Teaching and Learning. *The Internet and Higher Education*, *10*(1), 53–64. doi:10.1016/j.iheduc.2006.10.003

Glaser, B. G., & Strauss, A. L. (1967). *The discovery of grounded theory*. Hawthorne, NY: Aldine.

Gleich, U. (1996). Sind Fernsehpersonen die "Freunde" des Zuschauers? Ein Vergleich zwischen parasozialen und realen sozialen Beziehungen [Are media figures "friends" of the viewers? A comparsion between parasocial and real social relationships]. In P. Vorderer (Ed.), *Fernsehen als "Beziehingskiste": Parasoziale beziehungen und interaktionen mit TV-personen* (pp. 113–144). Opladen: Westdeutscher Verlag. doi:10.1007/978-3-322-83274-0_12

Globalwebindex. (2019). *The latest social media trends to know in 2019*. Retrieved from https://www.globalwebindex.com/reports/social

Godfray, H. C. J., & Garnett, T. (2014). Food security and sustainable intensification. *Philosophical Transactions of the Royal Society of London. Series B, Biological Sciences*, *369*(1639), 20120273. doi:10.1098/rstb.2012.0273 PMID:24535385

Goffman, E. (1959). *The presentation of self in everyday life*. Garden City, NY: Doubleday.

Goleman, D. (1999). *La práctica de la inteligencia emocional*. Barcelona: Kairos.

GoMRI. (2015). *Gulf of Mexico research initiative*. Retrieved May 21, 2018, from http://gulfresearchinitiative.org

Gonzales, A. L., & Hancock, J. T. (2011). Mirror, mirror on my Facebook wall: Effects of exposure to Facebook on self-esteem. *Cyberpsychology, Behavior, and Social Networking*, *14*(1-2), 79–83. doi:10.1089/cyber.2009.0411 PMID:21329447

Google Identity Platform. (2018). *Using OAuth 2.0 to Access Google APIs*. Retrieved from https://developers.google.com/identity/protocols/OAuth2

Gottesman, A. (1997, March 22). Arizona 85, Kansas 82: 'cats run wild, scald Jayhawks--superior speed ignites tournament shocker. *Chicago Tribune*. Retrieved from http://articles.chicagotribune.com/1997-03-22/sports/9703220149_1_jayhawks-fourth-seeded-arizona-southeast-regional-semifinal

Graesser, A. C., Chipman, P., Haynes, B. C., & Olney, A. (2005). AutoTutor: An intelligent tutoring system with mixed-initiative dialogue. *IEEE Transactions on Education*, *48*(4), 612–618. doi:10.1109/TE.2005.856149

Graesser, A., D'Mello, S., Craig, S., Witherspoon, A., Sullins, J., McDaniel, B., & Gholson, B. (2008). The relationship between affective states and dialogue patterns during interactions with AutoTutor. *Journal of Interactive Learning Research*, *19*(2), 293–302.

Gray, J. (2003). New audiences, new textualities: Anti-fans and non-fans. *International Journal of Cultural Studies*, *6*(1), 64–81. doi:10.1177/1367877903006001004

Griol, D., & Callejas, Z. (2013). An Architecture to Develop Multimodal Educative Applications with Chatbots. *International Journal of Advanced Robotic Systems*, *10*(3), 175. doi:10.5772/55791

Guijarro, L., Pla, V., Vidal, J. R., & Martinez-Bauset, J. (2011). *Analysis of price competition under peering and transit agreements in Internet Service provision to peer-to-peer users. In 2011 IEEE Consumer Communications and Networking Conference* (pp. 1145–1149). CCNC. doi:10.1109/CCNC.2011.5766356

Gunawardena, C. N., & Zittle, F. J. (1997). Social presence as a predictor of satisfaction within a computer-mediated conferencing environment. *American Journal of Distance Education*, *11*(3), 8–26. doi:10.1080/08923649709526970

Guzley, R. M. (1992). Organizational climate and communication climate: Predictors of commitment to the organization. *Management Communication Quarterly*, *5*(4), 379–402. doi:10.1177/0893318992005004001

Habak, K., Harras, K. A., & Youssef, M. (2015). Bandwidth aggregation techniques in heterogeneous multi-homed devices: A survey. *Computer Networks*, *92*, 168–188. doi:10.1016/j.comnet.2015.08.039

Haight, C. M. (2009). *One Wire Viewer and an iButton Quick Start Guide*. Retrieved March 11, 2009, from https://www.maximintegrated.com/en/design/technical-documents/tutorials/4/4373.html

Halawani, S. M. (2008). Arabic sign language translation system on mobile devices. *IJCSNS International Journal of Computer Science and Network Security*, *8*(1), 251–256.

Halawani, S. M., & Zaitun, A. B. (2012). An avatar based translation system from Arabic speech to Arabic sign language for deaf people. *International Journal of Information Science and Education*, *2*(1), 13–20.

Hall, A. A., Delello, J. A., & McWhorter, R. R. (2017). Using Facebook to supplement instruction in online and hybrid courses. *Journal of Innovation and Learning*, *22*(1), 87–104. doi:10.1504/IJIL.2017.085250

Hallberg, A. (2017). The Top 4 Countries That Have Developed eLearning. *eLearning Industry*. https://elearningindustry.com/countries-that-have-developed-elearning-top-4

Hall, J., & Matos, S. (2010). Incorporating impoverished communities in sustainable supply chains. *International Journal of Physical Distribution & Logistics Management*, *40*(1/2), 124–147. doi:10.1108/09600031011020368

Hammer. (2010). The OAuth 1.0 protocol. *The Internet Eng. Task Force RFC 5849*. Retrieved from https://tools.ietf.org/html/rfc5849

Hammer, M. (1996). *Beyond Reengineer*. Harper Business.

Hao, H., Vicky, S., & Wenliang, D. (2013). On the effectiveness of API-level access control using bytecode rewriting in Android. *8th ACM SIGSAC Symposium on Information, Computer and Communications Security*, 25–36. 10.1145/2484313.2484317

Hardt. (2011). The OAuth 2.0 authorization framework. *The Internet Eng. Task Force RFC 6749*. Retrieved from https://tools.ietf.org/html/rfc6749

Hardt. (Ed.) (2012). *The OAuth 2.0 authorization framework*. IETF Std., RFC6749. doi:10.17487/RFC6749

Harriet, T. (2015). Google's Chromebooks make up half of U.S. devices sold. *CNBC*. http://www.cnbc.com/2015/12/03/googles-chromebooks-make-up-half-of-us-classroom-devices.html

Hartig, O., & Langegger, A. (2010). A database perspective on consuming linked data on the web. *Datenbank-Spektrum, 10*(2), 57–66. doi:10.100713222-010-0021-7

Hartmann, T., Stuke, D., & Daschmann, G. (2008). Positive parasocial relationships with drivers affect suspense in racing sport spectators. *Journal of Media Psychology, 20*(1), 24–34. doi:10.1027/1864-1105.20.1.24

Hartsock, A. (2000, July 7). "I'm staying": Roy opts to remain at Kansas. *Lawrence Journal-World*. Retrieved from http://www2.kusports.com/news/2000/jul/07/im_staying/

Hausenblas, M., & Karnstedt, M. (2010). Understanding linked open data as a web-scale database. *Proceedings of the second conference on Advances in Databases Knowledge and Data Applications*, 56-61. Accessed 20 November 2018: https://aran.library.nuigalway.ie/handle/10379/1127

Hausenblas, M. (2009). Exploiting linked data to build web applications. *IEEE Internet Computing, 13*(4), 68–73. doi:10.1109/MIC.2009.79

Heath, T., & Bizer, C. (2011). Linked data: Evolving the web into a global data space. *Synthesis Lectures on the Semantic Web: Theory and Technology, 1*(1), 1-136. doi:10.2200/S00334ED1V01Y201102WBE001

Heiberger, G., & Junco, R. (2015). *Meet your students where they are: Social media*. National Education Association. Retrieved from http://www.nea.org/assets/docs/HE/HigherEdSocialMediaGuide.pdf

Heintz, F., Mannila, L., & Färnqvist, T. (2016). A review of models for introducing computational thinking, computer science and computing in K-12 education. In *Frontiers in Education Conference (FIE), 2016 IEEE* (pp. 1-9). IEEE.

Helvie-Mason, L. (2011). Facebook, friending, and faculty–student communication. In C. Wankel (Ed.), Teaching Arts and Science with the New Social Media. Emerald Group Publishing Limited. doi:10.1108/S2044-9968(2011)0000003007

Herpich, F., Voss, G. B., Nunes, F. B., & Medina, R. D. (2016). Three-Dimensional Virtual Environment and NPC: A Perspective about Intelligent Agents Ubiquitous. In Handbook of Research on 3-D Virtual Environments and Hypermedia for Ubiquitous Learning (pp. 510-536). IGI Global.

Herr, R. (2004). Telecommunications Relay Services: An Introduction to Communication Options. *Colorado Commission for the Deaf and Hard of Hearing E-Journal*, 1-6.

Hershkovitz, A., & Forkosh-Baruch, A. (2013). Student-teacher relationship in the Facebook era: The student perspective. *International Journal of Continuing Engineering Education and Lifelong Learning, 23*(1), 33–52. doi:10.1504/IJCEELL.2013.051765

Hiestand, M. (2008, April 7). CBS' Packer says honesty is still his policy. *USA Today*. Retrieved from https://www.usatoday.com/sports/columnist/hiestand-tv/2008-04-06-packer_N.htm

Hilsen, A. I., & Helvik, T. (2014). The construction of self in social media, such as Facebook. *AI & Society*, *29*(1), 3–10. doi:10.100700146-012-0426-y

History of ELearning & Interactive Books. (n.d.). *Interactive eBooks Articles*. The Institute of Progressive Education and Learning. Retrieved January 29, 2020, from http://institute-of-progressive-education-and-learning.org/elearning-i/history-of-elearning/

Hittle, S. (2009, July 23). In the world of online comments, there are plenty of opinions, but few names. *Lawrence Journal-World.* Retrieved from http://www2.ljworld.com/news/2009/jul/23/plenty-opinions-few-names/

Hitt, M. A., Ireland, R. D., Sirmon, D. G., & Trahms, C. A. (2011). Strategic Entrepreneurship: Creating Value for Individuals, Organizations, and Society. *The Academy of Management Perspectives*, *25*(2), 57–75.

Holland, J. H. (1992). *Adaptation in Natural and Artificial Systems : An Introductory Analysis with Applications to Biology, Control and Artificial Intelligence*. Cambridge, MA: MIT Press; doi:10.7551/mitpress/1090.001.0001

Horton, D., & Richard Wohl, R. (1956). Mass communication and parasocial interaction: Observations on intimacy at a distance. *Psychiatry*, *19*, 215–229. doi:10.1080/00332747.1956.11023049 PMID:13359569

Huang, H. C., Tsai, F. C., & Mao, T. F. (2011). Intelligent Embedded Control Warning System For Car Reversing. In *Measuring Technology and Mechatronics Automation* (pp.849-852). Third International Conference Shangshai. 10.1109/ICMTMA.2011.785

Huijboom, N., & Van den Broek, T. (2011). Open data: an international comparison of strategies. *European Journal of ePractice, 12*(1), 4-16.

Hu, M. (2015). The influence of a scandal on parasocial relationship, parasocial interaction, and parasocial breakup. *Psychology of Popular Media Culture*. doi:10.1037/ppm0000068

Hunt, K. A., Bristol, T., & Edward Bashaw, R. (1999). A conceptual approach to classifying sports fans. *Journal of Services Marketing*, *13*(6), 439–452. doi:10.1108/08876049910298720

Hussana, J. R., Priyanka, A., Revathi, A. M., Suchitha, K., & Sumana, K. J. (2014). Multiple sign language translation into voice. *International Journal of Engineering and Innovative Technology*, *3*(10).

InBIA. (2015). *International business innovation association*. Retrieved May 21, 2018, from https://www.inbia.org/resources/directories/soft-landings-members

Introduction to video relay services. (2005). Sorenson's Communications.

Introduction to VRS Interpreting. A Curriculum Guide. (2017). Retrieved in 17 July 2019, from http://www.vrsii.com/assets/pdf/curriculum/VRS_Curriculum.pdf

IP. (2015). *International programs at LSU*. Retrieved May 21, 2018, from http://international.lsu.edu

Irwin, C., Ball, L., Desbrow, B., & Leveritt, M. (2012). Students' perceptions of using Facebook as an interactive learning resource at university. *Australasian Journal of Educational Technology*, *28*(7), 1221–1232. doi:10.14742/ajet.798

Jaggars, S. S. (2014). Choosing between online and face-to-face courses: Community college student voices. *American Journal of Distance Education*, *28*(1), 27–38. doi:10.1080/08923647.2014.867697

Jaggernath, R., & Khan, Z. (2015). Green supply chain management. *World Journal of Entrepreneurship, Management and Sustainable Development*, *11*(1), 37–47. doi:10.1108/WJEMSD-06-2014-0018

Janssen, M., Charalabidis, Y., & Zuiderwijk, A. (2012). Benefits, adoption barriers and myths of open data and open government. *Information Systems Management, 29*(4), 258–268. doi:10.1080/10580530.2012.716740

Jason, B., Elie, B., Divij, G., & John, C. M. (2010). State of the Art: Automated Black-Box Web Application Vulnerability Testing. *IEEE Symposium on Security and Privacy, 31*(1), 332 - 345. doi:10.1109/SP.2010.27

Jean-Francois, E. (2018). *Transnational Perspectives on Innovation in Teaching and Learning Technologies.* Koninklijke Brill, NV. doi:10.1163/9789004366077

Jensen, T. (2008, April 5). Final Four 2008: Kansas vs North Carolina betting preview. *Point-Spreads.* Retrieved from http://www.point-spreads.com/college-basketball/040506-final-four-2008-kansas-vs-north-carolina-betting-preview.html

Joe-Wong, C., Ha, S., & Chiang, M. (2015). *Sponsoring mobile data: An economic analysis of the impact on users and content providers. In 2015 IEEE conference on computer communications (infocom)* (pp. 1499–1507). IEEE.

John, D., & Catherine, T. (2010). Social capital and self-presentation on social networking sites: A comparative study of Chinese and American young generations. *Chinese Journal of Communication, 3*(4), 402–420. doi:10.1080/17544750.2010.516575

Johnson, L., Adams Becker, S., Estrada, V., & Freeman, A. (2015). *NMC horizon report: 2015 K-12 edition.* Retrieved from http://cdn.nmc.org/media/2015-nmc-horizon-report-HE-EN.pdf

Johnson, K. (2011). The effect of Twitter posts on students' perceptions of instructor credibility. *Learning, Media and Technology, 36*(1), 21–38. doi:10.1080/17439884.2010.534798

Johnson, L., Adams Becker, S., & Cummins, M. (2012). *The NMC horizon report: 2012 higher education edition.* Austin, TX: The New Media Consortium.

Johnson, W. L., & Lester, J. C. (2018). Pedagogical Agents: Back to the Future. *AI Magazine, 39*(2), 33–44. doi:10.1609/aimag.v39i2.2793

Johnson, W. L., Rickel, J. W., & Lester, J. C. (2000). Animated pedagogical agents: Face-to-face interaction in interactive learning environments. *International Journal of Artificial Intelligence in Education, 11*(1), 47–78.

Jones & Hardt. (2012). *OAuth 2.0: Bearer token usage.* Available: https://tools.ietf.org/html/draft-ietf-oauth-v2-bearer-23

Jones, E. E. (1990). *Interpersonal perception.* New York: W.H. Freeman.

Junco, R., & Cotton, S. (2011). *A decade of distraction? How multitasking affects student outcomes.* Paper presented at A Decade in Internet Time Symposium on the Dynamics of the Internet and Society, Oxford Internet Institute, University of Oxford, Oxford, UK. 10.2139srn.1927049

Junco, R. (2012). Too much face and not enough books: The relationship between multiple indices of Facebook use and academic performance. *Computers in Human Behavior, 28*(1), 187–198. doi:10.1016/j.chb.2011.08.026

Junco, R., & Cotton, S. (2012). The relationship between multitasking and academic performance. *Computers & Education, 59*(4), 1–10. doi:10.1016/j.compedu.2011.12.023

Junco, R., Heiberger, G., & Loken, E. (2010). The effect of Twitter on college student engagement and grades. *Journal of Computer Assisted Learning, 27*(2), 119–132. doi:10.1111/j.1365-2729.2010.00387.x

Justis, R. T., & Judd, R. J. (2002). *Franchising.* DAME Publishing.

Kahan, J., Koivunen, M. R., Prud'Hommeaux, E., & Swick, R. R. (2002). Annotea: An open RDF infrastructure for shared Web annotations. *Computer Networks, 39*(5), 589–608. doi:10.1016/S1389-1286(02)00220-7

Kaplan, A. M., & Haenlein, M. (2010). Users of the world, unite! The challenges and opportunities of Social Media. *Business Horizons*, *53*(1), 59–68. doi:10.1016/j.bushor.2009.09.003

Kassing, J. W., & Sanderson, J. (2009). You're the kind of guy that we all want for a drinking buddy: Expressions of parasocial interaction on floydlandis.com. *Western Journal of Communication*, *73*(2), 182–203. doi:10.1080/10570310902856063

Kaufmann, H. (2011). Virtual environments for mathematics and geometry education. *Themes in Science and Technology Education, 2*(1-2), 131-152.

Kaushik, D., Prabhavathi, P., & Joy, B. (2017). Security Mechanism for Packaged Web Applications. *2017 IEEE International Conference on Web Services (ICWS).* doi: 10.1109/ICWS.2017.72

Kawachi, P. (2013). Online social presence and its correlation with learning. *International Journal of Social Media and Interactive Learning Environments*, *1*(1), 19–31. doi:10.1504/IJSMILE.2013.051653

Keiser Foundation. (2013). *Generation M²: Media in the lives of 8-to-18-year-olds*. Retrieved from https://kaiserfamilyfoundation.files.wordpress.com/2013/04/8010.pdf

Kelley, S. W., & Tian, K. (2004). Fanatical consumption: An investigation of the behavior of sports fans through textual data. In Sports marketing and the psychology of marketing communication (pp. 27–66). Mahwah, NJ: Lawrence Erlbaum Associates.

Kelly, A. E., & Seppala, M. (2016). Changing Policies Concerning Student Privacy and Ethics in Online Education. *International Journal of Education Technology*, *6*(8), 652–655. doi:10.7763/IJIET.2016.V6.768

Kelly, K., Schwarz, C. J., Gomez, R., & Marsh, K. (2018). Time required to move cash through slot machines. *Journal of Money Laundering Control*, *21*(3), 376–384. doi:10.1108/JMLC-09-2017-0049

Khanduja, J. (2013). *Shared tastes* (Doctoral dissertation). Columbia University, New York, NY.

Kim, H.-S., & Yoon, C.-H. (2004). Determinants of subscriber churn and customer loyalty in the Korean mobile telephony market. *Telecommunications Policy*, *28*(9), 751–765. doi:10.1016/j.telpol.2004.05.013

Kim, J., LaRose, R., & Peng, W. (2009). Loneliness as the cause and the effect of problematic Internet use: The relationship between Internet use and psychological well-being. *Cyberpsychology & Behavior*, *12*(4), 451–455. doi:10.1089/cpb.2008.0327 PMID:19514821

Kim, J., & Lee, J. E. R. (2011). The Facebook paths to happiness: Effects of the number of Facebook friends and self-presentation on subjective well-being. *Cyberpsychology, Behavior, and Social Networking*, *14*(6), 359–364. doi:10.1089/cyber.2010.0374 PMID:21117983

King, J. (2013, May 3). Roy vs. Kansas feels different this time. *ESPN.com*. Retrieved from http://espn.go.com/blog/north-carolina-basketball/post/_/id/13278/roy-vs-kansas-feels-different-this-time

King, T. (2019, July). Bitmojis, GIFs, and Snaps in the Classroom? Oh My! *Faculty Focus*. Retrieved from https://www.facultyfocus.com/articles/effective-classroom-management/bitmojis-GIFs-and-snaps-in-the-classroom/

Kirschner, P. A., & Karpinski, A. C. (2010). Facebook and academic performance. *Computers in Human Behavior*, *26*(6), 1237–1245. doi:10.1016/j.chb.2010.03.024

Kitchin, R. (2014). *The data revolution: Big data, open data, data infrastructures and their consequences*. Sage.

Kitura. (2018). *OAuth 2.0 authentication with Facebook/Google*. Retrieved from https://kitura.io/docs/authentication/fb-google-oauth2.html

Kleinrock, L. (1975). *Queueing Systems*. New York, NY: Wiley-Interscience.

Kline, C., Shah, N., & Rubright, H. (2014). Applying the positive theory of social entrepreneurship to understand food entrepreneurs and their operations. *Tourism Planning & Development*, *11*(3), 1–13. doi:10.1080/21568316.2014.890126

Klugman, M. (2009). Loves, suffering and identification: The passions of Australian Football League fans. *The International Journal of the History of Sport*, *26*(1), 21–44. doi:10.1080/09523360802500071

Knight-McCord, J., Cleary, D., Grant, N., Herron, A., Jumbo, S., Lacey, T., ... Smith, R. (2016). What social media sites do college students use most? *Journal of Undergraduate Ethnic Minority Psychology*, *2*(21), 21–26.

Konzen, A., Oliveira, O. L., Kist, L. S., Anjos, A. R., Moraes, L., Freitas, C. I. L., ... Mega Vitta, A. (2011). *Agente conversacional aplicado al juego de la educación*. XXII Simposio Brasileño de Informática en Educacion.

Koutsoupias, E., & Papadimitriou, C. (1999). Worst-case equilibria. In *Annual Symposium on Theoretical Aspects of Computer Science*. Springer.

Kramer, R. M. (1996). *Trust in organizations: Frontiers of theory and research*. Thousand Oaks: Sage Publications.

Kubler-Ross, E. (1969). *On death and dying*. New York: MacMillan.

Kubler-Ross, E. (1972). On death and dying. *Journal of the American Medical Association*, *221*(2), 174–179. doi:10.1001/jama.1972.03200150040010 PMID:5067627

Kumi-Yeboah, A., & Blankson, H. (2018). Social media and use of technology in higher education. In M. Khosrow-Pour (Ed.), *Teacher Training and Professional Development: Concepts, Methodologies, Tools, and Applications, 4* (Vol. 41, pp. 932–952). Information Resources Management Association. doi:10.4018/978-1-5225-5631-2.ch041

Labrinidis, A., & Jagadish, H. V. (2012). Challenges and opportunities with big data. *VLDB Endowment*, *5*(12), 2032–2033. doi:10.14778/2367502.2367572

LAFI. (2015). *LSU AgCenter food incubator*. Retrieved May 21, 2018, from https://www.lsuagcenter.com/en/our_offices/departments/food_science/extension_outreach/incubator

LAM. (2007). Alligator collagen: New source for medical, cosmetic uses. *Louisiana Agriculture Magazine*. Retrieved May 21, 2018, from https://www.lsuagcenter.com/en/communications/publications/agmag/archive/2007/winter/alligator+collagen+new+source+for+medical+cosmetic+uses.htm

LAM. (2012). Value-added from crawfish and catfish. *Louisiana Agriculture Magazine*. Retrieved May 21, 2018, from https://www.lsuagcenter.com/en/communications/publications/agmag/Archive/2002/Fall/Valueadded+from+Crawfish+and+Catfish.htm

Landed, J. (n.d.). *Language in Kuwait*. Retrieved in 17 July 2019, from https://www.justlanded.com/english/Kuwait/Kuwait-Guide/Language/Language-in-Kuwait

Lane, C., & Bachmann, R. (1998). *Trust within and between organizations: Conceptual issues and empirical applications*. New York: Oxford University Press.

Larson, R. R. (2010). Introduction to information retrieval. *Journal of the American Society for Information Science and Technology*, *61*(4), 852–853. doi:10.1002/asi.21234

Lasaulce, S., Debbah, M., & Altman, E. (2009). Methodologies for analyzing equilibria in wireless games. *IEEE Signal Processing Magazine*, *26*(5), 41–52. doi:10.1109/MSP.2009.933496

Lawrence Journal-World. (2006, June 24). *Tar Heels coach battling bad back.* Retrieved from http://www2.kusports.com/news/2006/jun/24/tar_heels_coach_battling_bad_back/

Lawrence Journal-World. (2008, January 12). *Williams Falls, Needs Stitches.* Retrieved from http://www2.kusports.com/news/2008/jan/12/williams_falls_needs_stitches/?mens_basketball

Lawrence Journal-World. (2009, July 8). *KUSports.com Jayhawk flashback: July 8, 1988, Roy Williams introductory press conference* [Video file]. Retrieved from https://www.youtube.com/watch?v=bnpT7AhUuwk

LBTC. (2015). *Louisiana business and technology center.* Retrieved May 21, 2018, from http://www.lbtc.lsu.edu

LCWRP. (1993). *Louisiana coastal wetlands restoration plan.* Retrieved May 21, 2018, from https://lacoast.gov/reports/cwcrp/1993/1993lcwrp-1executive_summary.pdf

LDWF. (2007). Nutria control program. *The Louisiana Department of Wildlife and Fisheries.* Retrieved May 21, 2018, from https://nutria.com/site9.php

Leal Filho, W., & Kovaleva, M. (2015). *Food waste and sustainable food waste management in the Baltic Sea Region.* Springer. doi:10.1007/978-3-319-10906-0

Leary, M. R. (1996). *Self-presentation: Impression management and interpersonal behavior.* Boulder, CO: Westview.

LED. (2018). International commerce. *Louisiana Economic Development.* Retrieved May 24, 2018, from https://www.opportunitylouisiana.com/international-commerce

Lee, Y. J., & Kim, J. S. (2012). Automatic web api composition for semantic data mashups. In *Proceedings of Fourth International Conference on Computational Intelligence and Communication Networks.* IEEE Computer Society. 10.1109/CICN.2012.56

Lee, A. R., & Horsley, J. S. (2017). The role of social media on positive youth development: An analysis of 4-H Facebook page and 4-H'ers' positive development. *Children and Youth Services Review, 77*, 127–138. doi:10.1016/j.childyouth.2017.04.014

Lee, A., & Wall, G. (2012). *Food clusters: Towards a creative rural economy.* Martin Prosperity Institute, Rotman School of Management, University of Toronto.

Lee, S. H., Simkins, T. J., Luster, S., & Chowdhury, S. A. (2018). Forgiving sports celebrities with ethical transgressions: The role of parasocial relationships, ethical intent and regulatory focus mindset. *Journal of Global Sport Management, 3*(2), 124–145. doi:10.1080/24704067.2018.1441737

Lee, S. J. (2009). Online communication and adolescent social ties: Who benefits more from Internet use? *Journal of Computer-Mediated Communication, 14*(3), 509–531. doi:10.1111/j.1083-6101.2009.01451.x

Lee, S. J., Quigley, B. M., Nesler, M. S., Corbett, A. B., & Tedeschi, J. T. (1999). Development of a selfpresentation tactics scale. *Personality and Individual Differences, 26*(4), 701–722. doi:10.1016/S0191-8869(98)00178-0

Lehmann, J., Isele, R., Jakob, M., Jentzsch, A., Kontokostas, D., Mendes, P. N., ... Bizer, C. (2015). DBpedia–a large-scale, multilingual knowledge base extracted from Wikipedia. *Semantic Web, 6*(2), 167–195. doi:10.3233/SW-140134

Lester, J., Converse, S., Kahler, S., Barlow, S., Stone, B., & Bhogal, R. (1997). The persona effect: Affective impact of animated pedagogical agents. In *Proceedings of the SIGCHI Conference on Human Factors in Computing Systems* (pp. 359-366). New York, NY: ACM. 10.1145/258549.258797

LETC. (2015). *Louisiana emerging technology center.* Retrieved May 21, 2018, from https://laetc.com/the-la-emerging-tech-center

Lewis, S., & Lallo, E. (2013). *Alligators are helping save Louisiana's coastal wetlands*. Retrieved May 21, 2018, from https://gulfseafoodnews.com/2013/09/18/the-business-of-alligator-part-four-the-thick-skinned-environmentalists

lhshaoren. (2010). *A vulnerability on weibo.com (sina.com)*. Retrieved from http://www.wooyun.org/bugs/wooyun-2010-039727

Liebers, N., & Schramm, H. (2019). Parasocial interactions and relationship with media characters—An inventory of 60 years of research. *Communication Research Trends*, *38*(2), 4–31.

Lincoln, Y. S., Lynham, S. A., & Guba, E. G. (2000). Paradigmatic controversies contradictions and emerging confluences. In The handbook of qualitative research (4th ed., pp. 163–188). Beverly Hills, CA: Sage Publications.

Lin, E. I., Eskicioglu, A. M., Lagendijk, R. L., & Delp, E. J. (2005). Advances in Digital Video Content Protection. *Proceedings of the IEEE*, *93*(1), 171–183. doi:10.1109/JPROC.2004.839623

Linke, B. (2009). *One Wire Viewer User's Guide*. Retrieved September 25, 2009, from https://pdfserv.maximintegrated.com/en/an/AN3358.pdf

Lin, L., Ginns, P., Wang, T., & Zhang, P. (2020). Using a pedagogical agent to deliver conversational style instruction: What beneðts can you obtain? *Computers & Education*, *143*, 103658. doi:10.1016/j.compedu.2019.103658

LIP. (2015). *LSU innovation park*. Retrieved May 21, 2018, from https://sites01.lsu.edu/wp/innovationpark

Lister, J. J., Nower, L., & Wohl, M. J. A. (2016). Gambling goals predict chasing behavior during slot machine play. *Addictive Behaviors*, *62*, 129–134. doi:10.1016/j.addbeh.2016.06.018 PMID:27347654

Liu, Y., Gong, W., & Shenoy, P. (2001). On the impact of concurrent downloads. In *Simulation Conference, 2001. Proceedings of the Winter*. IEEE.

Lodderstedt, McGloin, & Hunt. (2013). *OAuth 2.0 Threat Model and Security Considerations*. IETF Std., RFC6819. doi:10.17487/RFC6819

Lopez Ludeña, V., San Segundo Hernández, R., González Morcillo, C., López López, J. C., & Pardo Muñoz, J. M. (2012). *Methodology for developing a Speech into Sign Language Translation System in a New Semantic Domain*. Academic Press.

Lorkowski, J., & Kreinovich, V. (2018). *Bounded Rationality in Decision Making Under Uncertainty : Towards Optimal Granularity*. Consulté à l'adresse https://www.springer.com/gp/book/9783319622132

Loss, J., Lindacher, V., & Curbach, J. (2013). Do social networking sites enhance the attractiveness of risky health behavior? Impression management in adolescents' communication on Facebook and its ethical implications. *Public Health Ethics*, *7*(1), 1–12.

LS. (2015). *The economy of seafood in Louisiana*. Retrieved May 21, 2018, from https://www.louisianaseafood.com/industry

LSU. (2014). *Gov. Jindal announces funding hike for higher education & new workforce incentive fund*. Retrieved May 21, 2018, from https://sites01.lsu.edu/wp/lovepurple/2014/01/21/gov-jindal-announces-funding-hike-for-higher-education-new-workforce-incentive-fund

LSUOREC. (2012). The constant pursuit of discovery. *LSU office of research & economic development*. Retrieved May 21, 2018, from https://sites01.lsu.edu/wp/lsu2015/files/2013/04/20130409ResearhMtgOREDStrategicPlan.pdf

Lu, F., Wang, Z., Liao, X., & Jin, H. (2013). Concurrent Transmission Method for Mobile Applications using Multihoming Devices. *2013 IEEE 9th International Conference on Mobile Ad-hoc and Sensor Networks*, 240–247 doi:10.1109/MSN.2013.114

Maciejewski, P. K., Zhang, B., Block, S. D., & Prigerson, H. G. (2007). An empirical examination of the stage theory of grief. *Journal of the American Medical Association*, *297*(7), 716–723. doi:10.1001/jama.297.7.716 PMID:17312291

Maillé, P., & Tuffin, B. (2019). Analysis of Sponsored Data Practices in the Case of Competing Wireless Service Providers. In H. Peng, C. Deng, Z. Wu, & Y. Liu (Eds.), *Computational Intelligence and Intelligent Systems* (pp. 55–70). Springer.

Manogaran, G., & Lopez, D. (2018). Spatial cumulative sum algorithm with big data analytics for climate change detection. *Computers & Electrical Engineering*, *65*, 207–221. doi:10.1016/j.compeleceng.2017.04.006

Marian, H., Markus, H., Susanne, W., & Matthew, S. (2014). Using personal examples to improve risk communication for security & privacy decisions. *32nd Annual ACM Conference on Human Factors in Computing Systems,* 2647–2656. doi: 10.1145/2556288.2556978

Markham, A. N. (2004). Representation in online ethnographies: A matter of context sensitivity. In Online Social Research: Methods, Issues, and Ethics (pp. 141–155). New York: Peter Lang.

Mars, P., Chen, J. R., & Nambiar, R. (2018). Learning Algorithms: Theory and Applications in Signal Processing, Control and Communications. In Stochastic Learning Automata (pp.6-9). CRC Press Taylor & Francis.

Marschark, M., Pelz, J. B., Convertino, C., Sapere, P., Arndt, M. E., & Seewagen, R. (2005). Classroom interpreting and visual information processing in mainstream education for deaf students: Live or Memorex®? *American Educational Research Journal*, *42*(4), 727–761. doi:10.3102/00028312042004727 PMID:16628250

Martins, P., Rodrigues, H., Rocha, T., Francisco, M., & Morgado, L. (2015). Accessible options for deaf people in e-learning platforms: Technology solutions for sign language translation. *Procedia Computer Science*, *67*, 263–272. doi:10.1016/j.procs.2015.09.270

Masse, M. (2011). *REST API Design Rulebook: Designing Consistent RESTful Web Service Interfaces*. Sebastopol: O'Reilly Media, Inc.

Mastromartino, B., Chou, W. W., & Zhang, J. J. (2017). The passion that unites us all - The culture and consumption of sport fans. In C. L. Wang (Ed.), *Exploring the rise of fandom in contemporary consumer culture* (pp. 52–70). Hershey, PA: IGI Global.

MC. (2015). *Mobile classroom*. Retrieved May 21, 2018, from https://sites01.lsu.edu/wp/innovationpark/about-lbtc/mobile-classroom

McEvily, B., Perrone, V., & Zaheer, A. (2003). Introduction to the special issue on trust in an organizational context. *Organization Science*, *14*(1), 1–4. doi:10.1287/orsc.14.1.1.12812

McFadden, D. T., & Marshall, M. I. (2014). Local food systems and interactions with entrepreneurship. *Journal of Food Distribution Research*, *45*(3).

McGloin & Hunt. (2011). *OAuth 2.0 Threat Model and Security Considerations*. Retrieved from http://tools.ietf.org/id/draft-ietf-oauth-v2-threatmodel-00. txt/

McKay, J. (2005, Feb.). Employers complain about communication skills. *Pittsburgh Post-Gazette*.

MD. (2015). *Marsh dog*. Retrieved May 21, 2018, from www.marshdog.com

ME. (2015). *Martin ecosystems*. Retrieved May 21, 2018, from www.martinecosystems.com

Mearns, J. (1991). Coping with a breakup: Negative mood regulation expectancies and depression following the end of a romantic relationship. *Journal of Personality and Social Psychology*, *60*(2), 327–334. doi:10.1037/0022-3514.60.2.327 PMID:2016673

Mehdizadeh, S. (2010). Self-presentation 2.0: Narcissism and self-esteem on Facebook. *Cyberpsychology, Behavior, and Social Networking*, *13*(4), 357–364. doi:10.1089/cyber.2009.0257 PMID:20712493

Mellinger, S. (2013, May 3). Roy Williams may be here, but this is Bill Self's time at KU. *The Kansas City Star*. Retrieved from https://www.kansascity.com/2013/03/21/4135991/carolina-may-be-here-but-this.html

Mencin, O., & Erikson, C. (2009). Silicon valley's US market access center: The incubator as a soft landing zone. *International Journal of Entrepreneurship and Innovation*, *10*(3), 233–241. doi:10.5367/000000009789067842

Metzger, A. H., Finley, K. N., Ulbrich, T. R., & McAuley, J. W. (2010). Pharmacy faculty members' perspectives on the student/faculty relationship in online social networks. *American Journal of Pharmaceutical Education*, *74*(10), 188. doi:10.5688/aj7410188 PMID:21436929

Metzler, A., & Scheithauer, H. (2017). The Long-Term Benefits of Positive Self-Presentation via Profile Pictures, Number of Friends and the Initiation of Relationships on Facebook for Adolescents' Self-Esteem and the Initiation of Offline Relationships. *Frontiers in Psychology*, *8*, 1981. doi:10.3389/fpsyg.2017.01981 PMID:29187827

Michalewicz, Z. (1996). *Genetic Algorithms + Data Structures = Evolution Programs* (3e éd.). Consulté à l'adresse https://www.springer.com/gp/book/9783540606765

Microsoft Corporation. (2013). *Administrator's Guide for Microsoft Application Virtualization (App-V) 5.0 MDOP Information Experience Team*. TechNet Library.

Mike, T. L., & Venkat, V. (2009). Blueprint: Robust Prevention of Cross-site Scripting Attacks for Existing Browsers. *IEEE Symposium on Security and Privacy, 30*(1), 331-346. doi: 10.1109/SP.2009.33

Miles, M. B., & Huberman, A. M. (1984). *Qualitative data analysis* (Vol. 16). Newbury Park, CA: Sage.

Millar, D. (2012). Louisiana alligator industry stages a comeback, thanks to growing demand for skins. *The Times-Picayune*. Retrieved May 21, 2018, from https://www.nola.com/business/index.ssf/2012/06/louisianas_alligator_industry.html

Miller, H. (1995). *The presentation of self in electronic life: Goffman on the Internet. In Embodied knowledge and virtual space conference* (Vol. 9). Retrieved from: http://www.dourish.com/classes/ics234cw04/ miller2.pdf

Miller, A., Roberts, S., & LaPoe, V. (2014). *Oil and water: Media lessons from hurricane Katrina and the deepwater hrizon disaster*. Univ. Press of Mississippi.

Mitchell, B. (2015). Making cities resilient: New Orleans/Gothenburg city-to-city exchange. *Stephenson Disaster Management Institute*. Retrieved May 21, 2018, from http://www.sdmi-resilient-cities.com

Mitrano, J. R. (1999). 'Sudden death' of hockey in Hartford: Sports fans and franchise relocation. *Sociology of Sport Journal*, *16*(2), 134–154. doi:10.1123sj.16.2.134

Modi, K., & More, A. (2013). Translation of Sign Language Finger-Spelling to Text using Image Processing. *International Journal of Computers and Applications*, *77*(11).

Mohamed, S., & Fadi, M. (2014). Towards Enhancing the Security of OAuth Implementations in Smart Phones. *IEEE Third International Conference on Mobile Services*, 9–46. doi: 10.1109/MobServ.2014.15

Moody, E. J. (2001). Internet use and its relationship to loneliness. *Cyberpsychology & Behavior*, *4*(3), 393–401. doi:10.1089/109493101300210303 PMID:11710265

Morales-Urrutia, E. K., Ocaña, J. M., Pérez-Marín, D. P., & Tamayo-Moreno, S. (to appear). Interfaz de usuario enfocado en el co-diseño con niños. *Revista Ibérica de Sistemas y Tecnologías de la Información.*

Morales-Urrutia, E. K., Ocaña, J. M., Pérez-Marín, D., & Tamayo, S. (2017). *A first proposal of Pedagogic Conversational Agents to develop Computational Thinking in children.* TEEM. doi:10.1145/3144826.3145350

Moran, M., Seaman, J., & Tinti-Kane, H. (2011). *Teaching, learning, and sharing: How Today's Higher Education Faculty Use Social Media.* Pearson Learning Solutions and Babson Survey Research Group.

Moran, M., Seaman, J., & Tinti-Kane, H. (2012). *Social Media in Higher Education.* Pearson Learning Solutions and Babson Survey Research Group.

Morgan, B. J. T. (1974). On Luce's choice axiom. *Journal of Mathematical Psychology, 11*(2), 107–123. doi:10.1016/0022-2496(74)90002-9

Muchinsky, P. M. (1977). A Comparison of within- and Across-Subjects Analyses of the Expectancy-Valence Model for Predicting Effort. *Academy of Management Journal, 20*(1), 154–158.

Mueller, J. P. (2006). *Mining Google? Web Services: Building Applications with the Google? API.* New York: John Wiley & Sons.

Mujal, I., & Colilles, E. (2010). *Estudi d'un amplificadorclasse D.* Retrieved May 10, 2010 fromhttps://upcommons.upc.edu/bitstream/handle/2099.1/9546/memoria.pdf

Myatt, B. M., & Kennette, L. N. (2017, Jan.). *Towards a positive u.* Faculty Focus. Retrieved from http://www.facultyfocus.com/articles/effective-classroom-management/towards-positive-u/

Nadkarni, S., & Narayanan, V. (2004). Strategy frames, strategic flexibility and firm performance: the moderating role of industry clockspeed. *Academy of Management Proceedings.* 10.5465/ambpp.2004.13863821

Nardi, B. A., & Whittaker, S. (2002). The place of face-to-face communication in distributed work. *Distributed Work*, 83-110.

Nayef, N. (2017). *7 thousands deaf in Kuwait and requests for generalization of sign language.* Localities.

Nazmul, H., Md, A. H., Md, Z. H., Md, H. I. S., & Shawon, R. (2018). OAuth-SSO: A Framework to Secure the OAuth-Based SSO Service for Packaged Web Applications. *IEEE International Conference on Big Data Science and Engineering (TrustCom/BigDataSE), 12*(1), 1575-1578. doi:10.1109/TrustCom/BigDataSE.2018.00227

Nguyen, S. C., Zhang, X., Nguyen, T. M. T., & Pujolle, G. (2011). Evaluation of throughput optimization and load sharing of multipath TCP in heterogeneous networks. In *Wireless and optical communications networks (WOCN), 2011 eighth international conference on.* IEEE. doi:10.1109/WOCN.2011.5872966

Nielson. (2016). *2016 Nielson social media report. Social studies: A look at the social landscape.* Retrieved from https://www.nielsen.com/content/dam/corporate/us/en/reports-downloads/2017-reports/2016-nielsen-social-media-report.pdf

Niimi, M., Wakes, S. J., & McGuire, M. H. (2014). Design for sustainability: Addressing food waste behaviour through social practice approaches. *Food Design on the Edge*, 29.

NLAFI. (2014). To market, to market…. *News - LSU AgCenter Food Incubator.* Retrieved May 21, 2018, from https://www.lsu.edu/departments/nfs/outreach/incubator.htm

NOGE. (2015). *Making cities resilient: New Orleans / Gothenburg Exchange.* Retrieved May 21, 2018, from http://sdmi-resilient-cities.com

O'Donnell, S., & Epstein, L. H. (2019). Smartphones are more reinforcing than food for students. *Addictive Behaviors, 90*, 124–133. doi:10.1016/j.addbeh.2018.10.018 PMID:30390436

O'Neil, D. (2008, March 16). Kansas-Carolina, Ivy revenge and mid-major matchups are 'serendipity.' *ESPN.com*. Retrieved from: http://sports.espn.go.com/ncb/ncaatourney08/columns/story?columnist=oneil_dana&id=3297190

OB. (2015). *Oyster Bed*. Retrieved September 29, 2019, from https://www.theoysterbed.com

Ocak, M. A. (2011). Why are faculty members not teaching blended courses? Insights from faculty members. *Computers & Education*, *56*(3), 689–699. doi:10.1016/j.compedu.2010.10.011

Ocaya, R. O. (2006). An experiment to profile the voltage, current and temperature behaviour of a P–N diode. In *European Journal of Physics* (Vol. 27). IOP Publishing Ltd. doi:10.1088/0143-0807/27/3/015

Omar, D. A., Baslam, M., Nachaoui, M., & Fakir, M. (2018). Towards a New Approach to Rationalize the Operators' Choice in Telecommunication Networks. *International Journal of Control and Automation*, *11*(8), 39–50. doi:10.14257/ijca.2018.11.8.04

Omar, D. A., Garmani, H., El Amrani, M., Baslam, M., & Fakir, M. (2019a). Analysis of Bargaining Game Policy in the Internet Content Distribution Chain. *International Journal of Mobile Computing and Multimedia Communications*, *10*(3), 47–73. doi:10.4018/IJMCMC.2019070103

Omar, D. A., Garmani, H., El Amrani, M., Baslam, M., & Fakir, M. (2019b). Analyzing the Customers' Dynamic Confusion in Telecommunication Networks Share Game. *International Journal of Business Data Communications and Networking*, *15*(2), 15–34. doi:10.4018/IJBDCN.2019070102

OpenID Authentication 2.0 - Final. (2007). Retrieved from https://openid.net/specs/openid-authentication-2_0.html

OREA. (2015). *ORA Estuaries*. Retrieved September 29, 2019, from https://www.oraestuaries.com

Osterman, L. L., & Hecmanczuk, T. A. (2019). Parasocial forgiveness: The roles of parasocial closeness and offense perceptions. *Journal of Social and Personal Relationships*, *§§§*, 0265407519879511.

Outanoute, M., Garmani, H., Baslam, M., Ayachi, R. E., & Bouikhalene, B. (2019). A Non-Cooperative Game Analysis of Competition between Content Providers in the Internet Market. *International Journal of Business Data Communications and Networking*, *15*(1), 88–104. doi:10.4018/IJBDCN.2019010106

Oxford University Press. (2015). *Word of the year 2015*. Retrieved from https://languages.oup.com/word-of-the-year/word-of-the-year-2015

Özpolat, E., & Akar, G. B. (2009). Automatic Detection of Learning Styles for an e-Learning System. *Computers & Education*, *53*(2), 355–367. doi:10.1016/j.compedu.2009.02.018

Paloviita, A., & Järvelä, M. (2015). *Climate change adaptation and food supply chain management*. Routledge Advances in Climate Change Research. doi:10.4324/9781315757728

Papacharissi, Z. (2002). The presentation of self in virtual life: Characteristics of personal home pages. *Journalism & Mass Communication Quarterly*, *79*(3), 643–660. doi:10.1177/107769900207900307

Papert, S. (1980). *Mindstorms: Children, computers, and powerful ideas*. New York, NY: Basic Books.

Pappano, L. (2018). The iGen shift: Colleges must change to reach the next generation. *New York Times*. Retrieved from https://www.nytimes.com/2018/08/02/education/learning/generationz-igen-students-colleges.html

Parker, G. G., Van Alstyne, M. W., & Choudary, S. P. (2016). *Platform Revolution: How Networked Markets Are Transforming the Economy and How to Make Them Work for You*. WW Norton & Company.

Park, S. Y. (2009). An Analysis of the Technology Acceptance Model in Understanding University Students' Behavioral Intention to Use e-Learning. *Journal of Educational Technology & Society*, *12*(3), 150–162. www.jstor.org/stable/jeductechsoci.12.3.150

Park, Y. (2011). A pedagogical framework for mobile learning: Categorizing educational applications of mobile technologies into four types. *International Review of Research in Open and Distance Learning*, *12*(2), 78–102. doi:10.19173/irrodl.v12i2.791

Parton, B. S. (2005). Sign language recognition and translation: A multidisciplined approach from the field of artificial intelligence. *Journal of Deaf Studies and Deaf Education*, *11*(1), 94–101. doi:10.1093/deafed/enj003 PMID:16192405

Pascarella, E. (1985). College environmental influences on learning and cognitive development: A 32 2g critical review and synthesis. In J. Smart (Ed.), *Higher education: Handbook of theory and research* (Vol. I). New York: Agathon.

PBTI. (2013). *Pennington biotech initiative*. Retrieved May 21, 2018, from https://www.businessreport.com/business/budding-ecosystem

Pearson. (2018). *Beyond millennials: The next generation of learners*. Retrieved from https://www.pearson.com/content/dam/one-dot-com/one-dot-com/global/Files/news/news-annoucements/2018/The-Next-Generation-of-Learners_final.pdf

Pempek, T. A., Yermolayeva, Y. A., & Calvert, S. L. (2009). College students' social networking experiences on Facebook. *Journal of Applied Developmental Psychology*, *30*(3), 227–238. doi:10.1016/j.appdev.2008.12.010

Pérez, J., Arenas, M., & Gutierrez, C. (2006). Semantics and Complexity of SPARQL. *Lecture Notes in Computer Science, 4273*, 30-43. 10.1007/11926078_3

Pérez-Marín, D., & Caballero, C. (2013). Using Conceptual Models to Implement Natural Language Pedagogic Agent Student Conversations. *International Journal of Interactive Communication Systems and Technologies*, *3*(2), 29–47. doi:10.4018/ijicst.2013070103

Pérez-Marín, D., Hijón-Neira, R., & Martín-Lope, M. (2017). Propuesta de Metodología basada en Metáforas para la enseñanza de la Programación a Niños. *IEEE Revista Iberoamericana de Tecnologias del Aprendizaje*, *6*(1). doi:10.1109/RITA.2018.2809944

Pérez-Marín, D., Hijón, R., Bacelo, A., & Pizarro, C. (2018). *Can computational thinking be improved by using a methodology based on metaphors and Scratch to teach computer programming to children?* Computers in Human Behaviour. doi:10.1016/j.chb.2018.12.027

Perrin, A., & Anderson, M. (2019). *Share of U.S. adults using social media, including Facebook, is mostly unchanged since 2018*. Retrieved from https://www.pewresearch.org/fact-tank/2019/04/10/share-of-u-s-adults-using-social-media-including-facebook-is-mostly-unchanged-since-2018/

Peters, N. J., Hofstetter, J. S., & Hoffmann, V. H. (2011). Institutional entrepreneurship capabilities for interorganizational sustainable supply chain strategies. *International Journal of Logistics Management*, *22*(1), 52–86. doi:10.1108/09574091111127552

Picciano, A. G., Dziuban, C. D., & Graham, C. R. (2013). Blended learning: Research perspectives, Volume 2. Routledge.

Picó Martín, I. (2011). *Sistema Microcomputador per a l'Enregistrament de Dades GPS d'una Bicicleta Elèctrica*. Projecte/Treball Final de Carrera. UniversitatPolitècnica de Catalunya.

Pilkey, O. H., & Young, R. (2011). *The rising seas*. Island Press.

Pivato, S., Misani, N., & Tencati, A. (2008). The impact of corporate social responsibility on consumer trust: The case of organic food. *Business Ethics (Oxford, England), 17*(1), 3–12. doi:10.1111/j.1467-8608.2008.00515.x

Planalp, S., & Rivers, M. (1996). Changes in knowledge of personal relationships. In Knowledge structures in close relationships: A social psychological approach (pp. 299–324). Mahwah, NJ: Lawrence Erlbaum Associates.

PN. (2016). *Water Challenge Winner Recycles Water Bottles to Protect Coasts & Reduce Pollutants: Martin Ecosystems wins $15,000 at New Orleans Entrepreneur Week.* Retrieved September 29, 2019, from http://gopropeller.org/news/1583

Pomerantz, J., Hank, C., & Sugimoto, C. R. (2015). The state of social media policies in higher education. *Plos.* Retrieved from http://journals.plos.org/plosone/article?id=10.1371/journal.pone.0127485

Prateek, S., David, M., & Livshits, B. (2011). ScriptGard: Automatic Context-Sensitive Sanitization for Large-Scale Legacy Web Applications. *ACM Conference on Computer and Communications Security, 18*(2), 601-614. doi: 10.1145/2046707.2046776

Prigerson, H. G., & Maciejewski, P. K. (2008). Grief and acceptance as opposite sides of the same coin: Setting a research agenda to study peaceful acceptance of loss. *The British Journal of Psychiatry, 193*(6), 435–437. doi:10.1192/bjp.bp.108.053157 PMID:19043142

Przybylski, A. K., & Weinstein, N. (2013). Can you connect with me now? How the presence of mobile communication technology influences face-to-face conversation quality. *Journal of Social and Personal Relationships, 30*(3), 237–246. doi:10.1177/0265407512453827

PwC. (2014). Sustainable supply chains: Making value the priority. *PwC.* Retrieved May 21, 2018, from https://www.pwc.com/us/en/operations-management/publications/sustainable-supply-chain.jhtml

Qi, J., Zhang, Y., Zhang, Y., & Shi, S. (2006). TreeLogit Model for Customer Churn Prediction. *2006 IEEE Asia-Pacific Conference on Services Computing (APSCC'06)*, 70-75. doi:10.1109/APSCC.2006.111

Quan-Haase, A., & Young, A. L. (2010). Uses and gratifications of social media: A comparison of facebook and instant messaging. *Bulletin of Science, Technology & Society, 30*(5), 350–361. doi:10.1177/0270467610380009

Quilitz, B., & Leser, U. (2008). Querying distributed RDF data sources with SPARQL. In *Lecture Notes in Computer Science: vol 5021. Proceedings of European Semantic Web Conference.* Springer. 10.1007/978-3-540-68234-9_39

Rakhmawati, N. A., Umbrich, J., Karnstedt, M., Hasnain, A., & Hausenblas, M. (2013). *Querying over Federated SPARQL Endpoints. A State of the Art Survey.* Technical Publication. Accessed 20 November 2018: https://arxiv.org/pdf/1306.1723

Ramaboli, A. L., Falowo, O. E., & Chan, A. H. (2012). Bandwidth aggregation in heterogeneous wireless networks: A survey of current approaches and issues. *Journal of Network and Computer Applications, 35*(6), 1674–1690. doi:10.1016/j.jnca.2012.05.015

Reagan, T. (2019). Sign Language and the Deaf-World: 'Listening without hearing'. In Linguistic Legitimacy and Social Justice, (pp. 135-174). Palgrave Macmillan.

Real, M. (2011). Theorizing the sports-television dream marriage: Why sports fit television so well. In Sports media: Transformation, integration, consumption (pp. 19-39). Academic Press.

Refuel Agency. (2015). *College explorer '15.* Retrieved from http://research.refuelagency.com/wp-content/uploads/2016/01/College%20Explorer%202015%20Final.pdf

Reiffers-Masson, A., Hayel, Y., & Altman, E. (2015). *Pricing Agreement between Service and Content Providers: A Net Neutrality Issue.* arXiv preprint arXiv:150503555

Resnick, M. (1996). New paradigms for computing, new paradigms for thinking. In Constructionism in practice: Designing, thinking, and learning in a digital world. Mahwah, NJ: Erlbaum.

Resnick, M., Maloney, J., Monroy-Hernandez, A., Rusk, N., Eastmond, E., Brennan, K., ... Kafai, Y. (2009). Scratch: Programming for all. *Communications of the ACM*, *52*(11), 60–67. doi:10.1145/1592761.1592779

RF. (2015). *Righteous fur*. Retrieved May 21, 2018, from http://www.righteousfur.com

RFIC. (2015). *Rutgers food innovation center*. Retrieved May 21, 2018, from http://foodinnovation.rutgers.edu

Richardson, J. C., Maeda, Y., Lv, J., & Caskurlu, S. (2017). Social presence in relation to students' satisfaction and learning in the online environment. *Computers in Human Behavior*, *71*(C), 402–417. doi:10.1016/j.chb.2017.02.001

Richer, Mills, & Tschofenig. (2012). *OAuth 2.0 message authentication code (MAC) tokens*. Retrieved from https://tools.ietf.org/html/draft-ietf-oauth-v2-http-mac-00

Rivlin, G. (2015). *Katrina: After the flood*. Simon & Schuster.

RobbK. (2015). Available at http://friedtechnology.blogspot.com/2015/11/use-this-feelings-wheel-by-kaitlin-robb.html

Roberts, J. A., Yaya, L. H., & Manolis, C. (2014). The invisible addiction: Cell-phone activities and addiction among male and female college students. *Journal of Behavioral Addictions*, *3*(4), 254–265. doi:10.1556/JBA.3.2014.015 PMID:25595966

Roberts, N., & Stockport, G. J. (2009). Defining strategic flexibility. *Global Journal of Flexible Systems Managment*, *10*(1), 27–32. doi:10.1007/BF03396553

Rodriguez, G. R., & Brebbia, C. A. (2015). *Coastal cities and their sustainable future*. Wit Pr/Computational Mechanics.

Rodriguez, N. J. (2016). Vaccine-Hesitant Justifications: 'Too Many, Too Soon,' Narrative Persuasion and the Conflation of Expertise. *Global Qualitative Nursing Research*, *3*, 3. doi:10.1177/2333393616663304 PMID:28508015

Roorda, D. L., Koomen, H. M. Y., Spilt, J. L., & Oort, F. J. (2011). The influence of affective teacher–student relationships on students' school engagement and achievement. *The Journal of Educational Research*, *81*(4), 493–529.

Rosenberg, M. J. (2001). *E-learning: Strategies for delivering knowledge in the digital age*. New York: McGraw-Hill.

Rosen, J. B. (1965). Existence and uniqueness of equilibrium points for concave n-person games. *Econometrica*, *33*(3), 520–534. doi:10.2307/1911749

Rothenberg, M. (2019). *Emoji tracker*. Retrieved from http://emojitracker.com/

Rousseau, D. M., Sitkin, S. B., Burt, R. S., & Camerer, C. (1998). Not so different after all: A cross-discipline view of trust. *Academy of Management. Academy of Management Review*, *23*(3), 393–404. doi:10.5465/amr.1998.926617

RT. (2015). *Rising Tide: Louisiana kids help build Floating Islands Restoration Project*. Retrieved September 29, 2019, from https://risingtide.joincca.org/news3

Rubin, R. B., & McHugh, M. P. (1987). Development of parasocial interaction relationships. *Journal of Broadcasting & Electronic Media*, *31*(3), 279–292. doi:10.1080/08838158709386664

Rui, J., & Stefanone, M. A. (2013). Strategic self-presentation online: A cross-cultural study. *Computers in Human Behavior*, *29*(1), 110–118. doi:10.1016/j.chb.2012.07.022

Rui, W., Shuo, C., & Xiaofeng, W. (2012). Signing Me onto Your Accounts through Facebook and Google: A Traffic-Guided Security Study of Commercially Deployed Single-Sign-On Web Services. *2012 IEEE Symposium on Security and Privacy*, 365-379. doi: 10.1109/SP.2012.30

Russell, M. A. (2013). *Mining the Social Web: Data Mining Facebook, Twitter, LinkedIn, Google+, GitHub, and More*. Sebastopol: O'Reilly Media, Inc.

Ryokai, K., Vaucelle, C., & Cassell, J. (2003). Virtual peers as partners in storytelling and literacy learning. *Journal of Computer Assisted Learning, 19*(2), 195–208. doi:10.1046/j.0266-4909.2003.00020.x

Salinger, J. H. (2013). *Economic development policies through business incubation and co-working: A study of San Francisco and New York City*. Academic Press.

Sanchez, V. M. (2015). Análisis de proyectos "Open Data" en España: propuestas de mejora. *Cuadernos de Gestión de Información, 5*(1), 53-72. Accessed 20 November 2018: https://revistas.um.es/gesinfo/article/view/232231

Sanchez, R. (1995). Strategic flexibility in product competition. *Strategic Management Journal, 16*(S1), 135–159. doi:10.1002mj.4250160921

Sanderson, J. (2013). From loving the hero to despising the villain: Sports fans, Facebook, and social identity threats. *Mass Communication & Society, 16*(4), 487–509. doi:10.1080/15205436.2012.730650

Sanderson, J., & Cheong, P. H. (2010). Tweeting prayers and communicating grief over Michael Jackson online. *Bulletin of Science, Technology & Society, 30*(5), 328–340. doi:10.1177/0270467610380010

Sanderson, J., & Emmons, B. (2014). Extending and withholding forgiveness to Josh Hamilton: Exploring forgiveness within parasocial interaction. *Communication & Sport, 2*(1), 24–47. doi:10.1177/2167479513482306

Sanger, L. (2010). Individual knowledge in the Internet age. *EDUCAUSE Review, 45*(2), 14–24.

San-Tsai, S., & Konstantin, B. (2012). The Devil is in the (Implementation) Details: An Empirical Analysis of OAuth SSO Systems. *CCS '12 Proceedings of the 2012 ACM conference on Computer and communications security*, 378-390. doi: 10.1145/2382196.2382238

San-Tsai, S., Kirstie, H., & Konstantin, B. (2012). Systematically breaking and fixing OpenID security: Formal analysis, semi-automated empirical evaluation, and practical countermeasures. *Computers & Security, 31*(4), 465–483. doi:10.1016/j.cose.2012.02.005

Sari, M. B., Sanjaya, Y., & Djamal, M. (2017). Pengembangan Spektrometer Cahaya Tampak Menggunakan LED RGB untuk Menentukan Konsentrasi Glukosa. In Risalah Fisika (vol. 1, pp.21-27). doi:10.35895/rf.v1i1.40

Schlenker, B. R., & Pontari, B. A. (2000). The strategic control of information: Impression management and self-presentation in daily life. In A. Tesser, R. B. Felson, & J. M. Suls (Eds.), *Perspectives on self and identity* (pp. 199–232). Washington, DC: American Psychological Association; doi:10.1037/10357-008

Schwartz, H. L. (2012). From the classroom to the coffee shop: Graduate students and professors effectively navigate interpersonal boundaries. *International Journal on Teaching and Learning in Higher Education, 23*(3), 363–372.

SDMI. (2015). *Stephenson disaster management institute*. Retrieved May 21, 2018, from http://sdmi.lsu.edu

Seaman, J., & Tinti-Kane, H. (2013). *Social media for teaching and learning*. Pearson Learning Solutions and Babson Survey Research Group. Retrieved from http://www.pearsonlearningsolutions.com/assets/downloads/reports/social-media-for-teaching-and-learning-2013-report.pdf#view=FitH,0

Secchi, D. (2010). *Extendable rationality: Understanding decision making in oganizations*. Springer.

SecurEnvoy. (2012). *66% of the population suffer from Nomophobia the fear of being without their phone.* Retrieved from http://www.securenvoy.com/blog/2012/02/16/66-of-the-population-suffer-fromnomophobia-the-fear-of-being-without-their-phone/

SEI. (2015). *Stephenson entrepreneurship institute.* Retrieved May 21, 2018, from https://www.business.lsu.edu/Stephenson-Entrepreneurship-Institute/Pages/SEI.aspx

senorpalmer. (2009, December 17). *Roy Williams could give a sh!t about North Carolina* [Video file]. Retrieved from https://www.youtube.com/watch?v=pymcjOpiII8

SG. (2015). *The Silverfin™ Group - Our Alternative Solution to the Nation's Asian Carp Crisis.* Retrieved May 21, 2018, from http://cantbeatemeatem.com

Shackleton, S. (2014). Impacts of climate change on food availability: Non-Timber forest products. In *Global Environmental Change* (pp. 695–700). Springer Netherlands. doi:10.1007/978-94-007-5784-4_117

Shadbolt, N., & Gibbins, N. (2010). Resource Description Framework (RDF). In *Encyclopedia of library and information sciences* (pp. 4539-4547). Accessed 20 November 2018: https://eprints.soton.ac.uk/268264/

Shah, J., Modi, B., & Singh, R. (2014, February). *Wireless home appliances controlling system.* Paper presented at theInternational Conference on Electronics and Communication System, Coimbatore, India.

Shank, M. D., & Lyberger, M. R. (2015). *Sports Marketing: A Strategic Perspective.* New York: Routledge.

Shapiro, J. (2001). Genetic Algorithms in Machine Learning. In G. Paliouras, V. Karkaletsis, & C. D. Spyropoulos (Eds.), *Machine Learning and Its Applications : Advanced Lectures* (pp. 146–168)., doi:10.1007/3-540-44673-7_7

Shawnee. (2018). *Sprint National Relay Services. TTY/ASCII/Voice/Spanish/Speech-to-Speech/IP Relay/Federal Relay.* Author.

Sheldon, P. (2015). Understanding students' reasons and gender differences in adding faculty as Facebook friends. *Computers in Human Behavior, 53*, 58–62. doi:10.1016/j.chb.2015.06.043

SI. (2015). *LSU student incubator.* Retrieved May 21, 2018, from https://sites01.lsu.edu/wp/innovationpark/about-lbtc/lsu-student-incubator

Siegel, D. J., & Bryson, T. P. (2012). *The whole-brain child: 12 revolutionary strategies to nurture your child's developing mind.* Bantam.

Simion, C., & Baciu, A. M. (2018). *Legislation related to the field of gambling.* Retrieved November 20, 2018from https://iclg.com/practice-areas/gambling-laws-and-regulations/romania

Simon, H. A. (1993). Altruism and Economics. *The American Economic Review, 83*(2), 156–161.

Simpson, J. A. (1987). The dissolution of romantic relationships: Factors involved in relationship stability and emotional distress. *Journal of Personality and Social Psychology, 53*(4), 683–692. doi:10.1037/0022-3514.53.4.683

Sirmon, D. G., Hitt, M. A., Ireland, R. D., & Gilbert, B. A. (2011). Resource orchestration to create competitive advantage breadth, depth, and life cycle effects. *Journal of Management, 37*(5), 1390–1412. doi:10.1177/0149206310385695

Sitkin, S., Sutcliffe, K., & Barrios-Choplin, J. (1992). A Dual-Capacity Model of Communication Media Choice in Organizations. *Human Communication Research, 18*(4), 563–598. doi:10.1111/j.1468-2958.1992.tb00572.x

SL. (2015). *SHORE|LINKS.* Retrieved May 21, 2018, from http://www.shore-links.com

Slack, N. (1983). Flexibility as a manufacturing objective. *International Journal of Operations & Production Management, 3*(3), 4–13. doi:10.1108/eb054696

Smith & Anderson. (2018). *Social Media Use in 2018*. Retrieved from https://www.pewinternet.org/2018/03/01/social-media-use-in-2018/

Smock, A. (2010, June). *Self-presentation on Facebook: Managing content created by the users and others*. Paper presented at the annual conference of the International Communication Association, Singapore.

Sović, A., Jagušt, T., & Seršić, D. (2014). How to teach basic university-level programming concepts to first graders? *In Integrated STEM Education Conference (ISEC),* 2014 *IEEE* (pp. 1-6). IEEE. 10.1109/ISECon.2014.6891050

Srivastava, B., & Koehler, J. (2003). Web service composition-current solutions and open problems. *Proceedings of ICAPS 2003 workshop on Planning for Web Services*, 28-35.

Stein, D., Dreuw, P., Ney, H., Morrissey, S., & Way, A. (2007). *Hand in hand: automatic sign language to English translation*. Academic Press.

Sturgeon, C. M., & Walker, C. (2009). *Faculty on Facebook: Confirm or deny*. Paper presented at the Annual Instructional Technology Conference, Murfreesboro, TN.

Sun, T., Youn, S., & Wells, W. D. (2004). Exploration of consumption and communication communities in sports marketing. In Sports Marketing and the Psychology of Marketing Communication (pp. 3–26). Mahwah, NJ: Lawrence Erlbaum Associates.

Sun, T., & Wu, G. (2012). Influence of personality traits on parasocial relationship with sports celebrities: A hierarchical approach. *Journal of Consumer Behaviour, 11*(2), 136–146. doi:10.1002/cb.1378

Suresh, C., Charanjit, J., & Arnab, R. (2011). Universally Composable Security Analysis of OAuth v2.0. *IACR Cryptology ePrint Archive*. Retrieved from https://eprint.iacr.org/2011/526

Tai, A. P., Martin, M. V., & Heald, C. L. (2014). Threat to future global food security from climate change and ozone air pollution. *Nature Climate Change, 4*(9), 817–821. doi:10.1038/nclimate2317

Tamayo-Moreno, S. (2017). *Propuesta de Metodología para el Diseño e Integración en el Aula de un Agente Conversacional Pedagógico desde Educación Secundaria hasta Educación Infantil* (Doctoral Dissertation). Universidad Rey Juan Carlos. Available on-line at http://hdl.handle.net/10115/14691

Tara, W., & Kori, I. Q. (2005). Gathering evidence: use of visual security cues in web browsers. *Proceedings of the Graphics Interface 2005 Conference,* 137–144. doi: 10.1145/1089508.1089532

Tarr, N. W. (2011). Food entrepreneurs and food safety regulation. *J. Food L. & Pol'y, 7*, 35.

Technology Usage for Deaf Community. (2017). *Communication*. Retrieved in October 25, 2017 at https://elbdie.nfshost.com/mod/page/view.php?id=145&lang=tr

Tenor. (2017). *10 Years with the iPhone: Communication is now visual*. Retrieved from https://blog.tenor.com/10-years-with-the-iphone-communication-is-now-visual-c2b7dae6cf07

Thalluri, J., & Penman, J. (2015). Social media for learning and teaching undergraduate sciences: Good practices guidelines from intervention. *Electronic Journal of E-Learning, 13*(6), 455–465.

The Institute of Politics at Harvard Kennedy School. (2017). *Use of social networking technology*. Retrieved from https://iop.harvard.edu/use-social-networking-technology

The U.S. Census Bureau. (2010). *Americans with disabilities: 2010*. Retrieved in May 1, 2010, from https://www.census.gov/prod/2012pubs/p70-131.pdf

Tice, D. M., Butler, J. L., Muraven, M. B., & Stillwell, A. M. (1995). When modesty prevails: Differential favorability of self-presentation to friends and strangers. *Journal of Personality and Social Psychology*, *69*(6), 1120–1138. doi:10.1037/0022-3514.69.6.1120

Toma, C. L., & Carlson, C. L. (2012, January). *I'm so much cooler online: An examination of self-presentation in Facebook profiles*. Paper presented at the annual conference of the International Communication Association, Phoenix, AZ.

TR. (2015). *Tierra resources*. Retrieved September 29, 2019, from https://tierraresourcesllc.com

Trammell, K. D., & Keshelashvili, A. (2005). Examining the new influencers: A self-presentation study of Alist blog. *Journalism & Mass Communication Quarterly*, *82*(4), 968–982. doi:10.1177/107769900508200413

Trine, E. (2013). ةمجرتم: *a case study of an Arabic/Jordanian sign language (liu) interpreter in Jordan*. Academic Press.

Tripathi, R., & Agarwal, S. (2014). An empirical study of marketing for guava and its sub-products by farmers in Allahabad: An approach towards agripreneurship through food processing units. *International Journal of Business and Globalisation*, *13*(1), 69–75. doi:10.1504/IJBG.2014.063395

Trivedi, A., & Awad, M. (2015). ARM Based Handy and Portable Oscilloscope Using Graphical Display. *Journal of Engineering Research and Applications*, *5*, 106–109.

Tsuladze, M. (2015). *Teaching and Learning of Deaf Students in Ordinary Vocational Education Setting-The Georgian Case* (Master's thesis).

U.S. Commercial Service. (2009). *India: Legal and regulatory aspects of doing business in the franchising industry*. Retrieved May 21, 2018 from https://www.franchise.org/uploadedFiles/Franchise_Industry/International_Development/Country_Profiles/India_Franchise%20Sector_July%202009.pdf

U.S. Commercial Service. (2010). *Doing business in South Africa - 2010 country commercial guide for U.S. companies*. Retrieved May 21, 2018 from https://www.franchise.org/uploadedFiles/Franchise_Industry/International_Development/Country_Profiles/South%20Africa_2010.pdf

U.S. Commercial Service. (2011). *Brazil: Franchising*. Retrieved May 21, 2018 from https://www.franchise.org/uploadedFiles/FranchisingReportBrazilCCGFeb2011.doc

U.S. Commercial Service. (2012). *Doing business in Russia: 2012 country commercial guide for U.S. companies*. Retrieved May 21, 2018 from https://photos.state.gov/libraries/russia/231771/PDFs/2012-ccg-russia.pdf

UNFCCC. (2015). Closing address by Ms. Christiana Figueres, *United Nations Framework Convention on Climate Change Executive Secretary*. Retrieved May 21, 2018, from https://unfccc.int/files/press/statements/application/pdf/20151212_closingplenary.pdf

UNISDR. (2018). *Making cities resilient: My city is getting ready*. United Nations International Strategy for Disaster Reduction. Retrieved May 24, 2018, from https://www.unisdr.org/campaign/resilientcities

UNSC. (2015). Security Council calls for eradicating ISIL safe havens in Syria and Iraq. *UN News Centre*. Retrieved May 21, 2018, from http://www.un.org/apps/news/story.asp?NewsID=52623#.VnYWI1Xnbm4

UNSDG. (2015). *Sustainable development goals: 17 goals to transform our world*. Retrieved May 21, 2018 from https://www.un.org/sustainabledevelopment/sustainable-development-goals

USCS. (2009). *The gold key matching service*. Retrieved May 21, 2018 from http://www.export.gov/salesandmarketing/eg_main_018195.asp

USDS. (2014). *The United States and Pacific Alliance Advance Cooperation on entrepreneurship*. Retrieved May 21, 2018 from https://www.state.gov/r/pa/prs/ps/2014/12/235517.htm

Verborgh, R., Steiner, T., Van Deursen, D., Coppens, S., Vallés, J. G., & Van de Walle, R. (2012). Functional descriptions as the bridge between hypermedia APIs and the Semantic Web. In *Proceedings of the third international workshop on restful design*. ACM. 10.1145/2307819.2307828

Vervloed, M. P., Van Dijk, R. J., Knoors, H., & Van Dijk, J. P. (2006). Interaction between the teacher and the congenitally deafblind child. *American Annals of the Deaf, 151*(3), 336–344. doi:10.1353/aad.2006.0040 PMID:17087444

Victoria, B. (2016). Characterization of web single sign-on protocols. *IEEE Communications Magazine, 54*(7). doi:10.1109/MCOM.2016.7514160

Vitak, J. (2012). The impact of context collapse and privacy on social network site disclosures. *Journal of Broadcasting & Electronic Media, 56*(4), 451–470. doi:10.1080/08838151.2012.732140

Vrandečić, D. (2012). Wikidata: A new platform for collaborative data collection. In *Proceedings of the 21st International Conference on World Wide Web*. ACM.

Wakefield, K. (2012). Should professors use facebook to communicate with students? *Faculty Focus*. Retrieved from http://www.facultyfocus.com/articles/edtech-news-and-trends/should-professors-use-facebook-to-communicate-with-students/

Walther, J. B., Slovacek, C. L., & Tidwell, L. C. (2001). Is a picture worth a thousand words? Photographic images in long-term and short-term computer-mediated communication. *Communication Research, 28*(1), 105–134. doi:10.1177/009365001028001004

Wang, Q., Chen, W., & Liang, Y. (2011). The effects of social media on college student (Paper 5). *MBA Student Scholarship*. Retrieved from https://scholarsarchive.jwu.edu/mba_student/5

Wang, Q., Quek, C. L., & Hu, X. (2017). Designing and improving a blended synchronous learning environment: An educational design research. *The International Review of Research in Open and Distributed Learning, 18*(3). doi:10.19173/irrodl.v18i3.3034

Wang, W., Xiong, Z., Niyato, D., Wang, P., & Han, Z. (2019). A Hierarchical Game With Strategy Evolution for Mobile Sponsored Content and Service Markets. *IEEE Transactions on Communications, 67*(1), 472–488. doi:10.1109/TCOMM.2018.2869820

Wann, D. L. (2006). The causes and consequences of sport team identification. In Handbook of sports and media (pp. 331–352). Mahwah, NJ: Lawrence Erlbaum Associates.

Wann, D. L., Peterson, R. R., Cothran, C., & Dykes, M. (1999). Sport fan aggression and anonymity: The importance of team identification. *Social Behavior and Personality, 27*(6), 597–602. doi:10.2224bp.1999.27.6.597

Wanpeng, L., & Chris, J. M. (2014). Security Issues in OAuth 2.0 SSO Implementations. *International Conference on Information Security, 8783*(1), 529-541. doi: 10.1007/978-3-319-13257-0_34

Ware, B. L., & Linkugel, W. A. (1973). They spoke in defense of themselves: On the generic criticism of apologia. *The Quarterly Journal of Speech, 59*(3), 273–283. doi:10.1080/00335637309383176

Waycott, J., Thompson, C., Sheard, J., & Clerehan, R. (2017). A virtual panopticon in the community of practice: Students' experiences of being visible on social media. *The Internet and Higher Education*, *35*, 12–20. doi:10.1016/j.iheduc.2017.07.001

Web Runtime – Tizen developer. (2018). Retrieved from https://developer.tizen.org/development/training/web-application/understanding-tizen- programming/web-runtime

Wei, C., Chen, N., & Kinshuk. (2012). A model for social presence in online classrooms. *Educational Technology Research and Development*, *60*(3), 529–545. doi:10.100711423-012-9234-9

Wenner, L. A. (1989). Media, sports, & society: The research agenda. In L. A. Wenner (Ed.), *Media, sports, & society* (pp. 13–48). Newbury Park, CA: Sage Publications.

What is video relay service. (2009). *Convo and Convo Relay are registered trademarks*. The Federal Communication Commission.

Wikipedia Contributors. (2018, July 12). Video relay service. In *Wikipedia, The Free Encyclopedia*. Retrieved in January 2019 from https://en.wikipedia.org/w/index.php?title=Video_relay_service&oldid=905944879

Will, M. (2017). Memes, emojis, and GIFs, oh my! Teachers tell how they use social media. *Education Week*. Retrieved from http://blogs.edweek.org/teachers/teaching_now/2017/08/most_teachers_think_social_media_are_bad_for_grammar_and_spelling_but_still_use_it_in_class.html

William, K., & Robertson, G. V. (2009). Static Enforcement of Web Application Integrity Through Strong Typing. *USENIX Security Symposium, 18*(6), 283-298.

WIotG. (2015). *The water institute of the Gulf*. Retrieved May 21, 2018 from https://thewaterinstitute.org/about/our-commitment

Wong, W. K. W. (2012). Faces on Facebook: A study of self-presentation and social support on Facebook. *Discovery-SS Student E-Journal*, *1*, 184–214.

Woods, T. A. (2014). Innovation in agri-food clusters: Theory and case studies. *American Journal of Agricultural Economics*, *96*(2), aat082. doi:10.1093/ajae/aat082

Xiong, Z., Feng, S., Niyato, D., Wang, P., & Zhang, Y. (2017). Economic analysis of network effects on sponsored content: a hierarchical game theoretic approach. In *GLOBECOM 2017-2017 IEEE Global Communications Conference*. IEEE. 10.1109/GLOCOM.2017.8254738

Xiong, Z., Feng, S., Niyato, D., Wang, P., Leshem, A., & Han, Z. (2019). Joint Sponsored and Edge Caching Content Service Market: A Game-Theoretic Approach. *IEEE Transactions on Wireless Communications*, *18*(2), 1166–1181. doi:10.1109/TWC.2018.2890469

Xiong, Z., Feng, S., Niyato, D., Wang, P., Leshem, A., & Zhang, Y. (2018a). Game Theoretic Analysis for Joint Sponsored and Edge Caching Content Service Market. In *2018 IEEE Global Communications Conference (GLOBECOM)*. IEEE. 10.1109/GLOCOM.2018.8647436

Xiong, Z., Feng, S., Niyato, D., Wang, P., & Zhang, Y. (2018b). *Competition and cooperation analysis for data sponsored market: A network effects model. In 2018 IEEE Wireless Communications and Networking Conference (WCNC)* (pp. 1–6). Barcelona: IEEE.

Xu, J. (2017). *Sign Language Translation Using Kinect And Dynamic Time Warping*. Available online at https://web.stanford.edu/class/cs231m/projects/final-report-xu.pdf

Yacin, N., Prateek, S., & Dawn, S. (2009). Document structure integrity: A robust basis for cross-site scripting defense. *Proceedings of the Network and Distributed System Security Symposium (NDSS), 16*(1), 1-20.

Yee, N., & Bailenson, J. (2007). The Proteus effect: The effect of transformed self-representation on behaviour. *Human Communication Research, 33*(3), 271–290. doi:10.1111/j.1468-2958.2007.00299.x

YiJie, Z., Xiangjun, Z., & Weibo, L. (2010, December). *Power supply's design and research for high voltage circuit of active ECT*. Paper presented at the Conference on Power System Technology, Hangzhou, China.

Yoshida, M. (2008). *Barriers for telecommunication accessibility and needs assessment of video relay services (VRS): Utilization of VRS for the deaf community* (Thesis). Rochester Institute of Technology.

Zhang, H., Towsley, D., & Gong, W. (2005). Tcp connection game: A study on the selfish behavior of tcp users. In *Network Protocols, 2005. ICNP 2005. 13th IEEE International Conference on*. IEEE.

Zhang, L., Wu, W., & Wang, D. (2016). TDS: Time-dependent sponsored data plan for wireless data traffic market. In *IEEE INFOCOM 2016-The 35th Annual IEEE International Conference on Computer Communications*. IEEE.

Zhao, Y., & Breslow, L. (2013). Literature review on hybrid/blended learning. *Teaching and Learning Laboratory*. Available at: https://tll. mit. edu/sites/default/files/library/Blended_Learning_Lit_Reveiw. pdf

Zmikly, J. (2016). Remix: How to use Snapchat in the classroom. *Mediashift*. Retrieved from http://mediashift.org/2016/09/remix-snapchat-classroom/

About the Contributors

Muhammad Sarfraz is a Professor at Kuwait University, Kuwait. His research interests include Computer Graphics, Pattern Recognition, Computer Vision, Image Processing, Soft Computing, Machine Learning, Data Science, Information Technology, Intelligent Systems and Information Systems. He is currently working on various projects related to academia and industry. Prof. Sarfraz has been keynote/invited speaker at various platforms around the globe. He has advised/supervised more than 85 students for their MSc and PhD theses. He has published more than 370 publications as journal and conference papers. His publications include around 55 Books as Author and Editor. Prof. Sarfraz is member of various professional societies including IEEE, ACM. He is a Chair, member of the International Advisory Committees and Organizing Committees of various international conferences, Symposiums and Workshops. He is also Editor-in-Chief, Editor and Guest Editor of various International Journals. He is the reviewer, for many international Journals, Conferences, meetings, and workshops around the world. He has achieved various awards in education, research, and administrative services.

* * *

Azza A. Ahmed, Ph.D. is a full professor of Mass Communication and the coordinator of the Arabic Media Program at Zayed University in the United Arab Emirates. Before that, she worked as a full professor and assistant dean of the College of Arts and Sciences at Abu Dhabi University. She worked as an associate professor at the United Arab Emirates University. Azza also keeps her position as a full professor in the Faculty of Mass Communication, Cairo University in Egypt. Azza has many academic and administrative accomplishments. She established two Mass Communication Programs, in English and Arabic, for ADU. She received Khalifa Award for the Distinguished University Faculty in Teaching on Emirates and Arab World Levels for the academic year 2015-2016. She also got the distinguished Professor Award from Abu Dhabi University and the distinguished service award from UAEU. Moreover, Azza participated in many international refereed conferences in UK, USA, Bahrain, Singapore, Canada, Japan, Kuwait, Malaysia, China, and Spain. She got four international research awards from Virginia, Georgia State Universities in the USA, Morocco, and Paris. She is an active member of many International Media associations such as the Arab-US Association for Communication Educators AUSACE in which she served as a vice president of the association until 2017. Azza was the Arabic editor of the Journal of Middle East Media 2005-2016. She currently serves as a reviewer for journals such as Journalism & Mass Communication, Journal of New Media & Society; and conferences, such as AEJMC and ICA. Her research interest includes media effects, television studies, and new media usages and impacts. Her researches appeared in Cross-Cultural Communication Journal, the International E-Politics Journal

(IGI), and the International Journal of Interactive Communication System and Technologies (IJICT), International Journal of Media Watch (Scopus), Arab Media and Society and Contemporary Review of Middle East (Sage Pub.) among others.

Driss Ait Omar received the B.Sc. degree in computer science from the University of Moulay Ismail, Errachidia, Morocco, in 2006 and the M.Sc. degree in business intelligence from the University of Sultan Moulay Slimane in 2016. Currently he preparing his Ph.D in computer science from the Université of Sultan Moulay Slimane, Beni Mellal, Morocco. Her research interests include Network Economics, Network Performance Evaluation and Interference Management in Cellular Networks using game theory framework.

Aisha Al-Ajeel is a graduate student pursuing her Master program in Information Technology, Kuwait University, Kuwait. She graduated from Kuwait University with a bachelor's degree. Her research focuses on Information Technology. Al-Ajeel likes to participate in conferences and meetings related to her area of research.

Lujain Al-Anjari is a graduate student pursuing her Master program in Information Technology, Kuwait University, Kuwait. She graduated from Kuwait University with a bachelor's degree. Her research focuses on Information Technology. Al- Anjari likes to participate in conferences and meetings related to her area of research.

Wadhah AlKhaldi is Master of Science in Information Technology from Information Science Department, Kuwait University, Kuwait. Her research focuses on Information Technology and Language Translation Systems. AlKhaldi likes to participate in conferences and meetings related to her area of research.

Michael Anthony Brown Sr. earned his PhD in Public Administration and Urban Policy, International Business, from Old Dominion University (ODU) in May 2011. He is teaching online social media, public relations and communication courses for Florida Institute of Technology. His most recent publications are Solutions for High-Touch Communications in a High-Tech World and Social Media Performance Evaluation and Success Measurements. He is a Navy civil servant working as the Public Affairs Officer for the Navy Warfare Development Command (NWDC) in Norfolk, Va. This PR professional has 40-plus years of military and civilian experience combined, and is an Air Force retiree who served 24 years in uniform.

Ye-Sho Chen received his Ph.D. degree in Operations Research from Purdue University in 1985. He is a Professor of Management Information Systems in the Department of Information Systems and Decision Sciences, E. J. Ourso College of Business, Louisiana State University. He is the Director of Globlization in the College. He also holds the James C. & Cherie H. Flores Professor of MBA Studies #2. He received The Tiger Athletic Foundation Tenured Faculty Teaching Award in 2011 & 2016, Tenured Faculty Teaching Award in 2007, and The Erich Sternberg Foundation Excellence in Teaching Award in 1992. He has published more than 120 papers in journals and conference proceedings. Dr. Chen's major interest of teaching, research, and consulting is to help companies grow globally through the "Flying High, Landing Soft" platform of global entrepreneurship.

Julie Delello is an Associate Professor in the College of Education and Psychology at The University of Texas at Tyler. She also serves as the Director of the Center for Excellence in Teaching and Learning at the university. She received her Ph.D. in Curriculum and Instruction with a specialization in science and technology from Texas A&M University. She has extensive experience in curriculum and technology implementation and faculty training and development. She currently serves in a leadership role for the International Society for Technology in Education (ISTE) Teacher Education Network (TEN) professional learning network (PLN). She recently served as an expert panelist for the EDUCAUSE 2019 and 2020 Higher Education Horizon Report and she is a Microsoft Innovative Educator Expert. She has authored numerous publications and her professional interests focus on teaching pedagogies, academic innovations, visual media technologies, artificial intelligence and STEM explorations, gerontechnology, and social media platforms for authentic learning.

Mohamed Fakir obtained a master degree in electronic and electric system engineering from Nagaoka university of technology in 1991 and a PHD degree from University Cadi Ayyad in 2001 in computer sciences. He is currently with the department of computer sciences, University Sultan Moulay Slimane. His interest includes image processing, pattern recognition and intelligence artificielle.

José Luis Fernández-Vindel graduated in Physical C. (first cycle, 3 years) by the UNED. Licensed by the UAM, specializing in Theoretical Physics. Doctor of Science from the UNED (Artificial Intelligence: Artificial Vision), in the Department of Computer Science and Automation. Professor of the Department of Artificial Intelligence of the UNED since its foundation. The years of the period [2006-2010] were marked by management activities, already completed: first as deputy director of the Higher Technical School of Computer Engineering and later as Vice Chancellor of Technology of the UNED.

Hamid Garmani is a PhD student on Computer Science in Faculty of Sciences and Techniques, Sultan Moulay Slimane University, Beni Mellal, Morocco.

Md. Alam Hossain (Ph.D, 2018, Jashore University of Science and Technology) is a dedicated research supervisor with experience in supporting a large BSc and MSc student team's thesis. Have excellent communication skills and the ability to work in collaboration with a number of organisations and show initiative, versatility and attention to detail. Currently he works at the Department of Computer Science and Engineering, Jashore University of Science and Technology. Dr. Md. Alam does research in IoT, Distributed Computing, Computer Communications (Networks) and Computer Security and Reliability.

Md. Zobayer Hossain was a B.Sc. student of the Department of Computer Science and Engineering at Jessore University of Science and Technology. Getting excellent communication skills and the ability to work in collaboration with a number of organizations and being able to demonstrate effort, versatility and information. Currently he works at the Jamuna Group as a Network Engineer, IT department. He wants to work in such an atmosphere that is appropriate for self-assessment and improvement in both individual and group-based research that often faces multiple critical challenges and offers individual skills to the team.

Nazmul Hossain (M.Sc, 2014, Jashore University of Science and Technology) is a committed research coordinator with experience of supporting the research of a large team of BSc and MSc students. Having excellent communication skills and the ability to work in partnership with a range of organizations and can demonstrate initiative, flexibility and have high attention to detail. Wants to work in such an environment what is suitable to self-assessment and development in both individual and group-based research that frequently faces various critical challenges and serves the community with individual skills.

Kouider Mokhtari is the Anderson-Vukelja-Wright Endowed Professor of Literacy Education in the School of Education at the University of Texas at Tyler.

Antoanela Naaji is an associate professor of Computer Science at the Faculty of Economics, Engineering and Computer Science within "Vasile Goldis" Western University of Arad, Romania. She is also a director of International Relations Department at the same university. She received the M.S. degree in 1987 from the Faculty of Electrotechnics in Timisoara, Romania, with a major in Automation and Computers and Ph.D. degree from "Politehnica" University of Timisoara, Romania, in 2005 with a doctoral thesis, "Contributions to the Modelling Techniques of the Human Upper Limb", tackling an interdisciplinary field. Her research interest includes web technologies (e-learning), biomechanics (modeling of biological systems) and medical informatics. In 2009 she received a LL.M degree in Administrative Law and the Statute of Public Officers from "Vasile Goldis" Western University of Arad, Romania. She published 98 articles in national and international peer reviewed journals or conference volumes, 6 books as author or co-author and 7 books as editor or co-editor. Her preoccupations in the above-mentioned fields are reflected in a number of 30 projects, as manager or member, funded by EU, national or international research authorities or county council. She is a senior member of IEEE, member of EAEEIE (European Association for Electrical Engineer and Information Engineering), AGIR (General Association of Romanian Engineers) and SRIM (Romanian Society for Medical Informatics).

Diana Pérez-Marín received the Ph.D. degree in Computer Science and Engineering from the Universidad Autonoma de Madrid in 2007. She is currently a lecturer and researcher at the Universidad Rey Juan Carlos. She has published more than 50 papers in journals such as Journal of Educational Technology and Society, or Computers and Education. She has edited the book "Conversational Agents and Natural Language Interaction: Techniques and Effective Practices", and authored the book "ICT in education" (in Spanish). She is currently involved in several research projects regarding Pedagogic Conversational Agents, and the use of new technologies for education.

Marius Popescu (b. May 19, 1965) received his MSc in Automation and Computers (1990) and in Informatics (1995), PhD in Electrical Engineering (2002) from University of Craiova, Romania. Now she is professor of computer science in Department of Engineering and Computer Science, Faculty of Economics, Engineering and Computer Science, "Vasile Goldis" Western University of Arad, Romania. He has (co-) authored and (co-) edited 22 books and more than 250 papers. His research interests involve signal processing, modelling, optimization and simulation processes. His preoccupations in the above-mentioned fields are reflected in a number of over 25 projects, as manager or member, funded by national or international research authorities or county council.

Shawon S. M. Rahman (PhD, Software Engineering, North Dakota State University) is an Associate professor in the Department of Computer Science and Engineering at the University of Hawai'i Hilo (UH-Hilo), Hawaii, USA. He is serving as the Faculty advisor of Pre-engineering program and the program director of Cybersecurity and Digital Forensics certificate program at UH-Hilo. Dr. Rahman is serving as the editor-in-chief in the International Journal on Cryptography and Information Security (IJCIS) and guest editors to a few journals and also serving more than 15 conferences/journals as a reviewer and/or program committee member. Dr. Rahman is serving as the Member-at-large and Academic Advocate: Information Systems Audit and Control Association (ISACA) at the University of Hawai'i at Hilo and Academic Advocate of the IBM Academic Initiative.

Nathan J. Rodriguez, Ph.D., is an Assistant Professor in the Department of Communication at Weber State University.

Antonio Sarasa-Cabezuelo has a MS. degree on Mathematics of Computer Sciences from the Complutense University of Madrid, BS. degree on Computer Sciences from the National University of Distance Education, MS. degree on Computer Sciences from the Open University of Catalonia and CS PhD from the Complutense University of Madrid. Currently, he is an associate professor in the Computer Science School at Complutense University of Madrid, and a member of the research group ILSA (Implementation of Language-Driven Software and Applications, http://ilsa.fdi.ucm.es).

Md. Hasan Imam Sohag (BSc, 2018, Jessore University of Science and Technology) was a B.Sc. student of the Department of Computer Science and Engineering at Jessore University of Science and Technology. Currently he works at the Infobizsoft as a Graphics Designer. He has some excellent skills in Graphics Design section. He wants to work in such an environment what is suitable to self-assessment and development in both individual and group-based research related his skilled field that frequently faces various critical challenges and serves the community with individual skills.

Index

S

T

V

W

Ensure Quality Research is Introduced to the Academic Community

Become an IGI Global Reviewer for Authored Book Projects

The overall success of an authored book project is dependent on quality and timely reviews.

In this competitive age of scholarly publishing, constructive and timely feedback significantly expedites the turnaround time of manuscripts from submission to acceptance, allowing the publication and discovery of forward-thinking research at a much more expeditious rate. Several IGI Global authored book projects are currently seeking highly-qualified experts in the field to fill vacancies on their respective editorial review boards:

Applications and Inquiries may be sent to:
development@igi-global.com

Applicants must have a doctorate (or an equivalent degree) as well as publishing and reviewing experience. Reviewers are asked to complete the open-ended evaluation questions with as much detail as possible in a timely, collegial, and constructive manner. All reviewers' tenures run for one-year terms on the editorial review boards and are expected to complete at least three reviews per term. Upon successful completion of this term, reviewers can be considered for an additional term.

If you have a colleague that may be interested in this opportunity, we encourage you to share this information with them.

www.ingramcontent.com/pod-product-compliance
Lightning Source LLC
LaVergne TN
LVHW062139300926
849161LV00010B/2576